W9-AMX-922

# ENVIRONMENTAL MANAGEMENT ACCOUNTING: INFORMATIONAL AND INSTITUTIONAL DEVELOPMENTS

# ECO-EFFICIENCY IN INDUSTRY AND SCIENCE

VOLUME 9

**Series Editor: Arnold Tukker, TNO-STB, Delft, The Netherlands**

**Editorial Advisory Board:**

**Martin Charter,** *Centre for Sustainable Design, The Surrey Institute of Art & Design*, Farnham, United Kingdom
**John Ehrenfeld,** *International Society for Industrial Ecology, New Haven, U.S.A.*
**Gjalt Huppes,** *Centre of Environmental Science, Leiden University, Leiden, The Netherlands*
**Reid Lifset,** *Yale University School of Forestry and Environmental Studies; New Haven, U.S.A.*
**Theo de Bruijn,** *Center for Clean Technology and Environmental Policy (CSTM), University of Twente, Enschede, The Netherlands*

# Environmental Management Accounting: Informational and Institutional Developments

Edited by

Martin Bennett
*University of Gloucestershire Business School, Cheltenham, U.K.*

Jan Jaap Bouma
*Erasmus Centre for Sustainable Development and Management,*
*Erasmus University, Rotterdam, The Netherlands*
*and Economics Faculty of the University of Ghent, Belgium*

and

Teun Wolters
*ISCOM Institute for Sustainable Commodities,*
*Utrecht, The Netherlands*

KLUWER ACADEMIC PUBLISHERS
DORDRECHT / BOSTON / LONDON

A C.I.P. Catalogue record for this book is available from the Library of Congress.

ISBN 1-4020-0552-0

---

Published by Kluwer Academic Publishers,
P.O. Box 17, 3300 AA Dordrecht, The Netherlands.

Sold and distributed in North, Central and South America
by Kluwer Academic Publishers,
101 Philip Drive, Norwell, MA 02061, U.S.A.

In all other countries, sold and distributed
by Kluwer Academic Publishers,
P.O. Box 322, 3300 AH Dordrecht, The Netherlands.

BUSINESS
TD
194.7
.E598
2002

*Printed on acid-free paper*

All Rights Reserved
© 2002 Kluwer Academic Publishers
No part of this work may be reproduced, stored in a retrieval system, or transmitted
in any form or by any means, electronic, mechanical, photocopying, microfilming, recording
or otherwise, without written permission from the Publisher, with the exception
of any material supplied specifically for the purpose of being entered
and executed on a computer system, for exclusive use by the purchaser of the work.

Printed in the Netherlands.

# Contents

## PART III
EMA POLICIES



## PART IV
DIFFERENT EMA PERSPECTIVES

# Preface

The Environmental Management Accounting Network (EMAN) first emerged in 1997, not as part of any well-funded major corporate or governmental programme, but from the bottom-up initiative of a small group of like-minded individuals. This group had worked together on a research project into EMA that was then concluding and wished to maintain and continue their association in at least a semi-structured way, and to bring into this network others who might share their interest. From this modest initial ambition EMAN was formed and held its first conference in December 1997.

The growth of EMAN since then reflects the increasing level of interest in the application of management accounting in the context of environmental management. Its membership is comprehensive and includes academics, professional researchers, practitioners from industry (in particular environmental managers), governments, NGOs, and consultants. EMAN's aim is to facilitate an exchange of knowledge and ideas, and the main activity supporting this has been the conference which has been organised annually since 1997, supplemented by special workshops which have been organised by EMAN at other conferences such as those of the Greening of Industry network (GIN).

EMAN's annual conferences offer an opportunity for those interested in the topic, from whatever background, to present and hear papers, to learn of projects in progress and planned, to develop new research projects, and to network with others in the field. Since 2000 these conferences have been supported as a high level scientific conference by a grant from the European Union towards both a part of their direct running costs and the travel and subsistence costs of participants attending. The 2002 conference has the benefit of further financial support from other sponsors.

This book of papers from recent EMAN conferences and workshops represents a further stage in the evolution of EMAN, and indicates the recognition by a major publishing house of the increasing interest in EMA and the demand for more information. It provides a unique overview of the state of the art within EMA at this stage. Several contributions refer to the role of stakeholders in the process of integrating EMA into organisations, and EMAN offers a network which can help to disseminate new tools, ideas and new examples of good practice

The definition and content of EMA has continued to evolve rapidly over the short life to date of EMAN. From an initial focus exclusively on the environmental issue, the broader issue of sustainability has become increasingly important, and it is likely that EMAN's focus will evolve to extend to include also the other aspects of the 'triple bottom-line' – economic and social, as well as environmental, sustainability.

Perhaps the most encouraging aspect of the increasing interest in EMA generally, and of EMAN as a vehicle to advance EMA, is the increasing interest across the globe. When first founded EMAN was inevitably mainly European-centred, since its original members had first met through their partnership in an EU-sponsored research project. Although in close contact with others with well-established expertise and interest, such as the US EPA's 'Environmental Accounting' project and the seminal work of the Tellus Institute, the original EMAN network has tended to remain mainly Europe-centred, with all conferences to date based (for sound practical organisational and economic reasons) in Europe. However we have still been very pleased to welcome to the conferences speakers and participants from other parts of the world, notably from North America and the Asia-Pacific region, although the time and cost overheads (and environmental impacts) of global travel have necessarily been a deterrent and constraint on this. Prompted in part by discussions at the EMAN conference at Erasmus University, Rotterdam in December 2000, and the EMAN workshop at the GIN conference, Bangkok in January 2001, a group of enthusiasts have recently set up an Asian-Pacific section of EMAN, in parallel to what we have now re-titled 'EMAN-Europe'. There are also discussions amongst another group in North and South America towards setting up an 'EMAN-Americas' section which we are currently hoping will also come to a successful fruition.

There are also further moves to develop EMAN-Europe, in particular to develop and extend the EMAN-Europe website.

EMAN-Europe would like to take this opportunity to thank all those who have contributed to its development over the period since 1997. Firstly, the work of EMAN is dependent on the research-efforts of individual EMAN-members. Without which the network could not exist. Each paper that is presented at an EMAN conference or workshops demonstrates a variety of financial and institutional support from a large number of organisations and individuals, and many thanks go to all of these supporters. With regard to the organisation of the EMAN-Europe conferences and workshops special thanks go to:

- the sponsors – firstly the European Union, who are supporting a series of EMAN-Europe conferences. Additional support has come from the Dutch scientific organisation NWO [Nederlandse organisatie voor Wetenschappelijk Onderzoek] with its research program 'Environment and Economy', and in the United Kingdom from the Association of Chartered Certified Accountants and the Environment Agency.
- those who have been generous with their time and often other resources too in actively organising EMAN-Europe as a network, in particular the members of the Steering Committee.
- the organisers of the conferences at which the papers in this book were first published:
  - Eberhard Seifert and Martin Kreeb (1999 conference at the Wuppertal Institute, Germany);
  - Jan Jaap Bouma (2000 conference at Erasmus University, Rotterdam);
  - Teun Wolters (special workshop on EMA at the 2001 Greening of Industry Network conference, Thailand);
  - and in particular, all the contributors of papers to this book.

We would also like to take this opportunity to invite anyone interested to join the EMAN network. Further information can be obtained from the EMAN-Europe Co-ordinator Jan Jaap Bouma (bouma@fsw.eur.nl), the Chairperson Martin Bennett (mbennett@chelt.ac.uk), or directly from the EMAN-Europe website (www.eman-eu.net). Further information on EMAN Asia-Pacific can be obtained from Katsushiko Kokubu at kokubu@rokkodai.kobe-u.ac.jp; further information on EMAN-Americas can be obtained from Deborah Savage at dsavage@tellus.org or Terri Goldberg at tgoldberg@newmoa.org.

## EMAN-Europe Steering Committee

The steering committee of EMAN-Europe consists of the following members:

Martin Bennett, University of Gloucestershire, Business School, Cheltenham, UK (Chairperson)

Jan Jaap Bouma, Erasmus University, Rotterdam, Netherlands and Ghent University, Belgium (Co-ordinator)
Martin Kreeb, University of Hohenheim, Germany
Torsti Loikkanen, VTT, Espoo, Finland
Pall Rikhardsson, Aarhus School of Business, Denmark
Stefan Schaltegger, University of Lueneburg, Germany
Eberhard Seifert, Wuppertal Institute for Climate, Environment and Energy, Wuppertal, Germany
Giorgio Vicini, Fondazione Eni Enrico Mattei (FEEM), Milan, Italy
Teun Wolters, Institute for Sustainable Commodities (ISCOM), Utrecht, The Netherlands (Immediate Past Chairperson).

We hope that this book, whose various contributions introduce the reader to individual researchers, governmental organisations and representatives from industry, will encourage both individuals and organisations to become interested and involved in EMA; the Steering Committee is open to all suggestions.

"Supported by the European Commission, Research DG, Human Potential Programme, High-Level Scientific Conferences, Contract No. HPCF-CT-1999-00102".

"The papers which have been collected together in this publication were first presented at various EMAN conferences and workshops. These include the 4th annual EMAN-Europe conference at Erasmus University, Rotterdam, 13–14 December 2000, which was generously supported by the European Commission, Research DG, Human Potential Programme, High-Level Scientific Conferences, Contract No. HPCF-CT-1999-00102".

*The editors:*
Martin Bennett
Jan Jaap Bouma
Teun Wolters

# 1. The Development of Environmental Management Accounting: General Introduction and Critical Review

*Martin Bennett, Jan Jaap Bouma and Teun Wolters*

## 1.1. Introduction

This volume represents a further stage in the development of Environmental Management Accounting (EMA). It brings together a selection of papers from recent EMAN and other conferences by an international range of authors in the vanguard of the development and promotion of EMA, and represents some of the leading thinking and practice in what is still a very young and developing area. The papers which are collected here reflect both new informational developments, in that environmental costs are increasingly becoming integrated into corporate information systems; and institutional developments, in particular in respect of programs by governments and international organisations to promote EMA as a means to help to integrate environmental management into the decision-making structures of companies and other organisations.[1]

## 1.2. The rise of EMA and key issues

Environmental Management Accounting (EMA) can be defined as the generation, analysis and use of financial and non-financial information in order to optimise corporate environmental and economic performance and to achieve sustainable business (Bennett and James, 1998). Like conventional management accounting, EMA defines itself primarily in terms of its main audience, its main purpose being to provide relevant and useful information to the management of the organisation, as distinct from external stakeholders, in order to support the various responsibilities of management – planning, decision-making, controlling, etc. The breadth of the information encompassed by EMA and conventional management accounting is less clear, though it is generally accepted that this is not restricted solely to financial (i.e. monetary) information but includes also relevant non-financial data and information.

Although EMA has attracted increasing interest and recognition in recent years, it is still far from having achieved the position of conventional management accounting as a well-established function in business and management. The evident value of accounting in supporting management generally has however stimulated interest, particularly from environmental managers and professionals, in its potential to support environmental management in a similar way. At the same time, some accountants have recognised that if the environment is likely to become an increasingly important issue for their organisa-

[1] Like conventional management accounting, EMA can be relevant for not only private-sector profit-seeking companies but also other types of entity including government agencies and other not-for-profit organisations – any organisation for whom its environmental performance, and how best to manage this, matters. The term 'organisations' is, therefore, used here in order to indicate this wide applicability, except in those particular places where it is specifically commercial companies that are being referred to.



*M. Bennett et al. (eds.), Environmental Management Accounting: Informational and Institutional Developments,* 1–18.
© 2002 *Kluwer Academic Publishers. Printed in the Netherlands.*

tions, this will need to be managed and that their own expertise may be able to make a contribution to this.

### *EMA as interface*

One function of EMA is therefore to link together for their mutual benefit two organisational functions and areas of expertise which conventionally and historically are distinct and may initially not seem to have much natural interface: environmental management (in its broadest sense), and management accounting. The latter is better established, and with more consensus on what it encompasses and on the competencies needed to be an effective practitioner. The former is more general and less well-defined, and can range from organisations which set up specialist environmental management functions (these are usually in larger organisations, with environmental management set up as a staff function at corporate level) to those who simply recognise that their environmental performance and management may be a significant issue in the future success of their businesses, and take positive steps towards managing this.

The most obvious use of EMA is to support organisations' environmental management, but it is also an important issue for accounting and finance professionals, because, if the case for environmental concern is accepted, environment is likely also to become an increasingly impact factor in the business context[2] as a major strategic variable for businesses to recognise and respond to. The adequacy of an organisation's response will therefore be a major factor in its long-term survival and prosperity in conventional business terms too, and for accounting and finance functions to overlook environment would be to overlook a key factor.

The twin premises on which EMA depends are therefore:

- that environment is important and will gain in importance over time; in the first place for society as a whole, and therefore for governments and organisations too;
- Management accounting techniques and approaches are potentially one of a number of tools which organisations can apply in order to manage better their environmental performance.

### *Sustainability*

In the broader public debate, concern for business and the environment has rapidly developed recently into a wider concern for sustainability generally, encompassing the three elements of the 'triple bottom line' – environmental, social, and economic. EMA inherently encompasses the environmental and economic aspects of this, though the concept of sustainability does raise challenges for how these are defined and dealt with. How to bring the social aspect within the remit of EMA (which might then become Sustainability Management Accounting) is less obvious, partly because there is less consensus on what is represented by good corporate social performance, even before going on to raise the questions of how best to achieve and measure this. This is a live issue amongst those active in EMA, who are watching with interest the sustainability debate and looking for ways in which management accounting may best contribute.

---

[2] The term 'business context' is used instead of the more usual term amongst business strategists of 'business environment' to avoid the obvious ambiguity and potential confusion of the latter in a discussion of business and the (natural) environment.

### *Reporting*

Although some seminal publications can be traced back to the 1960s and 1970s it was only in the 1990s that concern for the relationship between business and the environment became generally recognised in most countries, at least in the developed world, as a mainstream issue for business. Initially the main concern was for compliance with a rapidly increasing body of 'command-and-control' environmental legislation and regulation, which required little of accountants and financial analysts. However, the issue of environmental performance has also raised issues concerning organisations' responsibilities to society and external stakeholders generally, more broadly than the traditional focus on the responsibility to investors for the organisation's financial performance. The main symptom has been a rapid growth in the number of organisations publishing regular (usually annual) reports on their environmental management and performance, usually modelled at least loosely on the financial Annual Reports. This trend has been encouraged by the accounting profession through environmental reporting awards schemes such as that organised annually by the UK Association of Chartered Certified Accountants (ACCA), and a move to standardise the frequently variable content and quality of corporate environmental reports through the Global Reporting Initiative is involving representatives of the accountancy profession together with environmental NGOs (non-governmental organisations) and others.

### *Cost savings*

The initial stimuli were therefore legal and regulatory compliance, and external accountability. It has also become recognised that good environmental management need not represent only a cost to businesses (even if one which often provides valuable intangible benefits), but can frequently also generate direct business benefits too. Some organisations have perceived opportunities to generate 'green revenues' by designing and marketing products, which are designed to be environmentally responsible, to environmentally aware consumers. More frequent, however, has been the perception that environmental audits, even if they are aimed in the first place only at improving environmental performance, can also help to identify opportunities to save costs, in many cases without requiring any major initial investment by the organisations implementing them or any fundamental changes in their methods of operation. These 'win-win' opportunities have usually been in the areas of energy efficiency and resources efficiency (the converse of which is the wastes produced in production and other operations – those inputs that fail to end up as good sale-able outputs).

Several governments and their agencies have therefore set up energy efficiency schemes to offer advice and technical support to businesses (and to other organisations, and consumers), where the main public policy motive has been environmental, and the incentive for businesses has been economic. Several such schemes have reported that major savings could be identified and achieved, which had not previously been apparent to the managements of those organisations – in defiance perhaps of the classic free-market purist's argument that 'you cannot base a business plan on picking up $50 bills off the floor', the response would seem to be that sometimes a surprisingly high number of bills that were not previously apparent seem to become visible if viewed for the first time through a green lens!

However, the potential for easily achieved gains is not indefinite, and many organisations have found that after the initial 'low-hanging fruit' has been harvested, diminishing returns have set in, so that more work has then been required to go on to identify further possibilities. In this, conventional accounting and other information systems have often been found to be a hindrance rather than a help, since they frequently fail to make transparent the extent of costs and revenues that are related to the less obvious aspects of environmental performance. For example, the costs of disposing of waste (which for most organisations is usually an external outlaid cost) are usually separately identifiable (though even this is not universal – some accounting systems might allocate this item in a larger general overhead account). However, there may be substantial further costs incurred through other activities in the processes through which waste is generated which are not so transparent, including the internal handling and storage of wastes before they reach the point of disposal, and the cost of the original inputs of raw materials and other resources up to the point when the waste is identified and removed from the process.

### *Better cost analysis*

Much early work on EMA consisted of attempts at better cost analysis by tracking resources and their associated costs in more detail in order to be able to identify their causes, with the frequent finding that a substantially higher proportion of total costs was in some way environment-related than had previously been recognised. In principle and method this is close to activity based costing, which is still a relatively recent innovation in mainstream management accounting.

Other parallels can be drawn between work in EMA (which is often initiated and carried out by people with no previous contact with management accounting), and topical developments in mainstream management accounting. For example, capital investment appraisal processes as they are conventionally applied have come under scrutiny for their alleged tendency towards too-short time horizons and excessive discount rates, and have therefore been criticised as inappropriate to evaluate some of the more positive forms of environment-related investment.[3] This is a similar criticism to that made in other areas of management, such as in the appraisal of investments in advanced manufacturing technology and major strategic information technology systems.

### *Parallel developments*

EMA's interest in product life-cycle costing is similar in purpose and focus to product-line costing, and its interest in supply chain management for environmental reasons can also be seen in the context of the aim of strategic management accounting to look outside the boundaries of the organisation at the factors in its wider business context that will influence strategy and determine its future prosperity, and to bring management accounting techniques to bear on them. Not least, the area of non-financial performance measurement, as reflected in Balanced Scorecard and similar models, has become a major issue in mainstream management accounting and is paralleled in EMA and environmental management generally, where advanced techniques of performance measurement have for several years been driven by other demands (e.g. regulatory) than only providing management with useful information. EMA is also advancing in other ways, such as a recent revival of interest in the potential use of external costs within organisations in order to inform policy, which may suggest a further tool which organisations can use to attempt to predict and plan for the future.

---

[3] Typically, these are often cleaner production and pollution prevention (P2) projects which offer more potential for fundamental improvement than do simpler 'end-of-pipe' pollution controls, but which may take longer for the benefits to justify the initial costs.

### *EMA and the accounting profession*

There are still few positions within organisations which carry the title of 'environmental management accountant', but there are several examples of work being done which can be described as EMA, whether this is done by environmental specialists, by accountants taking a particular interest in environment, or by those working in other areas such as production, product design, and strategy planning and formulation who recognise environment as an increasingly important strategic variable for their organisations and apply financial together with other techniques to address this. In many cases the initial interest has come from environmental specialists who see EMA as one way to re-position their roles as more than the necessary but unglamorous hygiene factor of compliance-oriented technical specialists for which their positions may originally have been established. For those in this position, EMA can offer a way to communicate the positive benefits for the business of environment-related projects, and to produce a business case in a form that will be comprehensible and attractive to senior managers and to their colleagues in other areas of their organisations.

The accounting profession has also recognised the issue of business and the environment as significant for its members. Its main initial focus has been on the issues of external corporate responsibility and accountability, but it is increasingly seeing EMA also as an opportunity for its members to apply their skills in a new and growing area of management. In particular, the rapidly growing area of non-financial performance measurement is one which crosses traditional professional boundaries, and the argument that accounting and finance functions are the natural custodians of performance measurement systems is far from universally supported by non-accountants. Several debates on EMA have raised questions over the boundaries of management accounting generally, and what the distinctive competencies are that management accounting, and management accounting professionals, have to offer to business and to the rest of the world.

### *Financial services sector*

Although some have criticised the financial services sector for allegedly being slow to recognise environment as a substantive business issue, there are signs that this is changing as lenders and investors come to appreciate the potential influence of environmental performance on future returns and risks. This is reflected in the UNEP Financial Initiative (UNEP, 1998), which invites financial institutions to sign a statement to signify their recognition that sustainable development depends on a positive interaction between economic and social development on the one hand, and environmental protection on the other, in order to balance the interests of this and future generations. This statement also recognises sustainable development as being the collective responsibility of government, business, and individuals, and the financial institutions that sign this thereby commit themselves to co-operate with these sectors within a market framework. Currently, about 179 banks have signed this statement (Bouma et al., 2001). The financial services sector as a whole is concerned less with environmental performance for reasons of corporate responsibility for its own sake, than with its implications for the future business and financial performance of the organisations in which it invests. Financial specialists need to translate environmental information into financial terms. As accountants within organisations are expected to generate such information, this creates a further stimulus for EMA within organisations.

### *EMA and public policy*

The most immediate markets for EMA are organisations and their managements, accountants and financiers, and it is not immediately obvious that EMA should have an attraction for governments, at least in their specifically governmental role.[4] However the case has been made that public policy environmental objectives may in many situations be most effectively achieved not through traditional 'command-and-control' approaches which seek mere compliance, but through the use by governments of financial instruments as a tool to implement policy, following the 'Polluter Pays' principle. These can include 'green taxes' in which the scale of environmental impacts is the driver of organisations' and individuals' tax liabilities; tradable emissions quotas; and a shift in law not merely towards strict liability in which the ownership of an industrial site or hazardous materials can be sufficient to create legal liability for negative environmental impacts, but even a responsibility for post-sale environmental impacts during use and financial disposal of a product under the concept of 'extended producer responsibility'.

Several government agencies and international bodies have therefore started to explore the possibilities of encouraging the design, use and dissemination of EMA systems and techniques. The United States Environmental Protection Agency (US EPA) is the pioneer in this, though other countries are following as indicated by recent projects in Japan and Korea (with the support of the World Bank), and in particular a project set up in 1999 by the United Nations Division for Sustainable Development to establish an Experts Working Group, including experts and government representatives from across the world to consider 'The Role of Governments in Promoting Environmental Management Accounting'.

### *Structure of the book*

The dialogues and debates in recent EMAN conferences have reflected these main strands of interest, and this has determined the structure of this book. Section 1 provides overviews of EMA from different angles, progressing to Section 2 which brings together several papers on the development of information systems and techniques to provide EMA information. Section 3 considers policies and projects, by governments and others, to disseminate and promote EMA, and finally Section 4 provides a number of other perspectives on EMA, in particular related to the wider issue of sustainability. The remainder of this chapter therefore introduces each section of the book and offers some reflections on the contributions presented.

## 1.3. Overviews (Section 1)

Section 1 holds a number of chapters that give an overview of either the entire field of EMA or a major subset of EMA. There are a number of critical issues that emerge from these chapters:

a. The relationship between financial and physical figures in EMA
b. The materials flow approach as a possible new paradigm

---

[4] Many government ministries and agencies, of course, are themselves large-scale organisations with significant environmental impacts caused by their own activities, which call for the use of relevant environmental management techniques including EMA in the same way as for any other form of organisation.

Ad a. Those who see EMA most of all as a subset of general Management Accounting, may be inclined to regard the compounding and internal reporting of financial environmental figures as its primary assignment. In this way EMA can indicate what a variety of environmentally-relevant activities contribute to the company's economic bottom line. In this view physical figures (quantities, volumes, and frequencies) are of secondary importance, although they can be very useful to interpret the financial figures. However, historically EMA has placed more weight on physical information than conventional management accounting usually does. This has to do with an interest in measuring the environmental effects of current economic activity in terms of emissions and wastes as a result of governmental Command & Control policies. For a greater part environmental affairs were seen as operational matters not affecting corporate strategies. Because of this, environmental figures were not usually integrated into the bottom-line oriented figures produced by management accountants. In fact, this is still the case in many companies. Only where environmental issues acquire a certain degree of strategic importance can EMA be expected to become part of the core accounting figures that will be presented to and discussed by senior managers and executives at corporate level. Then insight in the financial aspects will increase in importance and this will reverberate upon how EMA is shaped in the company's different departments. The needed strategic impetus will relate to increasing environmental costs because of stricter norms, increased levies, corporate image and reputation and more elaborate environmental programs. A higher profile of environmental management can also be used to reinforce internal drives for more efficiency and costs savings. In this way environmental objectives can also be related to financial performance. However, new areas such as the greening of product chains can rekindle an interest in new EMA models that are primarily based on physical environmental data.

Ad b. Different chapters in Section 1 deal with the materials flow approach. They prompt a fundamental question about the implications of this approach. It seems that the chapters deal with conventional approaches to EMA and the materials flow approach as complementary parts of the same overall approach. However, an alternative interpretation might be that the materials flow approach is based on a new paradigm that cannot be easily reconciled with conventional thinking. According to this paradigm, the entire operation of an organisation is seen as an ecological activity. The production function here signals the use and transformation of scarce and often non-renewable resources on the one hand, and the release of transformed substances into a fragile ecosystem on the other. The 'transformed substances' consist of both products (that people value for their own use) and non-product outputs (wastes and emissions). In any case non-product outputs have to be reduced wherever possible ('zero waste'), priorities being determined on the basis of both ecological and economic impacts. EMA is there to show what these impacts are. The environmental manager will primarily focus on input efficiency rather than on the volumes of output. The close links here between environment, materials efficiency and cost savings bring, as it seems, a fundamental ecological perspective and a quantitative analytic approach, which could provide a strong stimulus for environmental – or perhaps better called – ecological accounting. Whether a company produces the right product cannot be decided on the basis of short-term economic success alone but should also be based on the longer-term requirements of sustainability. The latter involves a kind of strategic decision-making for which EMA has to develop new tools and indicators. Here qualitative scenario-based types of accounting tools need to be developed. To have a kind of common understanding of how these tools should look like is a matter of public interest deserving the support of governmental policies.

Indeed, it seems that the materials flow approach has the potential of offering new policy perspectives on sustainable development. However, for the foreseeable future this approach cannot replace conventional accounting insofar as the latter relates to conventional concepts of profitability that are inherent in current financial reporting. Many organisations therefore will find it appropriate to adopt the materials flow approach only partially or incidentally, if at all, since given its substantial informational requirements, it could be very costly to have a materials flow accounting system implemented on a permanent basis at the same time as a conventional accounting approach. Large companies could be an exception to this general rule, where at a corporate level the materials flow approach offers an excellent tool of management control and strategic planning which may be sufficient to justify the investment in setting up and operating the system. A number of chapters in Section 2 discuss how the two approaches may be best combined when an organisation makes decision on its information systems.

The materials flow perspective appears to introduce a kind of dialectics into the debate, which questions the previously-mentioned importance of integrating EMA with conventional management accounting. The materials flow approach suggests that the conventional approach to EMA, as embodied in conventional management accounting, may be inadequate to bring the true factors of sustainability into focus – is it likely that simply to distinguish environmental from non-environmental elements of costs in conventional management accounting processes will be sufficient to lead an organisation to sustainability, or does this require the new paradigm of the materials flow approach (or indeed other new paradigms too) to ensure a credible path towards sustainability? A broadly supported consensus on this is unlikely in the near future, but whilst this is being debated, it would still be wise not over-hastily to reject partial and pragmatic solutions, or to reject what is still in incremental terms good practice in favour of an apparently superior solution that as yet for the greater part exists only on paper.

### *MEMA and PEMA*

Chapter 2 (Burritt, Hahn and Schaltegger) offers an integrative framework which propounds a broad set of EMA concepts. Is distinguishes between Monetary EMA (MEMA) and Physical EMA (PEMA), and also add other categories such as 'past-oriented' and 'future-oriented' EMA. This framework gives equal status to MEMA and PEMA, although – one could add – MEMA might be considered to be primary, with the role of PEMA being to give background to the financial quantities and to help to analyse and evaluate them, for example in terms of cost drivers, efficiency and economies of scale.

### *Materials flows*

Chapter 3 (Jasch) also distinguishes between MEMA and PEMA, but broadens the boundaries by also including external environmental reporting (both financial and non-financial), and application areas such as environmental management systems, eco-design, cleaner production and supply chain management. Jasch's view on EMA is based on the materials flow approach, though she also refers to a more conventional management accounting framework when she states that the most important role of EMA is to make sure that all relevant costs are considered when making business decisions, with 'environmental' costs being a subset of the wider cost universe that corporate decision-makers need to take into account. She then, however, goes on to argue that EMA should focus on materials flows, "which means that EMA is no longer meant to assess the total 'environmental' costs but to develop a different look at the production costs that takes an organisation's environmental effects seriously". Through this, EMA can be an attention-director to encourage managerial decision-makers to take a different look at familiar processes in order to reflect new priorities.

### *A guideline to distinguish environmental costs*

The range of different approaches is demonstrated further in Chapter 4 (Kim), which is in the domain of conventional past-oriented environmental accounting. In most conventional financial and management accounting systems, little if any distinction is made between environmental and non-environmental costs. This raises the question of how this might be accomplished in a uniform way across industry, so that the figures generated by different organisations are genuinely comparable. Kim's paper proposes a set of guidelines, which would provide a classification of environmental costs, and a uniform way to separate environmental costs from existing accumulated figures. The guidelines are targeted primarily at large companies, and their success will of course be measured by the extent to which they are actually adopted. Moreover, inter-organisation comparability will remain a critical issue so long as organisations' accounting policies and structures differ, so it could be advisable to develop a core set of environmental figures whose meaning is clear and consistent irrespective of the particular situation in which they have been generated.

### *Flow cost accounting explained*

Chapter 5 (Redmann and Strobel) provides a comprehensive picture of flow cost accounting, which is based on the materials flow approach discussed above. The paper makes clear that materials flow accounting involves a new way of looking at an organisation. Flow cost accounting is a basic component of flow management, which aims to combine economic benefits with environmental benefits. The other two components are the flow model, which shows the materals flows running through the organisation and the flow organisation that channels the flows. If the materials flow approach takes precedence over other forms of institutionalisation, present organisational structures will have to be changed on the basis of what could be called a flow-oriented version of process engineering. In flow cost accounting, materials flows are distinguished between the cost categories of materials, system, and delivery and disposal. For these three cost categories the paper provides a systematic treatment of how quantities and costs are recorded and used in order to manage the organisation as a processor of materials flows.

### *Resource efficiency accounting*

Chapter 6 (Orbach and Liedtke) discusses a particular EMA approach that takes into account the ecological impact of the materials chains related to the inputs of an organisation. This 'resource efficiency accounting' is a combination of systematic and pragmatic thinking. On the one hand, organisations are held accountable for upstream materials flows in their entirety, irrespective of their power to implement improvements, and are therefore expected to make serious efforts to quantify them. On the other hand, this approach recognises the difficulties which are usually encountered when trying to trace materials, and makes some simplifying assumptions in order to address these. In particular, these refer to the idea that the accumulated materials inputs (in physical terms) are representative of the total ecological impact of a production process or product. The authors would probably not deny the inherent cause-and-effect and weighting problems, but rather than becoming involved in addressing these problems, resource efficiency accounting encourages the active collection of data. Comparison of products, services and infrastructure is done on the basis of their life-cycle-wide materials input per service-unit (on the basis that providing services to their users is the function of tangible products too).

The economic analysis is achieved by a cost calculation, based on the economic and ecological cost drivers, which have been generated by the preceding process analysis. Although the cost concept is broad (involving the costs of purchasing and handling of the materials that end up as waste), the system boundaries are now dictated not by the materials flows but by the limits of the organisation as a decision unit. The existence of external costs is not denied, but these are deliberately excluded since any attempt to quantify them would inevitably be controversial.

The resulting resource-efficiency portfolios at process and product level – taking into account both materials intensity and costs – are intriguing, but raise questions over the significance of what is indicated by the information that is generated. Is it only a 'crystal ball' that stimulates the creativity of those in the organisation, or can it be seen as a direct indicator that provides a straightforward message on what should be concluded from them and the appropriate action to be taken? The processes and products indicated as having the highest materials intensities and highest costs need to receive most attention, since the win-win opportunities are likely to be greatest there, but this raises two further questions. Firstly, should stakeholders be included? – A stakeholder analysis could lead to quite different priorities. Or should the resource efficiency approach be seen as an additional 'stakeholder' that indicates the priorities for the long term? And what is the significance of a high level of costs? In fact, what is important are the possibilities to save costs, and in general it can probably be safely assumed that these will usually be greatest where the costs are highest. However, this does seem rather approximate as an indicator, in particular when it is to be used by people who have a significant amount of inside information, for instance on the structure of the means of production. Resource efficiency accounting highlights the problems of dealing with the monetary and physical side of EMA, but its policy of avoiding certain controversies and complexities has brought to the fore other theoretical problems that cannot be easily resolved.

## 1.4. Contextual variables in information management

The considerable data demands of materials flow costing mean that this cannot be executed without automated information systems, which raises again the issue of integration. One approach would be a totally separate system, which runs in parallel to existing conventional systems. However, this means a duplication of effort, and leaves doubtful the real impact on decision-making, given its separate and therefore likely isolated position.

Integration with the physical flow data systems seems feasible and promising: only then can an organisation with different divisions and establishments be united in a common materials-flow oriented policy. Further integration of the materials flow approach into the area of cost accounting would be a more drastic step that appears as yet not to be feasible for most organisations. In fact, the entire materials flow approach still is in its 'laboratory phase', despite some very interesting practical cases. Although it may represent a paradigm that provides an improved perspective on the relationship between ecology and economy, it is still difficult to predict whether it will become generally accepted. For the time being we can expect to see a range of different approaches in practice, with differing degrees of novelty and pragmatism, attracted by on the one hand the established position of conventional management accounting and the influence that this offers, and on the other hand the more fundamental and perhaps ultimately more promising materials flow approach.

Section 2 provides several interesting chapters that are highly illuminating in showing what the implementation of comprehensive information systems may mean in practice.

### *Linking materials flows to decision making*

Chapter 7 (Scheide, Enzler and Dold) argues that an organisation's materials and energy flows can be central to environmental management only if these are integrated into the decision-making processes of organisational functions such as purchasing, product development, production and marketing. This integration requires information systems that are capable of linking materials and energy flows to decision-making processes. The paper presents the ECO-Integral Reference Model that comprehensively shows how integrated environmental management can achieve this. In a structured fashion it makes available the database concerned, and applies several different environmental management tools, such as materials and energy balances and environmental cost accounting. This model largely reflects the materials flow approach which has been discussed in previous chapters but is addressed more towards implementation. The model as presented is clearly intended primarily for large companies, though even for these the chapter's discussion of a number of actual applications seems to indicate a tendency to apply the model only partially. This then raises the question of whether the model is capable of stimulating an increasing integration of environmental management into the main management functions of organisations, or whether it is in fact likely to deter certain categories of organisations, considering the complexity of the model. However, even if its applicability at present is restricted to only a specific group of large companies, it offers a useful exercise and one that might perhaps be adapted for the benefit of SMEs.

Chapter 8 (Juergens) takes a similar approach to chapter 7 and also adds a number of new elements. It argues that the focus of current concepts and approaches for production planning and control on the mapping and planning of the core processes, i.e. products and materials, can have drawbacks since this means that auxiliary materials, spin-offs, energy consumption and waste materials are generally neglected. Because of this, cost saving potentials in the latter areas are obscured, leading to sub-optimal efficiency policies.

The comprehensive nature of the materials flow approach becomes apparent when Juergens suggests that an organisation's environmental manager could take over various service functions for other business processes by taking responsibility for the generation of regular internal reports on materials flows and related cost figures. This could be interpreted as suggesting that the environmental manager would then become the organisation's materials flow information manager, taking responsibility for co-ordinating the setting of priorities but still leaving to line management the responsibility for actually taking the necessary measures, including environment-related actions. To make materials flow management effective and to realise its potential environmental and cost-saving advantages, it is necessary to ensure the availability and quality of suitable data, with an adequate linkage between materials flow management, production planning and control, and financial accounting.

The case study that chapter 8 presents refers not to a process industry in which it would be possible to base materials flow balances on the immediate production data from SAP R/3, but to the production of complex investment goods that are ordered individually. This requires an additional concept of how to calculate materials flows from the SAP R/3 data in a way that is fairly straightforward and reliable. This is done by using three allocation principles (fixed, time-related and quantity-related). The resulting model can be used for EMA purposes, but at the same time also offers possibilities for stimulating the cost-effectiveness of new production plans.

The paper claims that this approach can also be extended to other software programs, in particular those that are used by SMEs, which could be worthy of further research into the opinions and information needs of SME managers.

### *Application in a large multinational company*

Chapter 9 (Thurm) presents a fascinating case on Siemens, which is a huge multinational company with research and development locations in over 30 countries and regional offices and agencies in over 190 countries. As the company operates in all parts of the world, it realises more than others the impact that world-wide population growth has on the world's natural resources and the conflicts that may result from this.

To cope with this situation, the chapter talks about a paradigm shift that is needed to secure a path towards sustainability. It involves the same new paradigm as discussed above, but it is defined more in terms of its practical consequences. Fundamental to this is a shift from eco-efficiency to eco-effectiveness, which is reflected also by several current trends such as the introduction of inherently clean technologies, a movement from emission-oriented controls to input-oriented controls, a movement from a site-oriented view to a value-chain-oriented network of actors, and a movement from a production-oriented to a product-oriented view.

This sets a new communication challenge, which for EMA means adaptation in cost accounting (Activity Based Costing, Full Cost Accounting and Life Cycle Costing).

Resource management accounting has to measure the materials and energy flows and to connect these with cost accounting, which Siemens is working on. Zero-waste management, that focuses on the residual materials (the materials which do not become part of the final product to be sold), takes a central place in this since it makes visible the residual materials flows and helps to indicate potential savings and to improve cost allocations.

The products themselves are central to the design process for which relevant cost information is highly relevant, with EMA providing a valuable tool here.

It appears that SAP R/3 can generate the data needed for the annual report on environmental protection, or for the priorities for waste management, but not a comprehensive balance of materials and energy flows for which other complementary software would be needed.

Cost figures show that over recent years Siemens' environmental management policies have led to reductions in environment-related operational and capital costs, even though external environmental costs such as levies and fees have increased.

### *Two integration options*

Chapter 10 (Rikhardsson and Vedsø) discusses the integration of EMA into corporate information systems, and identifies two options: EMA can either be integrated into an existing financial system, or can be integrated into a standard environmental management information system (EMIS).

The advantage of the first option is that an organisation's corporate environmental affairs are then treated on the same footing as other business functions such as purchasing, production, and sales. However, the financial system normally has only limited space for non-financial figures, while these often play an important role in analysing environmental performance.

The second option would be to use an available EMIS, which has the advantage of being capable of processing huge amounts of data. Normally these are physical (non-financial) in nature, but to include financial quantities also would not present a problem. However, a notable disadvantage of this option would be the inevitable need to transfer data between systems. For instance, relevant invoices will be entered by the accounting department but may have to be entered by the environmental management department too. Besides various technical problems that have to be solved, being at some distance from the financial system may make EMA less visible to the decision-makers and their immediate staff. Whatever solution is chosen, it is important that the system can produce overviews, as well as a number of standard analyses and reports for decision-makers. Current financial systems or EMISs have no special modules for EMA, and the authors suggest that implementing such modules can best be done incrementally rather than by starting an over-ambitious scheme – for instance, one may start with well-defined areas such as waste water or air emissions. The paper provides two interesting case studies on this.

### *Product design and development*

Chapter 11 (Krasowski) starts by recognising that the product design and development phase in the life cycle of a product frequently determines the majority of its environmental impacts over its whole lifecycle, and concludes that product design and development should take into account the whole lifecycle, including the possibility of reuse or recycling. At the same time it is necessary to consider the costs that are incurred not only during production but also later when the product is used.

The paper introduces Life Cycle Engineering (LCE) as a tool to support this. LCE integrates three methods: Life Cycle Analysis (LCA), Life Cycle Costing (LCC) and Product Structure Analysis (ProSA). In LCC, four categories of costs are distinguished: R&D costs, Production and Construction Costs, Operation and Maintenance Costs, and Retirement and Disposal Costs. The main requirement for an effective LCE is integration of the three LCE methods into the business processes and incorporation of the LCE supporting tools into the existing IT-landscape. Early involvement of the designers in implementing LCE is extremely important, since it is they who will have to work with it. At present not all the data that are needed can usually be obtained directly from available sources, which means that the designers will later have to add further information to the system. The relevance of the design process justifies the importance of a method like LCA, where EMA can also be helpful although the context for its application is in the hands of others. Where systems such as those discussed in this paper are integrated, it is important that those who apply them first understand the underlying concepts and are aware of the assumptions made in order to be able to use the integrated tool appropriately. Systems transparency therefore can be identified as a separate field of accountability, irrespective of whether it should be seen as an additional field of EMA or not.

## 1.5. EMA policies

### *Governmental initiatives*

Section 3 discusses several initiatives by governments and international bodies to develop and promote EMA. The pioneer here was the United States Environmental Policy Agency (US EPA) by setting up an environmental accounting programme in 1990 to inform industry about EMA and to encourage its adoption. This programme has largely focused on the potential profitability of pollution prevention and has advocated Total Cost Assessment as a key conceptual model. Some 5 years later the European Commission sponsored under its Environment and Climate Programme (DG XII, Human Dimensions of Environmental Change) a research project to explore the potentials of EMA (The Ecomac Project). However, still in 2000 European governments had not formulated a policy that explicitly promotes EMA as a tool of management. It was the United Nations Division for Sustainable Development (UN-DSD) that launched an initiative to enhance the role of governments in promoting EMA. This initiative stimulates European governments to consider EMA as a separate policy item.

### *Social and private perspectives*

The main source of external encouragement on business and other organisations to adopt EMA is likely to continue to be governments who see EMA as a means to encourage business and other organisations to take actions which will help to achieve environmental public policy objectives. Assessing the effectiveness of EMA will then depend upon which perspective is taken. EMA can be seen from a social perspective as well as from an organisational perspective, and can be assessed in the same way as other policy instruments, applying criteria such as (1) environmental effectiveness; (2) economic efficiency, including public administrative costs; and (3) conformity with the institutional framework (Opschoor, 1991, p. 172). If EMA were assessed from a private organisational perspective, then the criteria would differ from those appropriate for a social perspective. The effectiveness criterion would then be the extent to which corporate objectives are met; the efficiency criterion would be the minimisation of private costs that are incurred in order to achieve these corporate objectives; and conformity with the institutional framework would be interpreted differently, since institutions defined from a micro perspective will be of a different nature than if defined at a macro (societal) level. Here, for instance, a criterion could be whether an EMA tool fits with existing corporate principles of cost allocation or with corporate culture that defines certain roles of management accountants.

### *Environmental effectiveness*

From a social perspective, environmental effectiveness is defined by the success of policies in reaching a social optimum (that is also sustainable) by reconciling private and social interests. This may involve altering organisations' economic parameters through economic incentives, in particular by measures that promote the internalisation of external costs and reward the production of sustainable products. However, such incentives can be effective only to the extent that companies are receptive to them, which depends on the quality of their information systems in recognising and reacting to external stimuli such as government policies. A critical element of these for any organisation will be an adequate accounting structure that accurately records cost and benefit data and reports to the relevant decision-makers the information that is generated. However, it should be realised that EMA alone is insufficient to about significant change. EMA can open one's eyes for the wider impact of a company's activities on its environment, but in the final analysis it depends on the strategic orientation and behaviour of decision-makers whether EMA can make a difference.

### *Economic efficiency*

EMA can also support economic efficiency where environmental public policy is part of a broader approach to public policy generally which aims to bridge the gaps between private and social costs of economic activities. At private organisational level, costs would be incurred in the design and implementation of EMA systems and techniques, and in the continuing costs of running EMA. A typical cost at a governmental level could be the costs of disseminating best EMA practices.

### *Conformity with institutional framework*

Environmental policy instruments are regarded as in conformance with the institutional framework if they comply with the Polluter Pays Principle (Opschoor, 1991, p. 172). EMA supports this principle as it helps to identify where costs are made and the underlying cost drivers. However, other dimensions of the institutional framework can also be taken into account. In principle, EMA can be applied under different regimes of environmental policy (C&C, standards, economic instruments such as ecotaxes, voluntary agreements). Economic instruments have the advantage of directly and visibly relating environmental improvements to operating efficiently and being keen on cost savings. This increases the relevance of EMA.

Government agencies in other parts of the world have also become increasingly aware of the usefulness of EMA. Chapter 12 (Kokubu and Kurasaka) gives an example from Japan where EMA has developed rapidly during recent years. It argues that the Japanese Environment Agency has played a crucial role in this by publishing a guideline. Although issued by a governmental organisation, the guideline was the result of two 'non-governmental' study groups: one with the Japanese Institute of Certified Public Accountants, the other with practitioners from the business community. Although EMA is still very young in Japan, this country has already played a notable role in its development. The way accounting can address externalities, for instance, has been given special attention.

Chapter 13 (Lee, Jung and Chun) describes the situation in Korea, where several stakeholders (including financial institutions, local communities and governments) have been interested in corporate environmental performance and disclosure. A few leading Korean companies have since the mid 1990s started to introduce environmental accounting. A substantial increase in environmental costs has forced Korean companies to begin to integrate them into their decision-making procedures at different levels. However, the chapter makes clear that in Korea explicit attention for environmental performance and the reporting on it is in an early stage of development. The authors propose a number of policy options for the introduction and promotion of environmental accounting in Korea as well as in (other) developing countries.

Chapter 14 (Schaltegger, Hahn and Burritt) discusses the UN-DSD initiative, focusing on why many organisations have not yet implemented EMA in spite of the various advantages that might result from its application. The paper reviews a range of different approaches that governments can adopt to stimulate the use of EMA in organisations. It is argued that a strategic approach to policy-making in the field of promoting EMA could overcome the problems inherent in the 'muddling through' that is currently taking place.

However, it is recognised that among actors in the public sector the motivation to support the promotion of EMA by the government may vary strongly, depending on one's position in the institutional framework.

### *EMA applied to government organisations*

As well as encouraging the adoption of EMA in the private sector, some government agencies have also been active in using EMA in their own operations. Chapter 15 (Osborn) illustrates how local authorities and businesses alike can use an information system designed from a governmental perspective (environmental statistics in this case). Such a multiple use of available facilities is likely to improve the economic efficiency of the funds invested in them. It a sense the local government, its community and some external stakeholders participate together in a process of knowledge management.

The interest of government agencies in EMA is based on the assumption that EMA is indeed effective in helping organisations to achieve environmental goals. If EMA can also help organisations to become more aware of the business dimensions of sustainable development, it could even play a crucial role in establishing a process of change within society generally, although whether this may be possible is not yet clear and the mechanisms that shape such a process of change are not yet fully mapped and understood.

Chapter 12 gives some examples from Japan of other types of government initiatives to promote EMA, and, as has already been mentioned, also Chapter 15 is illustrative of a possible policy to promote EMA. Finally Chapter 16 (Reyes) shows how the Philippines Institute of Certified Public Accountants takes the greening of accounting on board. This organisation promotes environmental accounting as part of a larger universe of accounting tools necessary to make good business decisions. However, for EMA to be effective as a tool, it needs to be embedded in a wider institutional framework.

## 1.6. Further perspectives on EMA

This final section looks at a number of different perspectives on EMA, and considers the perspectives of a wider array of stakeholders.

Chapter 17 (Günther and Sturm) describes a model to measure, evaluate and assess environmental performance. This model sets out to determine the objectives (or even very concrete targets) that have to be reached. While objectives are to be set, it explicitly takes into account the interests of the organisation's major stakeholders.

Chapter 18 (Wolters and Danse) focuses on how to develop indicators that adequately measure corporate and chain-oriented performance in reaching sustainable development. The authors regard this perspective as fundamentally different from considering only the perspectives of environmental objectives, and argue that EMA may over time evolve into accounting for sustainability. Although the current status of EMA systems and techniques may not yet be classified as sustainability accounting, the paper presents a useful approach to develop sustainability indicators. Performance indicators developed from the perspective of management accounting flow from the perspective of the management's objectives. In the case of sustainability indicators, the consistency of the organisation's activities with a move towards sustainable development is crucial, which implies that the interests of future generations should be acknowledged. The authors therefore claim a need for corporate sustainability strategies that lead to necessary transformations and stress that the development of sustainability is a process in which a set of management systems (not limited to accounting alone) plays a role. It is this integration of EMA into the organisation as a whole that brings the environment into focus not only of those who are charged with specific environmental tasks, but also of those responsible for general management.

Chapter 19 (Loikannen and Katajajuuri) looks at EMA from the perspective of consumer choice, and present the results from a study of the production of environmental data based on life cycle assessment in the food sector. Although the consumer perspective is stressed, the authors state that the data which is generated is useful also for authorities, shareholders and other stakeholder groups. The authors show that the ultimate reason for producing environmental data within the food sector is related to the intense international competition in current and future markets. Clearly, the development of EMA in the case which is presented comes from a management perspective which is built upon a customer and stakeholder-oriented approach. In this case the stakeholders are closely involved in setting environmental objectives, as in the case of the model of Günther and Sturm. However, the stakeholders are crucial for producing environmental data through the use of the EMA technique, which is presented, which is based on Life Cycle Analysis (LCA) methodology. The paper focuses upon the Finnish foodstuff industry that produced the LCA-based data, which was considered to be a necessary element in the long-term development of its quality management and accounting systems. The article shows the importance of participation by the organisations from the whole business value chain, because the quality of the final product will determined by the weakest links in the business chains.

Chapter 20 (Montel) looks at EMA from the perspective of implementing an environmental management system in pig farms, where he argues that EMA can be a useful tool for implementing and evaluating environmental management systems. The author's choice of French pig farms as a context on which to focus is interesting since (at least as yet) there is little environmental concern over this sector. Nevertheless, the role of environmental management systems is perceived by the author as important in order to progress in the industrialisation process which the farms face. The author shows that the financial consequences of an EMS are very important in this context, although due to the lack of an environmental management accounting system the likely actual environmental costs and possible savings cannot be specified. The author underlines how these accounting systems may be useful as a support tool for the management.

Finally, Chapter 22 (Bouma and Van der Veen) deals with the difficulty of evaluating the effectiveness of EMA. One of the barriers in determining EMA's effectiveness is the difference in objectives that could be achieved by using EMA (what is the yardstick to measure effectiveness?). Another barrier discussed by the authors is the lack of an EMA theory and empirical insights in the process of development and adoption of EMA.

## 1.7. Conclusion

EMA as a new subject offers an interesting way of looking at ecological sustainability. It opens up the corporate gateway towards the continuous internalisation of external ecological effects. Until now, the main incentives to develop EMA have not originated in the business community itself, even though there are some interesting examples of companies implementing innovative forms of EMA. This means that there is room for additional governmental programmes that promote the adoption of EMA. However, it is of great importance to actively involve the business community in developing them. In this context, it is important to discuss whether mandatory external reporting of a company's environmental performance can be expected to lead to better – that means, more environmentally benign – internal decisions. Of course, this relates to developments such as the GRI and discussions about sustainable enterprise. In all of this, it is crucial to distinguish between what seems to be wishful and how companies behave in reality. In other words, there is a continuous need for both good policies and good research and a sound vision on how they interrelate.

# PART I

# OVERVIEWS

# 2. An Integrative Framework of Environmental Management Accounting – Consolidating the Different Approaches of EMA into a Common Framework and Terminology

*Roger L. Burritt*
*The Australian National University, Australia; E-mail: Roger.Burritt@anu.ed.au*

*Tobias Hahn*
*University of Lueneburg, Germany; E-mail: Tobias-Hahn@uni-lueneburg.de*

*Stefan Schaltegger*
*University of Lueneburg, Germany; E-mail: Schaltegger@uni-lueneburg.de*

## 2.1. Environmental Accounting and Environmental Management Accounting

With the emergence of environmental accounting in the last two decades, various perceptions of the concept of environmental accounting have been developed (Gray et al., 1993; Schaltegger and Stinson, 1994; EPA, 1995; Gray et al., 1996; Schaltegger et al., 1996; Schaltegger and Burritt, 2000). Given the variety of stakeholders that require company-related environmental information, numerous approaches and tools have been proposed within the field of environmental accounting in order to meet these information needs. However, there is a wide consensus that there are two main groups of company-related environmental impacts (Schaltegger and Burritt, 2000, p. 58):

- Environmentally induced impacts of companies on their economic systems; and
- Company-related impacts on environmental systems.

Each of these impacts can be reflected by its own category of company-related environmental information.

### 2.1.1. *Monetary and physical information and different addresses*

Environmentally induced impacts on economic systems are reflected through **monetary environmental information**. Monetary environmental information addresses all the impacts that company-related environmental impacts leave on past, present or future financial stocks and flows. Monetary environmental information is expressed in monetary units such as US$ or euros (e.g. expenditure on cleaner production; cost of fines for breaching environmental laws; dollar values of environmental assets).

Induced impacts of corporate activities on environmental systems are reflected in **physical environmental information**. Thus, at the corporate level, physical environmental information includes all past, present and future material and energy amounts that have an impact on ecological systems. Physical environmental information is always expressed in physical units – such as kilograms, cubic meters, or joules (e.g. kilograms of material per customer served; joules of energy used per unit of product). Physical environmental information has also been termed 'ecological information' in the past (see e.g. Schaltegger et al., 1996).



*M. Bennett et al. (eds.), Environmental Management Accounting: Informational and Institutional Developments,* 21–35.
© 2002 *Kluwer Academic Publishers. Printed in the Netherlands.*

In a similar way to the distinction introduced above for environmental information, conventional accounting also provides separate information about monetary and physical aspects of the company's activities. Conventional accounting systems expressed in monetary units include:

- conventional management accounting – designed to satisfy internal needs of corporate decision makers for short term cost and revenue, long term investment information and internal accountability;
- conventional financial accounting – which serves to provide external corporate stakeholders with information about the company's financial position at a specific point in time and changes in its financial position on a regular basis over specified periods of time;
- other accounting systems such as tax or bank regulatory accounting – intended to provide specific information, mostly for regulatory purposes.

Conventional accounting systems with information expressed in physical units include approaches such as production planning systems, inventory accounting systems and quality systems, as well as accounting systems which provide information about the amount of goods and materials being passed between the company and external third parties.

Conventional corporate accounting does not normally give explicit, separate recognition to company-related environmental impacts. Instead, it is designed mainly to satisfy the needs of different stakeholders who seek information about the economic performance of the company. Yet, from a pragmatic perspective, the critical test for any accounting system is whether it produces information that is useful to particular stakeholders for evaluating their own ends (Chambers, 1966, p. 54; Schaltegger and Burritt, 2000, p. 45). Hence, different accounting systems should be designed to satisfy the fact that various addressees require different information. Different conventional accounting systems can be distinguished according to the main target audiences.

Some stakeholders have a major concern with physical environmental impacts of corporate activities, whereas other stakeholders are interested mainly in monetary effects induced by the environmental impacts of the company. For instance, shareholders are primarily interested in the monetary bottom line and may be only partially interested in a separate report containing pollution information expressed in physical units, even if it is put into a clear context with its monetary consequences. Shareholders are interested in pecuniary information that shows material effects on shareholder value, including environmentally induced monetary impacts on the company. Environmental protection agencies on the other hand are interested in various waste and pollution figures expressed in physical units and may not have much direct interest in, for example, whether the costs of pollution abatement or waste reduction measures are capitalized or are considered as expenses in the monetary accounts.

Unfortunately, the conventional approach to accounting tends to neglect the fact that information interests vary very much between different stakeholders. It is, however, common to distinguish between at least two major target stakeholder groups in conventional accounting systems for companies: internal company addressees (e.g. management) and a fairly narrow range of external groups (e.g. shareholders, rating agencies and financial analysts). Internal and external accounting systems can be distinguished, depending on whether the main purpose of the accounting system is to satisfy the information needs of internal or external stakeholders.

Management accounting in general, for example, '. . . is the identification, measurement, accumulation, analysis, preparation, and interpretation of information that assists executives in fulfilling organizational objectives' (Horngren and Foster, 1987, p. 2) and thus focuses on internal accounting and reporting. On the other hand, financial accounting and reporting is the branch of accounting concerned with the classifying, measuring and recording of transactions, transformations and events external to the company for the purpose of external reporting. Financial accounting is concerned mainly with the provision of information to external company stakeholders, especially investors and shareholders. Other accounting systems such as regulatory accounting and reporting systems are dedicated to more specific audiences external to the company, such as tax agencies or other regulatory bodies. Apart from the distinction between monetary and physical information, this leads to a further possible criterion for structuring environmental accounting – internal vs. external stakeholders.

The main difference between conventional accounting and environmental accounting is that environmental accounting systems distinctly take into account environmental impacts related to company activities. Within the conventional approach this distinction between and environmental accounting is somewhat unclear. Figure 2.1 illustrates the scope and the limits of environmental accounting on the basis of a lowest common denominator. There is consensus that environmental accounting systems take company-related environmental aspects explicitly into account, expressed in both monetary and physical units. Such information can be provided for internal or for external addressees. Yet, differences in the units of measurement, in the data quality and its sources, as well as between the regulatory and very different market requirements, cannot simply be neglected if purpose-orientated information is to be provided. Thus, further differentiation of the concept of environmental accounting, as well as of the delineation from conventional accounting, is indispensable.

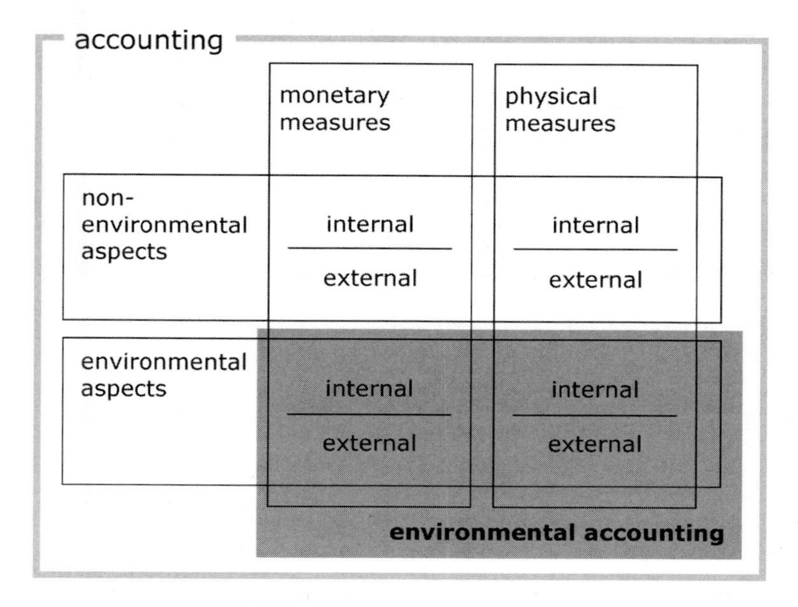


Figure 2.1. Fundamental scope and delineation of environmental accounting

## 2.1.2. *Environmental Accounting in monetary and physical units for different target audiences – Conceptual frameworks*

As with conventional accounting systems, both of the different classification criteria introduced above – accounting in monetary units vs. physical units, and internal vs. external accounting – also hold true for environmental accounting systems.

### *Early Framework of Environmental Accounting*

In an earlier work Schaltegger et al. (1996) introduced a framework of environmental accounting that emphasized very distinctly the separation of environmentally related economic aspects of corporate activity from company-related impacts on the environment. Accounting systems reflecting the former – expressed in monetary units – with their tight methodological linkages to conventional accounting systems, were labeled 'environmentally differentiated conventional accounting' (area shaded in dark grey in Figure 2.2). Being part of conventional accounting, they measure the environmentally induced impacts on the company in monetary terms. The remainder of the conventional accounting category, which does not address environmental issues, is shown in white. On the other side, accounting systems which refer to the physical impacts of a company on the environ-

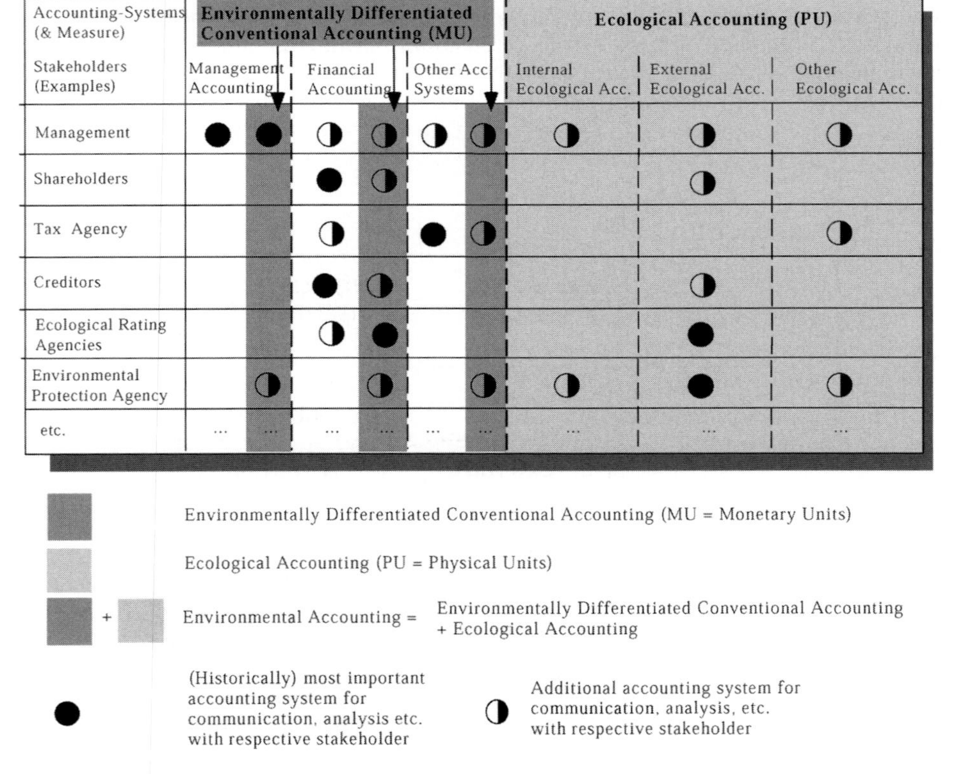


Figure 2.2. Early framework of environmental accounting (similar to Schaltegger et al., 1996; Schaltegger and Burritt, 2000, 58).

ment were considered under the umbrella of 'ecological accounting' (light grey area in Figure 2.2). Together, these two parts form corporate environmental accounting. Both environmentally differentiated conventional accounting and ecological accounting could further be distinguished according to their internal or external addressee focus and be assigned to different stakeholders' information requirements.

The strengths of this early framework are, *firstly*, in the clear distinction between environmentally induced economic impacts expressed in monetary units and the physical impacts of corporate activities on the natural environment. This distinction is important because physical environmental and monetary environmental information are often derived from different sources. In addition, physical environmental information has different measures of quality and quantity (e.g. kilograms) from monetary environmental information (e.g. monetary value added). Both kinds of information are often required for different purposes and by different stakeholders. All of these aspects make a clear distinction necessary between the accounting systems that deal with the two kinds of information.

*Secondly*, this framework underlines the methodological origin of environmental accounting systems dealing with environmentally related monetary impacts from conventional accounting.

However, there are two main shortcomings of this framework:

*Firstly*, conventional accounting – and within this its integral internal part, management accounting – is considered to deal with only the monetary aspects of running a company. As argued above, conventional accounting – as with environmental accounting – includes monetary *and* physical aspects. Production planning systems or stock accounting systems for example do not explicitly refer to environmental aspects but at the same time provide information expressed in physical units which is relevant for financial considerations. Thus, for the sake of consistency and completeness, there is a need to distinguish between monetary and physical aspects in any accounting system independently of whether they consider environmental aspects (see Figure 2.1).

This leads directly to a further requirement that any framework of environmental accounting should meet. Many aspects measured in physical units form the basis of monetary or monetarized measures. Thus, it is crucial to *integrate* physical and monetary issues and to combine them in one category, while maintaining the conceptual distinction between the two categories of corporate environmental aspects. Environmentally induced monetary impacts on a company are strongly interrelated with the environmental performance of a firm measured in physical units. A framework of environmental performance should therefore provide both a clear distinction between monetary and physical corporate environmental aspects, on the one hand, *and* an overarching structure which relates both aspects of environmental accounting.

The *second* shortcoming of the early framework is the quite unrelated and independent terminology of the two components that make up environmental accounting. In addition, from a semantic point of view both terms are often not understood or are confused. 'Ecological accounting' is to this extent somewhat cumbersome, as there is no real ecological notion contained in the conventional accounting tradition. That is to say, the term suggests a genuine basic ecological component instead of referring to physical corporate environmental aspects. In reality, the focus of accounting systems addressing with these aspects remains mainly within the corporate viewpoint rather than adopting an ecological view based on the natural environment and associated ecosystems. Accounting for environmentally induced monetary impacts on companies is referred to as 'environmentally differentiated conventional accounting' – thus it is seen as an integral part of conventional accounting. The question, however, is, what, apart from the methodological proximity with conventional accounting approaches, justifies considering these environmentally differentiated accounting systems to be thought of as 'conventional'. The current state of their use and acceptance throughout mainstream business practice does not justify the use of the term 'conventional' in the sense of being well established, i.e. representing a common convention.

### *Integrative framework of Environmental Accounting*

The effort and challenge for a more integrative framework of environmental accounting is to overcome these deficits, while maintaining the strengths of the early framework discussed above. Such an integrative framework has to be based on the common notion that corporate environmental accounting consists of monetary and physical environmental aspects of corporate activities and addresses both internal and external information needs using distinct accounting approaches (see Figure 2.1). While keeping these crucial distinctions, an integrative framework must provide an overarching structure to relate the different parts of environmental accounting and it should provide an intuitive and precise terminology.

According to the different types of information related to corporate environmental aspects, environmental accounting is considered to consist of monetary environmental accounting and physical environmental accounting. **Monetary environmental accounting** systems measure the environmentally induced economic impacts of the company in monetary terms. Monetary environmental accounting systems can be considered as a broadening of the scope, or a further development or refinement, of conventional accounting in monetary units as they are based on the methods of conventional accounting systems. **Physical environmental accounting** systems reflect the impacts of company-related activities on the environment. They are designed to satisfy the growing demands of various internal and external stakeholders for information about the company's environmental performance. Taken together, monetary and physical environmental accounting form environmental accounting (see the debate in e.g. Bennett and James, 1998; ECOMAC, 1996; IFAC, 1998; and Schaltegger et al., 2000). These two major parts of environmental accounting are reflected by the two halves (left and right) of Figure 2.3. On the basis of this fundamental understanding of environmental accounting the set of different environmental accounting systems can be positioned within the framework of environmental accounting.

Monetary environmental accounting covers:

- monetary environmental management accounting (MEMA) as internal environmental accounting expressed in monetary units;
- external monetary environmental accounting (EMEA) as external environmental accounting and reporting expressed in monetary units; and
- other monetary environmental accounting, such as environmental tax accounting.

Physical environmental accounting includes:

- physical environmental management accounting (PEMA) as internal environmental accounting expressed in physical units;
- external physical environmental accounting (EPEA) as external environmental accounting in physical units; and
- other physical environmental accounting, such as regulatory environmental accounting in physical units.

Figure 2.3 categorizes these accounting systems within the framework of environmental

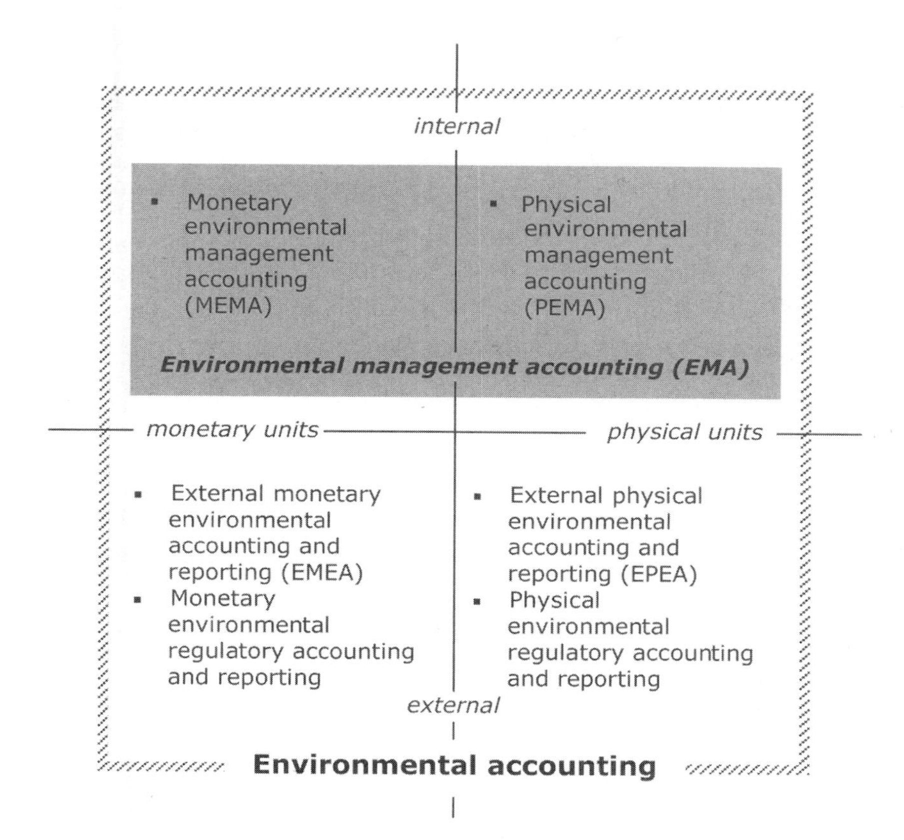


Figure 2.3. Integrative framework of environmental accounting (modified from Bartolomeo et al., 2000, p. 33).

accounting according to the two dimensions of (i) internal vs. external, and (ii) monetary vs. physical.

All the three major corporate environmental accounting systems:

- environmental management accounting,
- external environmental accounting, and
- other environmental accounting,

can be further divided according to their emphasis on monetary or physical aspects. This is also illustrated in Figure 2.2 by the area shaded in grey, which shows the scope of environmental management accounting (EMA).

A range of factors, outlined briefly below, supports the development of this conceptual framework of environmental accounting, including the framework of environmental management accounting.

*First*, given the assumption that the philosophy and tools associated with environmental accounting, including EMA, can assist the drive towards a sustainable society (Schaltegger and Burritt, 2000, p. 46), it is important to create a common understanding in order to facilitate its communication and promotion among managers and other stakeholders.

*Secondly*, a conceptual separation between internal and external accounting is based on the fact that the level of detail and aggregation of information and the extent of confidentiality differ between management and other stakeholder needs. It has also been argued that a separate focus on the accounting needs of management (rather than on the needs of external stakeholders) is to be encouraged because a focus on external reporting can lead to distortions in the collection and use of information for decision-making (Kaplan, 1984; see also Bennett and James, 1999, p. 32).

*Thirdly*, different types of managers rely on and have their performance assessed using either physical, or monetary, or both types of information. For example, managers in the corporate environmental department have various goals including:

- Identifying environmental improvement opportunities;
- Prioritizing environmental actions and measures;
- Environmental differentiation in product pricing, mix and development decisions;
- Transparency about environmentally relevant corporate activities;
- Meeting the claims and information demands of critical environmental stakeholders, to ensure resource provision and access;
- Justifying environmental management division and environmental protection measures.

Different forms of information are required to help environmental managers to meet their goals, including:

- Physical measures of material and energy flows and stocks and related processes and products, and their impacts upon the environment;
- Monetary measures about the economic impact of environmental initiatives (such as pay-back periods, return on capital/investment, etc.);
- Qualitative measures of stakeholder claims.

This contrasts with the needs of, for example, a production manager concerned with task control over operations, optimizing energy and material consumption, and reduction of environmentally induced risks and in need of physical measures of material and energy flows and process records.[1]

The need for *integrating* environmental with economic issues by combining them in one category – environmental accounting – provides a *fourth* driver. A major focus of environmental accounting is to raise management awareness about the potential importance, positive and negative, of environmental impacts on corporate economic performance. Environmentally induced monetary impacts of a company are strongly interrelated with corporate environmental performance measured in physical units. Integration can be typified through, for example, measures of eco-efficiency that combine measures of economic performance with measures of environmental impact in a ratio format (see e.g. Schaltegger and Sturm, 1992 and 1998).

Finally, conventional accounting in physical units, as noted above, exists independently of, and prior to, the development of environmental accounting systems (Horngren and Foster, 1987). Managers have always been concerned to improve materials and energy efficiency in order to improve economic results of their corporations. For example, productivity measures of efficiency, expressed in physical units, have long been derived in most conventional management accounting systems (e.g. material input per unit of

---

[1] See the typical goals of different types of managers outlined in Schaltegger et al. (2001): EMA-Links – The Promotion of Environmental Management Accounting and the Role of Government, Management and Stakeholders.

product). The derivation of physical material and energy flows is necessary information prior to their later expression in monetary units. Consequently, much physical information derived in conventional management accounting systems is of great use in environmental management accounting.

Taking all these factors into account, it is possible to provide greater insight into the development of a general framework for EMA. It is to this main purpose of the paper that attention is now directed.

## 2.2. General Framework for Environmental Management Accounting

Environmental management accounting – basically defined as internal corporate environmental accounting – has witnessed a range of different perceptions and conceptualizations developed in the environmental accounting literature. There have been two basic concepts, providing a narrow and a broader understanding of environmental management accounting (for detailed discussion see Schaltegger et al., 2000).

The narrower conception of environmental management accounting is based on the early framework introduced above (see section 1.2.1). From the perspective of this framework it is logical to see environmental management accounting as environmentally differentiated conventional management accounting, which means that environmental management accounting is an integral part of conventional accounting and deals only with the provision of corporate environmental aspects expressed in monetary terms for various different internal company decision makers. Physical environmental aspects of corporate activities are not covered by environmental management accounting under this narrow setting. Instead, physical aspects are seen as internal ecological accounting in accordance with the framework depicted in Figure 2.2. The general strengths and weaknesses of this framework have been outlined in the previous section. In principle these also hold true for the more specific case of internal environmental accounting. Therefore, in the following section, a detailed outline is provided of the application of the proposed integrative framework of environmental accounting to environmental management accounting.

### 2.2.1. *Monetary Environmental Management Accounting and Physical Environmental Management Accounting*

As already shown above, the strict bifurcation between monetary and physical information types is sometimes but not always a useful distinction in conventional management accounting where both coexist. Standard costing provides a case in point, where variance analysis includes price and quantity variances to be examined in tandem, not just price (monetary) variances for management control purposes. Internal tax planning provides another case where physical emissions of pollution are calculated first by management and then estimates of the cost of a specific environmental tax are made (e.g. a tax on carbon emissions).

However, because of the use of different measures the distinction between monetary and physical information must not be abandoned completely but should continue to be reflected in the accounting systems and in the nomenclature. Hence, it is proposed that EMA be defined as a generic term that includes both Monetary Environmental Management Accounting (MEMA) and Physical Environmental Management Accounting (PEMA). This situation is illustrated in Figure 2.3. The scope of EMA illustrated complements the views of authors who have canvassed the idea that companies should provide greater emphasis on the management and measurement of non-monetary aspects of corporate performance (Johnson and Kaplan, 1987; Kaplan and Norton, 1996) in order to encourage a mind set that takes the long term into account. Conceptions of a broader understanding of environmental management accounting, i.e. covering both monetary and physical environmental aspects for internal decision making have already been provided by other authors (see e.g. Bennett and James, 1998; ECOMAC, 1996; IFAC, 1998; UNDSD, 2000, p. 39). However, they lack an integrative framework and terminology to satisfy the needs for a comprehensive understanding of environmental accounting in general and environmental management accounting in particular.

*Monetary Environmental Management Accounting (MEMA)* deals with environmental aspects of corporate activities expressed in monetary units and generates information for internal management use. In terms of its methods MEMA is based on conventional management accounting that is extended and adapted for environmental aspects of company activities. It deals with the environmentally induced impacts on a company expressed in monetary terms (e.g. costs of fines for breaking environmental laws; investment in capital projects that improve the environment). It is the central, pervasive tool providing, as it does, the basis for most internal management decisions, as well as addressing the issue of how to track, trace, and treat costs and revenues that are incurred because of the company's impact on the environment (Schaltegger and Burritt, 2000, p. 59). MEMA is an accounting system with a focus on the economic impacts of environmentally induced corporate activities. It contributes to strategic and operational planning, provides the main basis for decisions about how to achieve desired goals or targets, and acts as a control and accountability device (Schaltegger and Burritt, 2000, section 6.1).

*Physical Environmental Management Accounting (PEMA)* also serves as an information tool for internal management decisions. However, in contrast with MEMA it focuses on a company's impact on the natural environment, expressed in terms of physical units such as kilograms. PEMA tools are designed to collect environmental impact information in physical units for internal use by management (Schaltegger and Burritt, 2000, pp. 61–63). According to Schaltegger & Burritt (2000, p. 261) PEMA as an internal physical environmental accounting approach serves as:

- an analytical tool designed to detect ecological strengths and weaknesses;
- a decision-support technique concerned with highlighting relative environmental quality;
- a measurement tool that is an integral part of other environmental measures such as eco-efficiency;
- a tool for direct and indirect control of environmental consequences;
- an accountability tool providing a neutral and transparent base for internal and, indirectly, external communication; and
- a tool with a close and complementary fit to the set of tools being developed to help promote ecologically sustainable development.

### 2.2.2. *Time: Frame, length and routineness*

Building on these arguments, which support the notions of MEMA and PEMA as core constructs in EMA, additional dimensions can also be seen as being a necessary, important part of environmental management accounting. In particular, three dimensions of environmental management accounting tools are emphasized below:

- time frame – the time frame being addressed by different tools (e.g. past vs. current vs. future time frames);

– length of time frame – how long is the time frame being addressed by the tool (e.g. tools addressing the short term vs. those with a focus on the long term); and
– routineness of information – how routinely is information gathered (e.g. ad hoc vs. routine gathering of information).

Figure 2.4 includes all of the five dimensions – internal vs. external; physical vs. monetary classifications, past and future time frames, short and long terms, and ad hoc vs. routine information gathering – in the proposed framework for EMA. Any specific EMA accounting tool can be assigned on the basis of the classification scheme drawn up by these five dimensions (see the detail in Figure 2.4 and Schaltegger and Burritt, 2000, Chapter 6 for a detailed description of different EMA tools).

| | | **Environmental Management Accounting (EMA)** | | | |
|---|---|---|---|---|---|
| | | Monetary Environmental Management Accounting (MEMA) | | Physical Environmental Management Accounting (PEMA) | |
| | | **Short-Term Focus** | **Long-Term Focus** | **Short-Term Focus** | **Long-Term Focus** |
| **Past Oriented** | Routinely generated information | Environmental cost accounting (e.g. variable costing, absorption costing, and activity based costing) | Environmentally induced capital expenditure and revenues | Material and energy flow accounting (short term impacts on the environment – product, site, division and company levels) | Environmental (or natural) capital impact accounting |
| **Past Oriented** | Ad hoc information | Ex post assessment of relevant environmental costing decisions | Environmental life cycle (and target) costing<br><br>Post investment assessment of individual projects | Ex post assessment of short term environmental impacts (e.g. of a site or product) | Life cycle inventories<br><br>Post investment assessment of physical environmental investment appraisal |
| **Future Oriented** | Routinely generated information | Monetary environmental operational budgeting (flows)<br><br>Monetary environmental capital budgeting (stocks) | Environmental long term financial planning | Physical environmental budgeting (flows and stocks) (e.g. material and energy flow activity based budgeting) | Long term physical environmental planning |
| **Future Oriented** | Ad hoc information | Relevant environmental costing (e.g. special orders, product mix with capacity constraint) | Monetary environmental project investment appraisal<br><br>Environmental life cycle budgeting and target pricing | Relevant environmental impacts (e.g. given short run constraints on activities) | Physical environmental investment appraisal<br><br>Life cycle analysis of specific project |

Figure 2.4. Proposed Framework of Environmental Management Accounting (EMA) (according to Schaltegger, Hahn and Burritt, 2000).

### *Time frame*

Accounting systems and associated tools of analysis, used to attach meaning to the signals produced by accounting tools, can be classified into those with a *focus* on the past, and those looking to the *future*. Rows headed 'past orientated' and 'future orientated', in Figure 2.4, distinguish between the MEMA and PEMA tools that are available to management for addressing environmental issues with a focus either on measurement of past transactions, transformations or events or the prediction of the impact of possible future transactions, transformations or events. For example, environmental cost accounting (in the top box in the third column) provides routinely generated short term information about the past environmental monetary impacts of activities, products, divisions, departments and the total economic entity, whereas monetary environmental operating budgeting (the third box down in the third column) projects this information into the short term future for planning and control purposes.

### *Length of time frame*

Environmental issues are generally considered to be *long term*; while management is frequently criticized for adopting a *short term* perspective, to appease the financial markets and one group of stakeholders in particular – shareholders. Columns headed 'short-term focus' and 'long-term focus', in Figure 2.4, distinguish between the MEMA and PEMA tools that are available to management for addressing environmental issues with either a short or long-term focus. The length of time frame associated with the discretion available to different levels of management has been highlighted by the need to emphasize length of planning periods, e.g. short run operational budgeting expressed in monetary terms (the third box down in the third column) vs long run financial planning (the third box down in the fourth column), and the span of control over physical actions, e.g. short span over tactical operational decisions in physical environmental budgeting (the third box down in the fifth column) vs a long span over strategic situations involving long term physical environmental planning (the third box down in the sixth column).

### *Routineness of information gathering*

From the viewpoint of internal management decision making and internal accountability, both past and future oriented approaches can be further distinguished into routinely generated information – general accounting systems that *routinely produce information* for management – and ad hoc information – specific accounting tools that produce information on a 'needs' basis for particular decisions. Rows headed 'ad hoc information' and 'routinely generated information', in Figure 2.4, distinguish between the MEMA and PEMA tools that are available to management for addressing environmental issues on a regular or irregular basis. For example, the PEMA tool 'environmental capital impact accounting' (the first box down in the final column) provides regular information about corporate impacts on natural capital (e.g. whether critical and non-critical environmental capital has been maintained, improved, or depleted), whereas PEMA information about 'life cycle inventories' (the second box down in the final column) is required only on an ad hoc basis for the purpose of conducting life cycle assessment of new products.

By combining all of these analytical factors this paper suggests a comprehensive conceptual framework for EMA within which the different tools of internal environmental accounting, MEMA as well as PEMA tools, can be placed and assigned according to the decision or internal accountability setting. Figure 2.4 shows that EMA encompasses a large range of different accounting approaches that serve different needs which depend on the decision context, purpose and management level. While detailed information about the EMA tools mentioned in the cells in Figure 2.4 are further discussed in the standard environmental accounting literature (see Schaltegger and Burritt, 2000), the question of choice of the most important EMA tools does need further consideration and is examined in the next section.

## 2.3. Choice of the most important EMA tools

To illustrate the benefits stemming from the general framework for EMA outlined above it is useful to outline the decision making and accountability context of some of the tools illustrated in Figure 2.4.

As shown in Figure 2.5 below, top management tends to be concerned with strategic, long term accounting information used to plan and control activities at the corporate level. Columns four and six display EMA tools that have a long-term focus and which may be of particular use to *top management* (e.g. when there is imposition of a carbon tax on an organization, or the introduction of a carbon trading scheme). Because top management need aggregate information, they look for measures that can be used to compare a range of diverse corporate activities. Hence they have a preference for monetary information that uses a common unit of account and facilitates comparison between different course of action. Hence, their emphasis is likely to be on MEMA tools that affect strategic decisions relating to monetary capital on a regular basis for the organization, as well as on an ad hoc basis for appraising the performance of individual projects with environmental impacts involving large amounts of monetary capital (column four boxes one and two). Top management are also responsible for steering their organization into the future

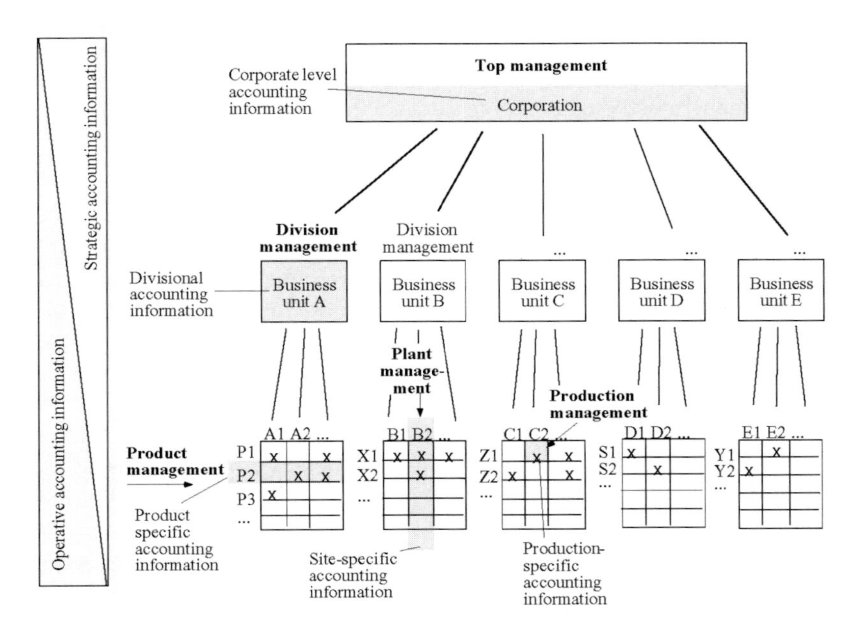


Figure 2.5. Accounting information relevant to different management (Schaltegger et al., 1996, p. 139; Schaltegger and Burritt, 2000, p. 92).

and would find routine long term MEMA planning tools of use, e.g. related to environmentally driven research and development plans for the company, and ad hoc monetary environmental investment appraisal tools such as NPV using growth options for large single investments where environmental considerations play a key role (e.g. the decision by producers of halons to cease production and to introduce substitutes because of the deleterious effects of halons on the ozone layer).

In contrast, *production managers* need production-specific accounting information. Such information, related to production activities in the value chain (Porter, 1980) has a tendency to be expressed in physical terms because production managers plan and control physical rather than monetary processes. Production management will tend towards the use of PEMA tools, especially short term PEMA tools because of their concern to keep production flowing, and to improve the technical efficiency with which production is carried out. Hence, use of materials and energy flow accounting information will be a routine requirement, relating to the past for control purposes (box one in column five) and projected through physical environmental budgeting (box three in column five) for production scheduling plans.

*Divisional management* are accountable to top management for their monetary performance and the performance of their divisions. Accountability implies the feedback of cost and revenue MEMA results about key divisional performance measures. The emphasis is likely to be on the type of short term routine EMA information represented in box one, column three of Figure 2.4.

To give another example, *product management* is mostly concerned with product specific information. Such information has to be expressed both in monetary and in physical units because decisions related to both pricing and environmental quality have to be made. Thus the ad hoc MEMA tool environmental life cycle (and target) costing represented in the second box down in column four of Figure 2.4 is of particular interest to product managers. However, they may also be interested in routinely generated environmental cost information, especially about material and energy flows. The most important PEMA tool for product managers is the past oriented ad hoc life cycle inventory information which covers all the physical environmental impacts of a product over all the stages of product life (see second box down in column six of Figure 2.4). In addition, product management might seek physical information on material and energy flows.

Further exploration of the full range of tools used for management decision making and accountability by different types of management, and in different organizations (e.g. manufacturing, service, knowledge, non-profit and government, small companies and companies in developing countries) demonstrates the effectiveness of the general EMA framework that has been developed.

## *2.4 Outlook*

At present there is still no precision in the terminology associated with EMA. Drawing upon the existing literature it has been argued above that there is scope for deriving an agreed, pragmatic general framework for EMA. Such an opportunity depends on the recognition of:

- monetary and physical accounting systems that, both separately and in combination, are of use to different types of managers in seeking to reduce environmental impacts from the activities of their organizations;
- a mapping of the tools available for EMA related to the time frame of impacts (impacts in the past, contemporary impacts, impacts in the future);

- a mapping of the tools available for EMA with the length of time frames used by different managers for analysis (the length of the time frame – short or long term; and
- a mapping of EMA information needs with the routineness of decisions and accountability processes faced by different managers.

Among the main advantages of the proposed new framework for EMA are:

- the movement towards a closure of the debate about what EMA is, or what it might be, is necessary for effective communication and research between academics as well as for the promotion and establishment of modern EMA approaches in practice;
- the recognition that EMA needs to include monetary and physical measures, albeit in systems that can be considered independently of each other, or in combination;
- the mapping of tools with EMA sub-systems that facilitate particular types of decisions and internal accountability processes; and
- the incorporation of time as a key element in the classification, in order to bring stronger focus on the links between short term and long term monetary considerations and short and long term ecological considerations in management decision making.

Furthermore, increased emphasis upon EMA systems that are largely required by management in need of information to help them achieve the goals of their various organizational segments (e.g. divisions or departments), may help to reduce the emphasis on manipulation of public environmental disclosures for political purposes which is evident in environmental financial accounting (see e.g. Gray et al., 1993) while, at the same time, stressing the need for improved environmental performance expressed in both physical and, where appropriate, monetary terms.

Finally, this development of a general framework of EMA is offered as a way forward for management seeking to adopt environmental management accounting systems. A major benefit that corporate managers will experience from the proposed general framework is that the framework considerably clarifies the concept and applicability of EMA and related tools. Once managers have a clear picture of the classification of MEMA and PEMA tools, promotion and adoption will be easier for them and therefore it will be more likely that they will adopt the appropriate tools in a particular decision making or internal accountability setting in which environmental aspects play a part, such as:

- the extent of subsidies from government that are environmentally damaging and which may be removed in the future;
- potential corporate impacts of environmental taxes and tightening regulations designed to bring corporations closer to tracking the full cost of their activities;
- divisional impacts on environmental capital such as biodiversity, land, water and air quality;
- corporate impacts on the goal of sustainable society; and
- product and production managers taking green opportunities when these are available.

# 3. Environmental Management Accounting Metrics: Procedures and Principles[1]

*Christine Jasch*
*Director of the Institute for Environmental Management and Economics, Vienna, Austria;*
*E-mail: info@ioew.at*

## 3.1. Introduction

This paper aims to define principles and procedures for environmental management accounting (EMA). It focuses on techniques for quantifying environmental expenditures or costs. These EMA metrics can be used by national governments as a starting point for developing EMA guidelines that are adapted to national circumstances and needs, and by companies and other organisations seeking to improve their control and benchmarking systems.

## 3.2. EMA metrics

### *Physical and financial*

EMA metrics include both **physical metrics** (materials and energy consumption, flows, and final disposal) and **financial metrics** (costs, savings, and revenues related to activities with a potential environmental impact).

Conventional **corporate monetary accounting** comprises:

- financial accounting (bookkeeping, balancing, consolidation, auditing of the financial statement and external reporting);
- management accounting (also called 'cost accounting');
- corporate statistics and indicators (past-oriented);
- budgeting (future-oriented);
- investment appraisal (future-oriented).

### *Conventional accounting systems*

**Management accounting** constitutes the central tool for **internal** management decisions such as product pricing, and is not regulated by law. This internal information system deals with questions such as 'what are the costs of production for different products, and what should be their selling prices?'. The main stakeholders in management and cost accounting are members of the managerial staff of the organisation, in various positions (e.g. executive, site, product and production managers).

Unfortunately, many companies do not have a separate cost accounting system, and therefore make calculations on the basis of the financial accounting data from bookkeeping. **Financial accounting** is designed mainly to satisfy the information needs of

[1] This chapter results from research done for the UN. Source: Environmental Management Accounting: Procedures and Principles, United Nations Division for Sustainable Development, Department of Economic and Social Affairs (United Nations publication, Sales No. 01.II.A.3). The website where this publication and others will be available is http://www.un.org/esa/sustdev/estema1.htm.



*M. Bennett et al. (eds.), Environmental Management Accounting: Informational and Institutional Developments,* 37–50.
© 2002 *Kluwer Academic Publishers. Printed in the Netherlands.*

**external** shareholders and financial authorities, both of whom have a strong interest in standardised, comparable data and in receiving true and fair information about the actual economic performance of the company. Therefore, financial accounting and reporting are dealt with in national laws and international accounting standards.

### *Environmental management information*

The core of **environmental information systems** consists of **materials flow (or mass) balances** measured in physical units of material, water and energy flows within the boundaries of a well-defined system. This can be at corporate level, but also at the level of cost centres and production processes or even of individual items of machinery and products. In the latter case, process technicians have to trace the data needed (see Figure 3.1).

| INPUT | | System boundaries | OUTPUT | |
|---|---|---|---|---|
| | | Nations | | |
| Materials | ⇒ | Regions | ⇒ | Products |
| Energy | ⇒ | Corporations | ⇒ | Waste |
| Water | ⇒ | Processes | ⇒ | Emissions |
| | | Products | | |

Figure 3.1. System boundaries for mass balances.

EMA derives its data from both financial accounting and cost accounting, and is instrumental in increasing materials efficiency, in reducing environmental impact and risk, and in reducing the costs of environmental protection. EMA uses both financial and physical data (see Figure 3.2).

| Accounting in Monetary Units | | Accounting in Physical Units | |
|---|---|---|---|
| Conventional Accounting | **Environmental management accounting** | | Other Assessment Tools |
| | **Financial EMA ('FEMA')** | **Physical EMA ('PEMA')** | |

Figure 3.2. EMA combines financial and physical data.

### *EMA applications*

Key application fields for the use of EMA data are:

- assessment of annual environmental costs/ expenditures;
- product pricing;
- budgeting;
- investment appraisal, calculating investment options;
- calculating costs and savings of environmental projects;
- design and implementation of environmental management systems;

- environmental performance evaluation, indicators and benchmarking;
- setting quantified performance targets;
- cleaner production and eco-design projects;
- external disclosure of environmental expenditures, investments and liabilities;
- external environmental or sustainability reporting;
- other reporting, such as the reporting of environmental data to statistical agencies and local authorities.

EMA data and its application can be divided into past-oriented and future-oriented tools (see Figure 3.3).

| EMA Environmental Management Accounting | | | |
|---|---|---|---|
| Monetary EMA ('MEMA') | | Physical EMA ('PEMA') | |
| Past-oriented tools | Future-oriented tools | Past-oriented tools | Future-oriented tools |
| Annual environmental expenditure or costs, tracing from book keeping and cost accounting | Monetary environmental budgeting and investment appraisal | Materials, energy and water flow balances | Physical environmental budgeting and investment appraisal |
| | Calculating costs, savings and benefits of projects | Environmental performance evaluation and indicators, benchmarking | Setting quantified performance targets |
| External disclosure of environmental expenditures, investments and liabilities | | External environmental reporting<br><br>Other reporting to agencies and authorities | Design and implementation of environmental management systems, cleaner production, pollution prevention, design for environment, supply chain management, etc. |

Figure 3.3. Past and future oriented EMA tools.

Figure 3.4 provides an overview of the different types of EMA.

### *Defining environmental costs*

The main problem in environmental management accounting is the lack of a standard **definition of environmental costs**. Depending on various interests, these can include a variety of costs, e.g. disposal costs or investment costs and, sometimes, also external costs (i.e. costs incurred outside the company and borne by others). Of course, also the

| **Accounting in Monetary Units** | | **Accounting in Physical Units** | |
| --- | --- | --- | --- |
| **Conventional Accounting** | **Environmental Management Accounting** | | **Other Assessment Tools** |
| | **Financial EMA (FEMA)** | **Physical EMA (PEMA)** | |
| ***Data at Corporate Level*** | | | |
| Conventional book-keeping | **Tracing of environmental costs from book-keeping and cost accounting** | **Materials flow balances on corporate level for mass, energy and water flows** | Production planning systems, stock accounting systems |
| ***Data at Process/Cost Centre and Product/Cost Carrier level*** | | | |
| Cost accounting | **Activity based costing, materials flow cost accounting** | **Materials flow balances at process and product level** | Other environmental assessments, measures and evaluation tools |
| ***Business application*** | | | |
| Internal use for statistics, indicators, calculating savings, budgeting and investment appraisal | **Internal use for statistics, indicators, calculating savings, budgeting and investment appraisal of environmental costs** | **Internal use for environmental management systems and performance evaluation, benchmarking** | Other internal use for cleaner production projects and eco-design |
| External financial reporting | External disclosure of environmental expenditures, investments and liabilities | External reporting (EMAS statement, corporate environmental report, sustainability report) | Other external reporting to statistical agencies, local governments, etc. |
| ***National application*** | | | |
| National income accounting by statistical agency | National accounting on investments and annual environmental costs of industry, Externalities costing | National resource accounting (mass balances for countries, regions and sectors) | |

Figure 3.4. What is EMA?

opposite occurs: benefits such as environmental cost savings may remain hidden. Moreover, many environmental costs that a company pays for are difficult to trace as they are not systematically allocated to appropriate processes and products, but simply allotted to general overheads.

### *Distorted calculations*

The fact that environmental costs are not recorded, or only poorly, often distorts calculations that could underpin improvement options. This, for instance, may mean that possible beneficial environmental projects that prevent emissions and waste at source (avoidance option) are not identified and therefore not implemented. The economic and ecological advantages that could be derived from such measures are foregone. Without the correct figures, those in charge often fail to realise that the costs related to producing waste and emissions can often be higher than the costs of disposing them.

### *Access to information*

Experience shows that environmental managers frequently do not have access to the actual cost accounting figures of the company. On the other hand however, accountants and financial controllers, who in principle have the information, are unable to distinguish which costs are environmentally relevant without further guidance. As well as the inadequacy of the figures, environmental staff and accountants/financial controllers tend to live in different worlds and find it hard to communicate with each other for want of a common language. EMA is intended to help to bridge this gap.

### *Hidden costs*

In conventional cost accounting, environmental and non-environmental costs are usually recorded together without differentiation in general overhead accounts, so that many environmental costs tend to remain 'hidden' from management. There is substantial evidence that management often tends to underestimate the extent and growth of such costs. By identifying, assessing and allocating environmental costs, EMA allows management to identify opportunities for cost savings. A prime example is the savings that can result from replacement of toxic organic solvents by non-toxic substitutes, thus eliminating the high and growing costs of regulatory reporting, hazardous waste handling and other costs associated with the use of toxic materials. Many other examples refer to more efficient use of materials, highlighting the fact that waste is expensive not only because of the costs of its disposal but also because of the purchase value of the wasted materials.

### *Cost internalisation*

**Environmental costs** comprise both internal and external costs and relate to all costs incurred because of environmental damage and protection. **Environmental protection costs** include costs of prevention, disposal, planning, control, shifting actions and damage repair (VDI, 2000)[2] as they occur in different types of organisation. This paper deals only

---

[2] VDI, the German Association of Technicians, together with German Industry representatives, have developed a guidance document on the definition of environmental protection costs and other terms of pollution prevention, VDI 2000.

with corporate environmental costs, and does not attempt to consider external costs, which result from corporate activities but are not internalised via regulations and prices. It is the role of governments to apply political instruments such as eco-taxes and emissions control regulations in order to enforce the 'polluter-pays' principle and thus integrate external costs into corporate calculations. When this happens, external costs are transformed into internal costs, i.e. they are 'internalised'.

## 3.3. How to define corporate environmental costs

What then are corporate environmental costs? Costs incurred to deal with contaminated sites, effluent control technologies and waste disposal may come first to mind.

**Measures for environmental protection** comprise all activities taken for legal compliance, compliance with own commitments, or voluntarily. The criterion is not economic effects, but the effect on the prevention or reduction of environmental impacts (VDI, 2000).

**Corporate environmental protection expenditure** includes all expenditure on measures for environmental protection of a company or on its behalf to prevent, reduce, control and document environmental aspects, impacts and hazards, as well as disposal, treatment, sanitation and clean up expenditure. The amount of corporate environmental protection expenditure is not directly related to the environmental performance of a company (VDI, 2000).

For calculating a company's internal environmental costs, one should look not only at expenditures on environmental protection. The concept of '**waste**' has a double meaning. Waste is a material which has been purchased and paid for, but which has not been converted into a marketable product. Waste therefore signals inefficiency. The costs of wasted materials, capital and labour have to be totalled in order to arrive at total corporate environmental costs as a sound basis for further calculations and decisions. 'Waste' is used in this context as a general term for solid wastes, waste water, and air emissions, and thus comprises all **non-product output**.

| | Environmental Protection Costs<br>(Emissions, Treatment and Pollution Prevention) |
|---|---|
| + | Costs of wasted materials (including water and energy) |
| + | Costs of wasted capital and labour |
| = | Total corporate environmental costs |

Figure 3.5. Total corporate environmental costs.

Adding the purchase value of non-materials output such as solid wastes and wastewater to the environmental costs increases the proportion of 'environmental' costs in relation to other costs.

### *EMA as an eye-opener*

The most important role of EMA is to make sure that all **relevant costs** are considered when making business decisions. In other words, 'environmental' costs are just a subset of the bigger cost universe that corporate decision-makers need to take into account.

'Environmental' costs are part of a company's usual materials and money flows. EMA can be an eye-opener to those in charge of accounting and management decisions in stimulating the company's processes to be viewed in a different way, to reflect new priorities. EMA's focus on materials flows is no longer intended to assess the total 'environmental' costs, but to develop a different look at the **production costs** that takes a company's environmental effects seriously.

### *Annual corporate figures as first step*

In the methodology adopted in this paper, the environmental cost assessment scheme is first used for the assessment of annual corporate environmental expenditure of the previous year, which can then be broken down to cost centres and processes. Total annual figures have limited value as such, but are a first step in a top-down approach of environmental cost management. Annual expenses are the best available data source; a further distinction into cost centres, processes, products and materials flow balances should then be done in a step-by-step procedure, gradually improving the information system. Calculating savings, investment options or estimating future price changes requires the consideration of future costs, and is dealt with separately.

### *From treatment to prevention*

The environmental cost categories that a company uses reflect the past process of awareness in this area.

The first block of environmental cost categories comprises conventional **waste disposal and emission treatment costs**, including related labour and maintenance materials costs. Insurance and provisions for environmental liabilities also reflect the spirit of treatment rather than prevention. This first block corresponds to the conventional definition of environmental costs and consists of all treatment, disposal and clean-up costs of existing wastes and emissions.

The second block is termed **prevention and environmental management** and adds labour costs and external services for good housekeeping as well as the 'environmental' share and extra costs of integrated technologies and green purchases, if significant. The main focus of the second block is on annual costs for prevention of waste and emissions, but without calculated cost savings. They include higher pro-rata costs for environment-friendly auxiliary and operating materials, low-emission process technologies and the development of environmentally benign products.

Conventionally, three production factors are distinguished: materials, capital (investments, related annual depreciation and financing cost) and labour. The next two blocks consider the costs of wasted materials, capital and labour due to inefficient production, generating waste and emissions.

In the third block, the **wasted materials purchase value** is added. All non-product output is assessed by a materials flow balance. Wasted materials are evaluated with their materials purchase value or materials consumed value in the case of stock management.

Lastly, the **production costs of non-product output** are added, including the respective production cost charges, which include labour hours, depreciation of machinery, and operating materials. In activity-based costing and flow cost accounting, the flows of residual materials are more precisely determined and allocated to cost centres and cost carriers.

**Environmental revenues** derived from sales of waste, grants and subsidies are accounted for in a separate block.

Costs that are incurred outside the company and borne by the general public (external costs) or that are relevant to suppliers and consumers (life cycle costs) are not dealt with.

Figure 3.6 shows the environmental cost assessment scheme developed for EMA.

### *Non-product output*

The basis of environmental performance improvements and for assessing the amounts and costs of non-product output (NPO) is the recording of materials flows in kilograms by an **Input-Output Analysis**. The system boundaries can be at the corporate level, or broken down further to sites, cost centres, processes and product levels. The materials flow balance is an equation based on 'what comes in must go out (or be stored)'. In a materials flow balance, information on both the materials used and the resulting amounts of product, waste and emissions are stated. All items are measured in physical units in terms of mass (kg, t) or energy (MJ, kWh). The purchased input is cross-checked with the amounts produced and sold as well as the resulting waste and emissions. The goal is to improve the efficient use of materials, from both an economic and an environmental point of view.

A materials flow balance can be made for a few selected materials or processes or for all materials and wastes of an organisation. The aim of process balances is to track materials on their way through the company. The starting point is often the corporate level, as much information is available on this system boundary, and this is also the level which is used for disclosure in corporate environmental reports.

During its first environmental review, most companies draw up a materials flow balance without going into much detail. This provides a basis on which to gain knowledge on where to achieve improvements in performance and information gathering. By improving the quality of the information and the information systems, a regular monitoring system can be established. This monitoring system shows resource input and production and waste output on a monthly basis. As a next step, the materials flows can be subdivided further according to processes and cost centres, and can also be subject to monetary evaluation.

### *Input-output balance*

Figure 3.7 shows the generally applicable structure of the input-output balance at a corporate level, which could also be used for environmental reporting. Specific subcategories will be needed for different sectors, but it should always be possible to aggregate in a standardised manner, in order to be able to compare them.

The input-output balance at corporate level is drawn up on an annual or a monthly basis and is linked to the bookkeeping, cost accounting, storage and purchase systems. All materials flows should be listed with their values and amounts per year. The assessment scheme for the materials flow balance should therefore record the amounts in kilograms, the values and the corresponding accounts. In addition, it should indicate whether materials are registered by materials stock number, and whether there is inventory management. It should also indicate whether there is consumption-based stock withdrawal according to cost centres. As the first step in setting up the materials input-output statement at corporate level, quantitative data are collected from the accounting and stock-keeping systems. The accounting system offers annual data on a company's inputs, and on the output in so far as it involves sales. All materials purchased will eventually leave the company as either product or waste; in between, they can also be stored. An annual input-output balance has to reflect these three possibilities.

| Environmental media / Environmental cost/expenditure categories | Air + Climate | Waste Water | Waste | Soil + Ground Water | Noise + Vibration | Biodiversity + Landscape | Radiation | Other | Total |
|---|---|---|---|---|---|---|---|---|---|
| **1. Waste and Emission treatment** | | | | | | | | | |
| 1.1. Depreciation for related equipment | | | | | | | | | |
| 1.2. Maintenance and operating materials and services | | | | | | | | | |
| 1.3. Related personnel costs | | | | | | | | | |
| 1.4. Fees, taxes, charges | | | | | | | | | |
| 1.5. Fines and penalties | | | | | | | | | |
| 1.6. Insurance for environmental liabilities | | | | | | | | | |
| 1.7. Provisions for clean up costs, remediation | | | | | | | | | |
| **2. Prevention and environmental management** | | | | | | | | | |
| 2.1. External services for environmental management | | | | | | | | | |
| 2.2. Personnel for general environmental management activities | | | | | | | | | |
| 2.3. Research and Development | | | | | | | | | |
| 2.4. Extra expenditure for cleaner technologies | | | | | | | | | |
| 2.5. Other environmental management costs | | | | | | | | | |
| **3. Materials Purchase Value of non-product output** | | | | | | | | | |
| 3.1. Raw materials | | | | | | | | | |
| 3.2. Packaging | | | | | | | | | |
| 3.3. Auxiliary materials | | | | | | | | | |
| 3.4. Operating materials | | | | | | | | | |
| 3.5. Energy | | | | | | | | | |
| 3.6. Water | | | | | | | | | |
| **4. Processing Costs of non-product output** | | | | | | | | | |
| **Σ Environmental Expenditure** | | | | | | | | | |
| **5. Environmental Revenues** | | | | | | | | | |
| 5.1. Subsidies, grants | | | | | | | | | |
| 5.2. Other earnings | | | | | | | | | |
| **Σ Environmental Revenues** | | | | | | | | | |

Figure 3.6. Environmental cost assessment scheme.

| INPUT in kg/kWh | **OUTPUT in kg** |
|---|---|
| **Raw materials** | **Product** |
| **Auxiliary materials** | Main Product |
| **Packaging** | By Products |
| **Operating materials** | **Waste** |
| **Merchandise** | Municipal waste |
| **Energy** | Recycled waste |
| Gas | Hazardous waste |
| Coal | **Waste Water** |
| Fuel Oil | Total volume |
| Other Fuels | Heavy metals |
| District heat | COD |
| Renewables (Biomass, Wood) | BOD |
| Solar, Wind, Water | **Air-Emissions** |
| Externally produced electricity | CO2 |
| Internally produced electricity | CO |

Figure 3.7. General Input/Output chart of accounts.

### *Gradual process*

Figure 3.8 shows the assessment scheme for the materials input-output statement. √ indicates the likely data source or the records which are likely to be of relevance. It serves the purpose of gradually improving the recording of materials flows. Given its complexity, the work to accomplish this is unlikely to be completed within a single year, so the method presented here is geared at step-by-step tracking and tracing the materials as completely and consistently as possible, in storage administration, cost centres and in production planning.

The next step, after environmental cost assessment and materials flow balances at a corporate level, is to allocate the data from the system boundary of the company to internal processes.

Process flow charts (see also Figure 3.9), which trace the inputs and outputs of materials flows on a technical process level, give insights into company-specific processes and allow the determination of losses, leakages and waste streams at the originating source. This requires a detailed examination of individual steps in production – again in the form of an input-output analysis, but sometimes linked to technical Sankey diagrams. The process flow charts combine technical information with cost accounting data. These are done not on an annual basis but for a specified production unit, machinery or cost centre. In total, they should aggregate to the annual amount.

Technicians will do this level of materials flow analysis, but the data gathered should be cross-checked to ensure consistency with the cost accounting system. Frequently, due to lack of inter-departmental communication such a harmonisation of technical data with data from financial bookkeeping is not undertaken, but experience has shown that such a consistency check has great optimisation potentials, and has thus become a major tool in environmental accounting. Therefore, a great advantage lies in having compatible technical and financial records.

| Assessment scheme for materials flow balances | Amount in kg, kWh, l | Purchase value | Account number | Materials stock number | Stock keeping | Production Planning Syst. | Direct costs | Overhead | Assigned to cost centre | Oth. records/measurements | Calculation/estimates |
|---|---|---|---|---|---|---|---|---|---|---|---|
| Raw materials | √ | √ | √ | √ | √ | √ | √ | | √ | √ | |
| Auxiliary materials | √ | √ | √ | √ | √ | √ | √ | | √ | √ | |
| Packaging | √ | √ | √ | √ | √ | √ | √ | | √ | √ | |
| Operating materials | √ | √ | √ | √ | | | | √ | √ | √ | |
| Energy | √ | √ | √ | | | | | √ | √ | √ | |
| Water | √ | √ | √ | | | | | √ | | √ | |
| Product | √ | √ | √ | | | | | | | √ | |
| Waste | √ | | √ | | | | | √ | √ | √ | |
| Waste Water | √ | | √ | | | | | √ | | √ | √ |
| Emissions to Air | √ | | | | | | | √ | | √ | √ |

Figure 3.8. Assessment Scheme for Materials flow balances.

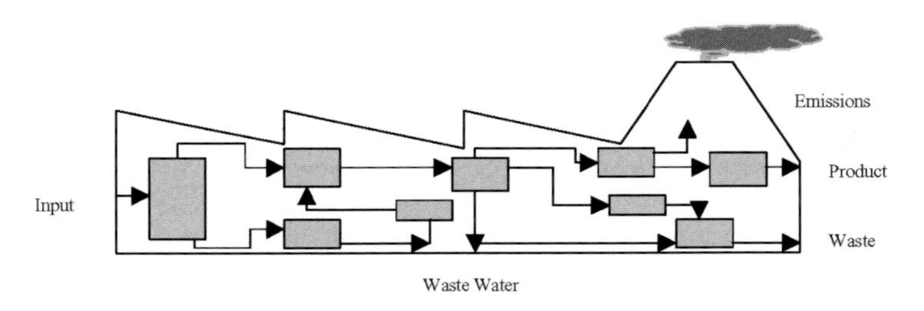


Figure 3.9. Process flow charts: Opening of the Black Box.

### *Detailed corporate flows*

Splitting up the corporate flows into cost centres, or even down to specific production equipment allows for more detailed investigation of technical improvement options, but also for tracing the sources of costs. Special attention should be drawn to the quantitative recording of materials on a consistent kilogram basis. The key questions answered by the approaches of **activity-based costin**g and **cost flow accounting** are:

- which cost centres have processed the materials, and how much?
- can materials input be further divided into production lines or specific equipment?
- how large were the resulting emissions, scrap and waste, preferably recorded separately for each cost centre, production line and machinery?
- what is the correct allocation of costs to products, thus reducing the amount of costs hidden in overhead cost categories?

## 3.4 Environmental indicators and investment appraisal

### *Indicators*

**Environmental performance indicators** condense extensive environmental data into critical information that allows monitoring, target setting, tracing performance improvements, benchmarking and reporting. Several publications and pilot projects highlight their relevance for supporting environmental management systems. As a general outline for generic indicators that can be applied throughout all sectors, the following items should be monitored. Sector-specific, more detailed indicators may be valuable, but aggregation

| | Absolute quantity | Relative quantity<br>Eco-intensity |
|---|---|---|
| Production output (PO) | kg, litres | |
| Raw materials input | kg | kg/PO |
| Auxiliary material | kg | kg(PO |
| Packaging | kg | kg/PO |
| Operating material | kg | kg/PO |
| Energy | kWh | kWh/PO |
| Water | $m^3$/litre | $m^3$/PO |
| Waste | kg | kg/PO |
| Waste water | $m^3$/litre | $m^3$/PO |
| Specific pollution loads | kg | kg/PO |
| Air emissions | $m^3$ | $m^3$/PO |
| Air emissions load | kg | kg/PO |
| | | |
| Other denominators | | |
| Number of employees | number | |
| Turnover | money value | |
| Earnings before Interest and Tax (EBIT ) | money value | |
| Production hours | time | |
| Workdays | days | |
| Building area | $m^2$ | |
| | | |
| Management performance indicators | | |
| Number of achieved objectives and targets | | |
| Number of non-compliances or degree of compliance with regulation | | |
| Number of sites with certified environmental management systems (EMS) | | |
| Number of sites with environmental reports | | |
| % turnover from EMS certified sites | | |
| % turnover of green products (e.g. organically grown versus conventional crops) | | |

Figure 3.10. Environmental Performance Indicator System.

to general categories should be possible. The indicator system should cover all major input and output categories (see Figure 3.10).

### *Investment appraisal*

Most methods of **investment appraisal** assume that all inputs and outputs of an investment decision can be expressed in money terms. The high risks, the difficulty of estimating acceptable monetary values, and the uncertainty that surrounds future environment-related costs and potential future cost savings make the assessment of long-term investments questionable. In certain cases end-of-life expenses (e.g. costs of dismantling, as with the Brent Spar) can be extremely high, but are discounted away by common discounting procedures. However, the methods are still widely used. The task of EMA is not so much to change the concept of discounting future cash flows, but to ensure that all relevant earnings and expenses (including reputational damage) are taken into account.

In addition to initial investment and annual operating expenditure,

- future liability costs and
- savings potentials

need consideration for investment appraisal.

**Initial investment costs** can comprise several items in addition to the purchased equipment.

**Annual operating costs** can relate to all the other cost categories of the assessment scheme. Therefore, annual assessment of total expenditure is vital as a starting point in environmental management accounting as it assures confidence that all relevant costs have been included in the analysis of costs to specific cost centres or equipment.

Measures for pollution prevention help to reduce disposal and emission treatment costs and to increase the efficient use of purchased materials. Often, when calculating investments, the reduced costs for materials and emission treatment are not completely calculated, which results in distorted investment decisions.

Additionally, future liability costs and less tangible benefits should be estimated.

**Future Liability Costs:** Two general forms of future liability costs can be distinguished: liability for personal injury or property damage (e.g. liability stemming from a leaking landfill), and penalties and fines for violation of environmental regulations. When calculating future risks and liabilities, an estimate of avoided future liability is also required.

**Saving potentials:** Less tangible benefits from pollution prevention investments, such as increased revenue from enhanced product quality, company reputation or product image, and the effects of improved health conditions on productivity are difficult to predict and quantify.

In addition to savings, other positive effects can arise from environmental management. These so-called soft factors, structured by stakeholder relations, can be:

- increased turnover, customer satisfaction, new markets, differentiation from competitors;
- image enhancement;
- better relations with authorities, reduced regulatory compliance costs;
- better creditworthiness, reduced insurance rates, good ratings by investment brokers and agencies;
- better public shareholder and community relations;
- increased job motivation and satisfaction, less absenteeism and illness.

## 3.5. Conclusion

This paper has highlighted the limitations of traditional accounting and has shown what EMA can do to overcome them, in the light of the need for companies to take the environment seriously.

Business risk can be defined as any chance that an organisation will not achieve its business objectives. Environmental aspects cannot be ignored when defining one's objectives, and therefore become part and parcel of any serious business activity. This is why environment is increasingly incorporated in corporate risk management and management control. EMA makes it possible to do this effectively.

# 4. A Guideline for the Measurement and Reporting of Environmental Costs

*Jong-Dae Kim*
*Associate Professor, College of Commerce & Business Administration, Chungbuk National University, and Chief Director, Research Institute for Industry and Management, Chungbuk National University*
*E-mail: jdkim@cbucc.chungbuk.c.kr*

## 4.1. Background

There has been an increasing demand for environmental cost information from various stakeholders, both internal and external, who need this for their own decision-making purposes. Many firms around the world have therefore found it critical to measure environmental costs accurately and report them to stakeholders in a timely manner.

In line with this movement, efforts to help firms with measuring and reporting environmental costs have been exerted by governmental and other organisations. 'A Guideline for Adopting an Environmental Accounting System' is one example, which was released by the Japanese Ministry of Environment in March 2000. It provides discussion and guidance on a wide range of issues concerning the measurement and reporting of environmental costs.

The World Bank has recently commissioned a project on 'Environmental Accounting Systems and Environmental Performance Indicators' from a group of academics and researchers in Korea. One of the key objectives of this project was to develop guidelines for measuring and reporting the environmental costs which are applicable to Korean firms. The guideline was developed by the author based on literature review and, mostly, on case studies in Korean firms. This chapter's main content reflects these research activities.

Although the project encompassed three case studies, this guideline was based mainly on the practices of LG Chemical Ltd., which is the largest chemical manufacturer in Korea. It has developed an environmental cost measuring system, and has implemented this for internal management purposes since 1999. The guidelines presented in this paper are therefore likely to be most suitable large firms, which is where the main concern to measure and report environmental costs to stakeholders still remains. Few if any SMEs in Korea have ever considered measuring, reporting and using environmental cost information for management purposes.

Although the target group for the guidelines presented in this paper are large firms in Korea, they could also be adopted by other large firms in Asia, which was the original intention of the World Bank. Since the major purpose of this chapter is to present the guidelines, the case study of LG Chemical Ltd. will be introduced only briefly.

## 4.2. Environmental costing project of LG Chemical Ltd.

In an effort to anticipate and prepare for future demand for environmental cost information for management purposes, the Environment and Safety Team of the Headquarters of the company initiated an environmental costing project, which aimed to standardise the measurement process of environmental costs. The focus of the project was placed on the classification of environmental costs, the segregation of environmental costs from



*M. Bennett et al. (eds.), Environmental Management Accounting: Informational and Institutional Developments,* 51–65.
© 2002 *Kluwer Academic Publishers. Printed in the Netherlands.*

non-environmental costs, and the calculation and systematic management of environmental costs. The Environment and Safety Team, which has 4–5 staff, is responsible for the design and implementation of environmental management for all local plants. The project was initiated by the team and started in 1996.

### 4.2.1. *Profile of the company*

Founded in 1947, LG Chemical is the largest chemical firm in Korea. Its product portfolio includes petrochemicals, industrial materials, auto parts, household goods, electronic materials and pharmaceuticals. It had 11,423 employees as of the end of 1999, in which year total sales revenue (non-consolidated) amounted to 4,546 billion Won. Although all of LG Chemical's eight plants nationwide joined the environmental accounting project, this case study concentrates on their Chongju Plant.

### 4.2.2. *Classification of environmental costs*

The first step was to classify environmental costs between *proactive* and *ex-post* costs. *Proactive costs* are those which are incurred in pollution prevention activities, which in turn are divided into costs for pollution prevention at source, treatment/disposal costs and stakeholder costs. LG Chemical's classification scheme focuses on management purpose, in that it classifies environmental costs so that pollution prevention costs are more efficient in terms of environmental improvements as well as cost savings than are treatment/disposal costs, and proactive costs are more efficient than ex-post costs. Stakeholder costs are incurred in an effort to develop and maintain better relations with outside stakeholders including governments, local communities, consumers, environmental activists, investors, etc.

*Ex-post* costs include fines and penalties for non-compliance with environmental regulations, and compensation to third parties as a result of loss or injury caused by past environmental pollution and damage. These costs differ from other types of environmental costs in that they do not provide any benefit or return to the enterprise. Ex-post costs also include taxes and charges.

#### 1. *Proactive environmental costs*

Proactive Environmental Costs are incurred in order to prevent pollution. They are divided into three categories:

a. *Costs for Pollution Prevention at Source*

These costs are related to environmental activities that are intended to stop pollutants from occurring at their source. Examples include R&D costs, the replacement of clean production process, utility replacement costs, and costs to establish an EMS.

b. *Pollution Treatment/Disposal Costs*

These costs are incurred in order to prevent pollutants that have already occurred from contaminating the environment external to the company. Examples include the purchase and installation of facilities, measurement costs, operation and maintenance of facilities, disposal of wastes, utilities, and operating costs related to environmental management.

c. *Stakeholder costs*

These are costs which are incurred in order to maintain good relations with outside stakeholders. Compliance costs, public relations costs, and advertising costs are included in this category.

2. *Ex-post costs*
These costs are incurred in order to remedy or restore environmental damages that have already occurred. Compensation paid to third parties for injury or property damages caused by environmental impacts, fines and penalties, taxes and charges, and opportunity costs, are some other examples of costs that belong to this category. Opportunity costs (loss of profit opportunities) in the cost sheet are used to represent the contingent loss that can result from environmental accidents or other events. However, inclusion of these costs in the cost sheet is inappropriate since the costs are not even captured by the traditional accounting system. Contingent losses are not normally recorded, since it is too difficult to make a reasonable estimate of the amount of the losses. In fact, this item did not show up in the costs sheets filled out by the staffs of local plants.

### 4.2.3. *Accounts in the existing accounting system and environmental Costs*

The main thrust of the measurement of environmental costs in LG Chemical is to segregate environmental components from the traditional cost accounts. The project team provided a table that indicates the classification of environmental costs corresponding to traditional cost accounts, in order to help local plant staffs to fill out the cost sheets. The project team, however, did not provide detailed criteria to distinguish between environmental and non-environmental costs.

### 4.2.4. *Collection of environmental costs*

After all environmental costs have been segregated from the traditional manufacturing costs (direct materials costs and direct labour costs as well as overhead costs), each environmental cost item was then assigned to the appropriate environmental impact area, such as air pollution treatment, liquid effluent treatment, solid wastes treatment, soil contamination restoration, etc. The project team provided the measurement sheets to LG Chemical's eight plants nationwide so that they could fill out the matrix forms as directed for 1998 and 1999.

Note that LG Chemical did not distinguish between environmental costs and investments – environmental investments are simply added to the other environmental expenses. Environmental investments in environmental facilities should not be treated as expenses, since investments are expensed through the process of depreciation. The data indicate that LG Chemical spends the largest amount on air pollution prevention and treatment costs, where environmental costs amount to 5.26 billion Won, compared to 0.37 billion Won on waste water treatment, and 0.69 billion Won on the disposal of solid wastes.

This case clearly reveals the critical problem of how to distinguish between environmental costs and non-environmental costs, which is not an easy task. LG Chemical realises that measuring environmental costs and investments is complicated, but after experiencing and solving the problems involved with this it will have come closer to achieving an accurate and useful environmental accounting system in the future.

### 4.2.5. *Summary*

LG Chemical did not go so far as to allocate environmental costs to each product, and the environmental cost data were not reported to the top management or utilised for management purposes. The internal use of environmental cost information is possible only with accurate product costing, without which management decisions such as product mix, choice from various investment options, product pricing, etc., cannot be made optimally.

Another problem for the project was the non-participation of accounting staff – the project team recruited engineers, but there was no accounting expert. It is crucial for top management to fully support the project financially and to empower the team members to design and implement the environmental accounting system in full scale. In doing so, outside consultants will be of great help.

## 4.3. A Guideline for measuring and reporting environmental costs

This section provides general concepts and tools for measuring and reporting environmental costs. Environmental accounting is a process of measuring the environmental impacts of companies, and the costs and benefits of reducing or eliminating environmental impacts. It records costs and benefits in appropriate accounts in an accounting system, and allocates them to production processes or products in an appropriate manner. An environmental accounting system is an information system that produces environmental cost information and conveys it to the stakeholders to support optimal decision-making.

Environmental accounting has two purposes: internal and external reporting respectively. The internal use of environmental costs improves management decision-making, including on product mix, product pricing, choice from various pollution prevention options, etc. Knowledge of environmental costs that were hidden in other cost items such as manufacturing overheads will facilitate near-optimal decision-making, as well as cost savings.

The external reporting of environmental cost information improves relations with outside stakeholders. Outside stakeholders include investors, creditors, governments, consumers, communities and environmental activists who are interested in an company's environmental costs and investments. These data will provide the stakeholders with information on the company's environmental impacts and environmental risks, which are essential to evaluating future values of the entity.

In internal reporting, each organisation can choose its methods of reporting. However, information provided to outside users needs to be consistent and comparable, which is one purpose of this guideline. Otherwise, it deals mostly with the measurement of environmental costs for internal use rather than with disclosure to outside stakeholders.

When a company is committed to environmental protection activities, it is essential to capture, accumulate, and analyse environmental investments and costs. Environmental cost/benefit analysis will enable management to make more efficient decisions, and understanding its own environmental costs and performance is essential to the establishment and maintenance of an effective environmental management system. In particular, in order to enhance environmental efficiency by saving energy and resources, by producing less wastes, or by developing better quality products, it is essential to measure and manage environmental costs as well as environmental impacts in terms of physical units. In addition, the environmental accounting system can be utilised to evaluate the effectiveness of environmental protection measures.

## 4.4. Definition and Scope

### 4.4.1. *Environmental costs*

Environmental costs are defined as those resources which are consumed, either voluntarily or in order to comply with regulation, in attaining environmental goals. Since resources are consumed by activities, environmental costs are defined and classified according to environmental activities. Organisations perform environmental activities in order to improve their environmental performance, that is, to reduce or eliminate their environmental impacts. However, costs referred to as 'environmental damage costs' do not contribute to the reduction of environmental impacts. Examples of these costs include fines and penalties for non-compliance with environmental regulations and compensation to third parties as a result of loss or injury caused by past environmental damage and pollution.

The terms 'costs' and 'expenses' are used interchangeably in most cases. Costs are a service potential that will provide economic benefits in the future, and when they are consumed in order to generate revenues, they become expenses. On the other hand, when costs are incurred without their making any contribution to the generation of revenues, they are termed 'losses' rather than expenses.

### 4.4.2. *Environmental goals and performances*

Since in many cases there is a trade-off between profit maximisation and environmental impact minimisation, each company should have decision-making systems and tools that will find optimal solutions which balance these two goals. Therefore, environmental goals cannot be defined in simple terms, but depend on each company's own strategic position. Generally speaking, the goal will be minimum environmental impacts within the constraint of available resources.

A company's environmental performance consists of the environmental impacts of its activities, so that an improved environmental performance therefore means less environmental impacts. Environmental activities are defined as all activities of a company which are undertaken with the purpose of improving its environmental performance, and resources which are used in these activities are environmental costs. It follows that the classification of environmental costs coincides with the classification of environmental activities.

### 4.4.3. *Environmental investments*

Environmental investments are expenditures that are made with a view to improving a company's environmental performance over a period of several years. Thus, the expenditures are first capitalised, and then expensed over time into the future. In the same way, environmental investments are environmental expenditures that are capitalised; the ensuing costs are deferred to the future through the process of depreciation. In most cases, investments involve the acquisition of plant and equipment, but can also include intangible assets such as R&D.

Consequently, current period expenditures are distinguished between expenses and assets., Environmental disclosures should therefore be made such that the following items will be separately reported:

- Total investments of current and past few years
- Environmental costs expensed during the current period (such as raw materials costs, labour costs, overhead costs, sales and administration expenses, etc.);
- Depreciation expenses.

It may be useful to calculate environmental investments as a percentage of total investments, and environmental costs as a percentage of total costs, for both internal and external reporting purposes. To see how the various costs develop, it is necessary to present environmental costs and investments for a number of previous years as well as for the current year, in order to facilitate comparisons. Since it is important to know the effects of environmental costs and investments, each company has to develop its own environmental performance indicators in order to assess them. If an company has a stake in environmental business activities, the costs involved do not constitute environmental costs. Separate disclosure on their magnitude and characteristics may be necessary as supplementary footnotes.

### 4.4.4. *Research and development costs*

Information on the environmental R&D activities of an company is important in making judgments on the company's environmental efforts and attitude. In principle, R&D expenditures are expensed, but the acquisition costs of depreciable assets which are purchased entirely for environmental purposes are regarded as environmental R&D investments. The acquisition costs of plant and equipment are not the only assets that are capitalised. Any costs that are incurred in relation with R&D activities that are expected to provide future economic benefits can be treated as environmental investments and capitalised.

It is recommended that for each year, the following information on R&D be presented:

- Total R&D expenditures;
- R&D expenditures that are capitalised (investments);
- R&D expenditures that are expensed (taken as costs for the period);
- Amortisation (depreciation) of past years' R&D investments.

### 4.4.5. *Social costs*

Social costs arise when society is compensated for the damage that a private organisation inflicts on it. They are also referred to as *externalities* because the costs are not yet internalised, that is, they are not borne by the company that caused them. Examples of externalities include damage to the health of neighbouring communities, crops or fisheries as a result of water contamination caused by the effluent released by an enterprise. Historically speaking, environmental costs are essentially internalised externalities. That is, environmental costs are charged not to the responsible entities but to society until those entities incur costs to reduce the environmental impacts of their activities. This guideline deals only with internalised environmental costs.

## 4.5. Classification of environmental costs

### 4.5.1. *Classification of environmental activities*

This guideline suggests a two-way classification of environmental costs. The first criterion relates to the types of the environmental activities throughout the business processes, so that environmental costs are first classified into:

1. Pollution Prevention Costs;
2. Pollution Treatment Costs;
3. Environmental System Costs;
4. Stakeholder Costs; and
5. Environmental Damage Costs.

Of the five categories of environmental costs, categories 1, 2, and 5 are directly related to environmental effects and/or impacts. Category 1 activities are likely to be more efficient than category 2, in the sense that a given amount of pollution prevention costs is more likely to result in greater environmental improvement or in greater cost reduction than is pollution treatment cost. Likewise, category 2 is more efficient than category 5.

On the other hand, categories 3 and 4 activities are related only rather indirectly to }pollution. It is not easy to identify direct connections between these activities and environmental improvement, and the benefits may take some time to be realised. However, just because they are not specifically related to environmental benefits in the short-term period, this does not mean that they are not efficient or important.

### *Pollution prevention costs*

These are costs which are incurred to prevent pollution at source. Examples include additional materials costs in order to switch to more environmentally friendly raw materials, additional energy costs to switch to cleaner energy, R&D costs to design and implement more environmentally friendly production process, costs for a greener supply chain, etc.

### *Pollution treatment costs*

Once generated, pollutants should be treated so that they will not contaminate the natural environment. If it is not possible to eliminate all the adverse impacts that pollutants might have on water, air, and soil, they should at least be minimised. Thus, the costs which are incurred in order to eliminate or reduce the environmental impacts of pollutants that are already being generated in the process of business activities, are referred to as pollution treatment costs. Examples are wastewater treatment costs, the operating costs of an incinerator, and depreciation expenses of treatment facilities.

### *Environmental management system costs (EMS costs)*

EMS Costs are indirect environmental costs which are incurred in order to design, establish, implement, monitor, evaluate and improve the environmental management system as a whole. All costs related to ISO certification belong to this category, which also includes costs incurred in order to carry out Life Cycle Assessments (LCA), Environmental Performance Evaluation (EPE), Environmental Audits (EA), and Environmental Labeling, and costs emanating from the environmental education of employees.

### *Stakeholder costs*

These are costs which are incurred in order to build up and maintain good relations with outside stakeholders such as investors, creditors, regulators, communities, consumers, and environmental activist groups. By maintaining good relations with its stakeholders, a company can avoid unnecessary conflicts. Building up an image of a 'green' enterprise can contribute to the creation of future profit opportunities. Examples are the costs of monitoring and reporting to the regulators and public, eco-marketing expenses, sponsoring the activities of environmentally-oriented NGOs, and financial support for community events.

### *Environmental damage costs*

When waste materials are discharged into the environment without adequate prior treatment, they are likely to cause damage to nature or people. Environmental damages have to be restored, and any damage to health, life and property should be compensated. Examples of these costs include fines and penalties for non-compliance with environmental regulations, and compensation to third parties as a result of loss or injury caused by past environmental damages and pollution.

### 4.5.2. *Classification by the types of environmental impacts*

The second criterion applied in classification is the type of environmental activities as related to environmental impacts. Thus, all environmental costs are classified into one of the following categories:

1. Air pollution prevention or treatment costs.
2. Water pollution prevention or treatment costs.
3. Solid waste reduction or disposal costs.
4. Soil contamination prevention or remedial costs.
5. Sustainable resource conservation costs.
6. Other costs including those to prevent noise, odors, etc.

Environmental cost information classified by the types of environmental impacts may not be so useful to information users as environmental cost information classified by activities, as mentioned previously, although the general public and regulators may be very interested in this type of information. If, however, the company implements LCA or EPE, and tries to apply the results of this to reduce environmental impacts, environmental cost information by types of environmental impacts will become significant.

## 4.6. Measurement of environmental costs

### 4.6.1. *Scope of environmental costing*

Each company may start with a plant, site, or office in measuring environmental costs, although it is preferable to expand this to the whole enterprise. The scope of environmental accounting need not necessarily be the same as that of ordinary accounting. Each company can choose whatever seems to be most appropriate for itself taking into account the goals, experience, and expected costs and benefits of the environmental costing project.

### 4.6.2. *Environmental costing period*

The measurement and reporting period of environmental costs can differ from company to company. The process should be performed on each occasion when management needs environmental cost information in order to make decisions on product mix, various waste management options, switching from one type of raw materials or energy source to others, investment plans, product pricing, etc. For the purpose of product costing and external reporting, however, environmental costs could be measured and reported at regular intervals, such as quarterly.

### 4.6.3. *Separation of environmental costs from traditional costs*

Distinguishing between environmental costs and non-environmental costs is a frequent procedure in environmental cost accounting. In a conventional accounting system, environmental costs are often hidden in raw material costs, labour costs or (most frequently) in overhead costs, without being separated from non-environmental costs. In order to separate environmental costs from non-environmental costs, we need to distinguish between environmental activities and non-environmental activities since costs are defined according to activities. That is, costs incurred by environmental activities are environmental costs, and costs incurred by non-environmental activities are non-environmental costs.

In practice, however, the separation is far from simple. In cases where activities are performed in order to comply with environmental laws and regulations, the expenditure related to those activities are clearly environmental costs. In many cases, however, environmental goals are mixed with other non-environmental goals, such as improving profits through realising energy savings or increased productivity. In such cases, the first criterion should be the primary motivation or purpose of the activities: if the costs are incurred primarily in order to improve environmental performance then they should be classified as environmental costs, regardless of the results. Environmental costs may not always lead to the improvement of environmental performance. Environmental costs that do not contribute to environmental performance improvement are called environmental losses; these belong to Environmental Damage Costs.

On the other hand, costs incurred to serve non-environmental purposes such as cost savings and quality improvement should be classified as non-environmental costs. Suppose that the change from low-quality to high-quality materials results in reducing solid wastes, thereby improving environmental performance. The costs incurred by the change are not classified as environmental costs so long as the change was driven by non-environmental motivation, which in this case is product quality improvement. The investment in the energy saving facilities for cost saving purpose may be another example of non-environmental costs that result in environmental performance improvement (i.e. reduction of $CO_2$ emissions) as by-products.

In most cases, however, economically motivated activities such as seeking cost savings by the use of lower grade materials and less clean energy will result in an increasing burden on the environment. Those actions may have the effect of lowering manufacturing costs in the short term, but will eventually increase costs in the future, thereby lowering the company's value. These additional costs will accrue in the future but are not usually reflected in current cost statements.

Costs incurred in order to comply with environmental laws and regulations are by definition classified as environmental, because the motivation of the expenditures is related to environmental protection. For example, if a company purchases a waste water treatment facility in order to comply with emission standards set by laws and regulations, the acquisition costs of the facilities are capitalised into environmental assets, and subsequently depreciated; and if the law requires the recycling of used products, the operating costs of recycling facilities are environmental costs. If, on the other hand, environmental activities are performed voluntarily by entities, then internal decision-makers (i.e. managers) are the only people who know the primary motivation of the expenditures. Managers should therefore try to disclose true and accurate information on environmental costs.

Specifically, compliance costs, remedial costs, operating costs of pollution control equipment, fines and penalties for violating laws, and costs related to environmental protection facilities, are some examples of environmental costs. Normally, costs related to end-of-pipe measures are environmental costs. But costs related to process improvement measures frequently have both environmental and environmental objectives.

These costs are referred to as 'complex costs'. One obvious way to break these down into their environmental and non-environmental components is to calculate exactly what portion of the activities are environmentally motivated. Since it is not easy to distinguish between the two different motivations, the following methods can be used in practice: the Differential Cost Method and the Proportional Cost Method.

### *Differential Cost Method*

Complex costs are composed of environmental and non-environmental costs. A facility that has a particular environmental quality is compared with a similar facility which does not currently have that particular environmental quality. The difference in price directly relates to the environment, and therefore gives the key to how to separate the environmental expenditure and environmental cost involved.

***Proportional Cost Method***

When it is improper or infeasible to apply the Differential Cost Method, total costs can be assigned to environmental and non-environmental costs in proportion to the purposes of the activities. In cases where the company must determine what fraction of an expenditure should be accounted for as environmental costs, the company may choose to divide the costs according to some predetermined proportions. For example, all complex costs may be assigned to one of the following three categories:

| | Environmental | Non-environmental |
|---|---|---|
| Category | A: 25% | 75% |
| | B: 50% | 50% |
| | C: 75% | 25% |

When a company finds it almost impossible to estimate what fraction of the costs are environmental or to what category they belong, it is recommended that the total costs are treated as non-environmental.

To illustrate a break-down of complex costs between environmental and non-environmental costs, consider labour costs and depreciation expenses. Staff who work in the environment and safety department normally also work for other departments too. Thus, labour costs for those staff can be allocated between environmental and non-environmental based on direct labour hours or on a different job analysis. Depreciation of purely environmental assets are charged to environmental costs, but if environmental protection equipment is retro-fitted on to the existing non-environmental assets, the amount of depreciation costs to be allocated as environmental should be computed in proportion to the costs of the added equipment.

### 4.6.4. *Matrix of traditional and environmental costs*

This guideline suggests the use of spreadsheets in measuring environmental costs. In an early stage of environmental cost accounting, environmental costs are measured by separating them from the total costs for each cost account. It therefore begins with the traditional statement of manufacturing costs and income, then for each cost account the environmental portion of the total amount is determined by using one of the above-mentioned procedures. The columns of the environmental cost sheet reflect the conventional cost accounts, while the rows reflect the environmental costs (see Appendix).

For each category of environmental costs, environmental costs are reclassified into six different types of environmental impacts: air pollution, water pollution, solid wastes, soil contamination, sustainable resource conservation, and others. As environmental accounting matures, all environmental costs will be identified from the outset with environmental and non-environmental costs being recorded as original entries in the accounting system and the environmental accounting system incorporated into the entire accounting system. In the meantime, however, the measurement of environmental costs is to be accomplished according to the procedures as suggested in this guideline.

## 4.7. Environmental investments

Environmental investments are to be presented in a separate form, distinct from environmental costs. Part of environmental expenditures is expensed in the current period, and the remainder is capitalised into assets and will be later expensed through the depreciation process. Information on environmental investments for the current and past periods should therefore be reported separate from the environmental cost information, so that information users can assess the company's environmental activities for the entire period.

### 4.7.1. *Classification of environmental investments*

It is recommended that investments in pollution prevention and treatment facilities and environmental R&D investments be reported separately. Investments in pollution prevention and treatment facilities correspond to pollution prevention costs/activities and pollution treatment costs/activities, and are measured by the acquisition costs of the facilities. On the other hand, R&D investments are expenditures related to R&D activities in order to develop clean products, design clean processes, and develop clean technologies. R&D investments can be divided into a few different categories such as R&D for clean products, R&D for clean processes, R&D for clean technology, R&D for sustainable resource conservation, etc., if necessary.

R&D investments are composed of not only the purchase of plant and equipment, but also ordinary costs such as materials costs, labour costs and other operating costs. The environmental investment table should present investments for both the current and the past few years, in order to indicate trends. Investments are also reported for each category of the types of environmental impacts.

### 4.7.2. *Environmental investments and expenses*

The relationship between environmental investments and environmental expenses hinges on depreciation expenses. Additional environmental investments increase environmental assets, which are capitalised investments, whereas depreciation decreases environmental assets. Thus, they must meet the following formula:

Beginning of Period Environmental Asset (Balance)
\+ Current Period Investments (Addition to Assets)
− Depreciation for Environmental Assets (Decrease in Assets)
= End of Period Environmental Assets (Balance).

For plant and equipment, the beginning-of-the-period amount, current period investments, current period depreciation and amortisation, and end -of-period amounts should be presented. In addition, total investments in plant and equipment are preferably presented for a number of previous years. It will be useful to the information users if all the environmental investments and related depreciation expenses are reported for each category of environmental impacts (air pollution, water pollution, solid waste, soil contamination, resource conservation, noise and odours).

If a company purchases or develops assets other than plant and equipment as a result of environmental investments (e.g. the purchase or development of patents), another table for the intangible assets will be necessary. In accounting, the term 'amortisation' is used for intangible assets instead of 'depreciation'.

## 4.8. Environmental benefits

When environmental costs result in the elimination or reduction of environmental impacts, or in improvements in the environmental performance of a company, the environmental benefits are to be matched with the corresponding costs to the extent possible. The scope and the reporting period of environmental costs can be applied to environmental benefits.

Environmental benefits can be measured in either physical units or monetary units. While environmental burdens and changes therein are best measured in physical units, economic benefits from environmental measures taken by a company, such as revenues and cost savings or avoidance, are better measured in monetary units.

Some environmental benefits, however, cannot be measured or presented in either physical or monetary terms. For example, an enterprise may be able to save future costs through pollution prevention investments because the investments will reduce or eliminate future environmental risks. Although these types of benefits are not always quantifiable, it is recommended that they be described in an appropriate way in the table whenever possible.

### 4.8.1. *Environmental benefits measured in physical units*

Examples of physical units include the emission of pollutants such as NOx and SOx, consumption of energy, amount of solid wastes generated, value of green purchases, amount of reuse and recycling, etc. Environmental benefits measured in physical units are preferably reported by environmental impact category (air pollution, water pollution, solid waste, resource conservation, use of toxic chemicals, and others). If physical units of pollution measurement are too numerous for inclusion in a single table, a separate table for each environmental impact category could be prepared, or disclosure could be restricted to only a selection of the most significant items.

### 4.8.2. *Environmental benefits measured in monetary terms*

Environmental benefits measured in monetary units include cost savings and increases in revenues. When environmental activities reduce either environmental or non-environmental costs, the cost savings are to be matched with the corresponding environmental costs (or activities) to the extent that these can be measured. Examples of cost savings include energy cost savings as a result of energy saving efforts and investments, and solid waste treatment costs reduced by recycling.

There are times when assumptions need to be made in assessing the economic benefits of environmental activities. This situation can arise, for instance, when there is an opportunity to avoid future environmental accidents through the successful implementation of an environmental management system. Pollution prevention measures may also help to avoid future contingent environmental losses. Examples of environmental losses which might be avoided by reducing environmental risks through environmental investments include savings of future remedial costs, and avoidance of the costs of stopped production lines and compensation paid to third parties for injury.

These potential benefits should be measured and reported wherever possible. Some cost savings, however, are so uncertain in terms of their timing and magnitude that they should not be reported until they are actually realised. However, even in such an uncertain case, mention of possible benefits in a non-quantitative form may be useful to information users.

Environmental activities can increase revenues in two ways – the first one is certain and realised in the short term, whereas the second one is uncertain. Examples of the first type include sales of recycled wastes, retrieved materials, retrieved electricity and retrieved containers and packages. These are all by-products of environmental activities. Sales of the by-products are reported in terms of weight and amount.

The second type of increased revenue, however, is uncertain. An example is the possibility of increasing sales in the future as a result of an improved image through successful environmental protection activities. This type of revenues can be ignored until they become visible.

### 4.9. Disclosures

For external reporting, environmental costs and benefits are disclosed in one table. The total amount of pollution prevention costs, pollution treatment costs, EMS costs, stakeholder costs, and environmental damage costs are summarised for each environmental impact category, and both sales of by-products from environmental activities and cost savings are matched with environmental costs.

Unlike the sales of by-products from environmental activities, cost savings are uncertain and sometimes impossible to estimate. Thus, as far as cost savings are concerned, environmental benefits are reported to the extent that they can be reasonably measured. Cost savings can be so uncertain that they cannot be reported in any credible manner. Then it could be counterproductive to allude to them, because external stakeholders may eventually feel misled or because management could become sceptical of the value of considering environmental costs at all.

Environmental investments are also summarised in a table. Investments in pollution prevention and treatment and R&D investments are disclosed for each environmental impact category. For environmental assets, beginning-of-year balances, current year investments, depreciation expenses, and end-of-year balances are disclosed in the same table for each environmental impact category.

Finally, environmental improvements measured in physical units are summarised in a table which is the same as Table 3-2 of Appendix 3. For external reporting, measurements of environmental improvements may be presented in less detail than for internal reporting.

### 4.10. Implementation of the Guideline

#### 4.10.1. *Incorporation of all other information available*

When evaluating a company's environmental performance, information users should consider not only the environmental cost information provided by the environmental accounting system but also other information from other sources such as environmental reports, annual reports, news releases, and so on.

#### 4.10.2. *Understanding the meaning of environmental cost figures*

Simply to compare environmental cost figures between two different enterprises may be meaningless or even reckless if the comparison made is between two entities in entirely different industries, or between entities in different stages of environmental management development. It should therefore be noted that:

- The structure of environmental costs varies from industry to industry, and depends on the stage of environmental management development which the particular company has reached;
- The same environmental cost amounts do not necessarily represent the same levels of environmental efforts or efficiency, because the specific procedures adopted by individual entities to measure environmental costs can be different from each other (note that this guideline provides only general principles and procedures in measuring and reporting environmental costs and benefits);
- Environmental costs tend to diminish as the company progresses towards a more advanced environmental management system, because the company has already improved its environmental performance in the past; and
- If the technology remains basically the same, at a certain point additional costs of continuously reducing certain emissions may rise sharply. Environmental cost accounting can bring this to light and may lead to a different strategy, such as adopting new technologies, or shifting investments to measures that are more eco-efficient.

### 4.10.3. *Evaluation of cost/benefits of environmental activities*

An environmental accounting system may be based on the conventional financial accounting system, but need not be restricted to it. Thus, when assessing environmental measures, it is not appropriate simply to go by conventional financial criteria as this tends to imply that certain costs and benefits related to the environmental measures are ignored. Such a procedure distorts both sound environmental and economic decision-making. The fact that the additional figures which are needed cannot be easily measured should not be readily accepted as a valid excuse. In general, a lack of information on environmental costs leads to an underestimation of the economic benefits of environmental investments: if the costs are not visible, the savings cannot be made visible either.

### 4.10.4. *Incremental nature of environmental costs*

Some environmental costs are defined as incremental costs. For example, environmental costs incurred by a change from cheap energy to more expensive clean energy can be calculated by subtracting the energy costs before the change from those after the change. The environmental costs incurred due to the change of energy source are therefore the incremental (additional) energy costs, and are incurred only when the energy source is changed. In other words, if the entity continues to use clean energy the following year and thereon, there will no longer be environmental costs related to the energy source. Another example would be a change of raw materials in order to reduce toxic wastes.

Most environmental costs are recorded as they are incurred. Unlike the example of the change of energy source, they will continue to occur so long as the related activities continue. If the entity ceases its environmental activities, there will no longer be any environmental costs. For example, environmental staff costs are environmental costs so long as the staff work in the environmental department, but if the environmental department were to be dissolved and the staff transferred to another department, then there would no longer be any environmental costs so far as the staff costs are concerned.

Depreciation expenses for environmental assets are recognised throughout the depreciation period. Other than that they are recognised over a long period of time, depreciation expenses are no different in principle from other environmental costs. If it were not for the investments in the environmental facilities, there would be no depreciation expenses to be classified as environmental costs.

It should be noted that since no environmental costs are recorded in respect of changes of raw materials or energy source in the year after the changes take place, therefore a lower level of environmental costs incurred in the current period compared with those incurred in the previous period does not necessarily indicate inferior environmental performance during the current period. This is why we need to take precaution in interpreting environmental costs, and to take into account environmental benefits as well as costs in order to evaluate an entity's environmental efforts and performances.

# 5. Flow Cost Accounting, an Accounting Approach Based on the Actual Flows of Materials

*Markus Strobel and Carsten Redmann*
*Institute for Management and the Environment, Augsburg, Germany; E-mail: redmann@imu-augsburg.de*

## 5.1. The basics of flow cost accounting

### 5.1.1. *Flow cost accounting as a basic element of flow management*

Flow cost accounting is an essential instrument in a new management approach known as flow management (see also Strobel, 2001; LfU,[1] 1999). The aim of **flow management** is to improve the management of production companies by overcoming compartmental thinking and instead to see one's organisation as a system that channels and transforms flows of materials[2] and information from beginning to end. These flows are to be structured in an efficient, goal-oriented manner.

Flow management, with the goal of having a company that is both profitable and environmentally sound, has three main components: the *'flow model'*, *'flow cost accounting'*, and *'flow organization'*.

**Flow modelling** makes the materials and informational flows transparent, while directing attention to materials losses. With the help of **flow cost accounting**, all costs within a company will be assigned to specific materials flows. To perform flow cost accounting successfully, existing information systems (such as SAP R/3) and available databases must be used.

Technical measures are frequently insufficient to reduce flow costs on a long-term basis because of existing organizational barriers. The flow-oriented restructuring of the organization that is needed for this – the **flow organization** – has links with a flow-oriented version of process-engineering. Co-ordination of the different elements requires effective **flow-oriented communication**.

### 5.1.2. *The purpose of flow cost accounting*

In addition to environmentally-oriented cost accounting systems, other different approaches have been developed to remedy the shortcomings of traditional cost accounting methods though without replacing them, such as activity-based costing, target costing, and logistics costing. This begs the questions: what can flow cost accounting add to the existing body of knowledge, and why is this approach increasingly adopted by industrial concerns?

In flow management, the company can be defined as a materials flow system[3] (see Figure 5.2). This includes, on the one hand, the classic materials flows from incoming

[1] LfU / Landesanstalt für Umweltschutz = Agency for the Environment of the German Federal State of Baden Württemberg.
[2] Energy flows can be thought of in the same way as materials flows, especially since it is often in a material form (in the full sense of the word, e.g. coal, oil, gas) that energy first enters a company. From now on in this paper, therefore, the word 'materials' is used as generic for materials and energy.
[3] A view purely in terms of inputs and outputs, as usually calculated in a company's in-house environmental audit, is not sufficient to define the materials flow system. See also LfU (1999), page 12 et seq.



*M. Bennett et al. (eds.), Environmental Management Accounting: Informational and Institutional Developments,* 67–82.
© 2002 *Kluwer Academic Publishers. Printed in the Netherlands.*

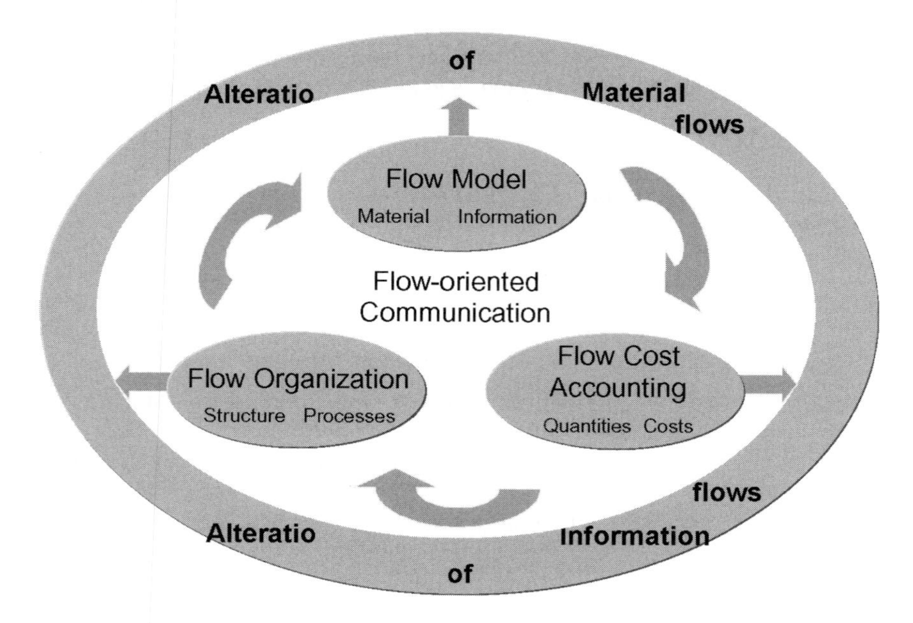


Figure 5.1. Basic components of flow management.

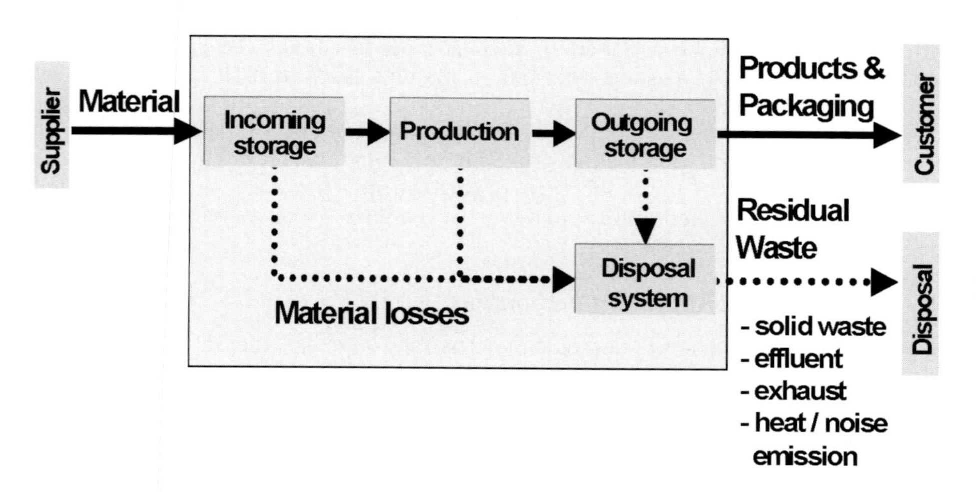


Figure 5.2. The company – seen as a (simplified) materials flow system.

goods, through various processing steps, up to delivery to the customer. On the other hand, it includes all the materials losses that occur along the process chain (e.g. rejects, scrap, cut-offs, chippings, outdated goods or damaged products), which leave as effluents and solid waste.

Flow cost accounting performs an important function within flow management, namely that of quantifying the cost factors in the materials flow system and improving intra-

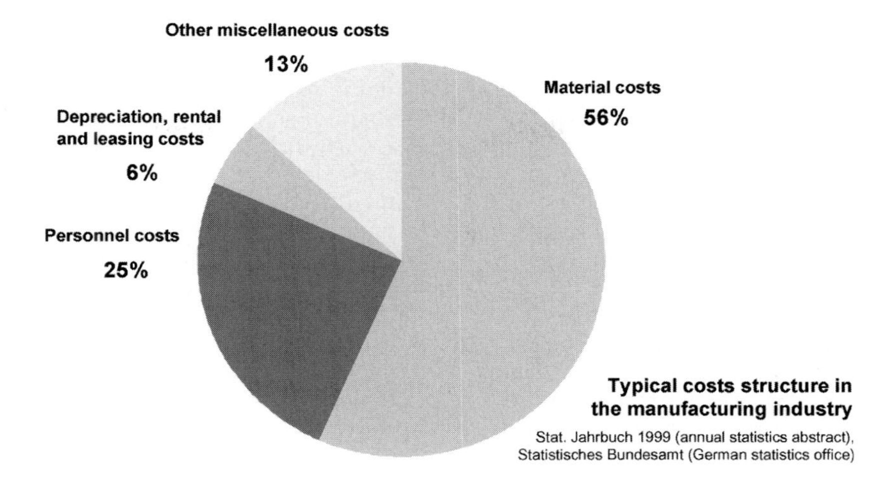


Figure 5.3. Typical proportion of costs in the manufacturing industry.[4]

company transparency, in order to lead to both economic and ecological improvements. Figure 5.3 shows a typical outcome of such an exercise.

Existing conventional accounting approaches are unable to provide sufficiently precise data on the costs of materials. Tracking the use and storage of materials along the materials flows, therefore, can be extremely difficult. Flow cost accounting makes a notable difference here by linking extended quantitative data to the materials flows.

Seen from an ecological perspective, flow cost accounting is beneficial because there is a systematic drive towards reducing the quantities of materials and energy used (avoidance of waste and emissions).

### 5.1.3. *From conventional environmental cost accounting to flow cost accounting*

Existing conventional cost accounting has been criticised for failing to provide sufficient insight into what happens with the various materials flows and what drives overheads. Moreover, it does not make transparent the costs which relate to ineffective use of materials and other environmental costs. In recent years, to address the ecological criticism in particular, a number of environmentally-oriented costing systems have been developed,[5] namely traditional environmental costing (or environmental protection costing), waste costing, and flow cost accounting.

As a way of describing and comparing these various approaches we have, below, divided up a company's costs on a flow-oriented basis into a total of six **cost segments** (cf. also Krcmar et al., 2000): materials costs, system costs (personnel, depreciation), end-of-pipe environmental costs and disposal costs.

The materials costs are divided up into two segments, namely those materials costs for materials which physically enter the product (including packaging) (cost segment 1), and

---

[4] This average costs structure in German production companies corresponds largely with the results of an American study (conducted by 'Business Week' magazine, 22 March 1993) which puts the part played by materials costs in US companies at 50–80%.

[5] A summary of the various approaches is given in Fichter et al. (1997), p. 34 et seq.

those for materials which are physically contained in materials losses (cost segment 2). The system costs[6] (e.g. personnel costs and depreciation) are divided into three segments. The first contains those system costs which are incurred in manufacturing the product (cost segment 3); the second contains those system costs used in handling material losses up to the point at which these losses are incurred as such (cost segment 4); and the third contains those system costs incurred in handling material losses after these have arisen (cost segment 5). This last segment of the system costs is frequently called the 'end-of-pipe' costs block. The disposal costs represent a single segment (cost segment 6).

**Traditional environmental costing** (cf. also US EPA, 1995; BMU and UBA, 1996) is used to report in detail the 'costs of environmental protection' (e.g. water treatment, waste segregation costs) and the 'costs of environmental damage' (e.g. fees for waste or effluent disposal), either as part of existing cost accounting or as a separate costing procedure. Such reporting of environmentally related costs remains limited essentially to end-of-pipe costs (cost segment 5) and disposal costs (cost segment 6). Much larger potential lies in reducing the costs of materials – but it is precisely this potential which is left untouched by traditional environmental costing.

The **waste costing** (cf. also Fischer and Blasius, 1995; Fischer, 1997) approach does go further. It considers not only the purely end-of-pipe costs and disposal costs, but also the materials costs involved in materials losses and the share of system costs connected with materials losses (i.e. cost segments 2, 4, 5, and 6). Waste costing thus places materials efficiency much more clearly in the foreground than does traditional environmental costing. However, in waste costing, the products and their packaging (and thus by far the largest materials quantity and the largest costs block) are still left untouched.

**Flow cost accounting** (cf. also Strobel, 2001; Wagner, B. and Strobel, 1999; Hessisches Ministerium für Wirtschaft,[7] 1999) aims to identify and analyse the entire system of materials flows as an essential cost driver. Not only the materials costs but also all the system costs are assigned to materials flows (cost segments 1 to 6). Flow cost accounting can thus be seen as a kind of total cost accounting, encouraging the following actions:

1. develop products that require less materials,
2. develop product packaging that requires less materials, and
3. reduce materials losses (e.g. rejects, scrap, cut-offs), and, as a result of this, reduce waste (i.e. solid waste, effluent, exhaust).

The following table, finally, compares the relevance in cost terms of traditional environmental costing, of waste costing, and of flow cost accounting.

### 5.1.4. *The basic idea in flow cost accounting*

The instrument of flow cost accounting (see also Wagner, B. and Strobel, 1999; Hessisches Ministerium für Wirtschaft, 1999; Strobel, 2001) shifts a company's in-house materials flows to the centre of the costs analysis.

In flow cost accounting, in order to achieve this vital transparency, the **values and costs of the materials flows** are divided up into the following categories: 1. materials, 2. system, and 3. delivery and disposal.

---

6 System costs by definition are those costs that are incurred in the course of in-house handling of the material flows. System costs are incurred by the company in its efforts to ensure that material movements can be made in the desired form.

7 Hessisches Ministerium für Wirtschaft = Hessen Ministry for the Economy, Traffic, and Development.

Table 5.1. Share of total production costs attributed to each approach.

| | % | Environmental costing | Waste costing | Flow cost accounting |
|---|---|---|---|---|
| **1) Material costs** in the product (product + packaging) | 57 | | | |
| **2) Material costs** in material losses | 6 | | | |
| **3) System costs** for products | 28 | | | |
| **4) System costs** before material losses occur | 6 | | | |
| **5) System costs** after material losses occur | 2 | | | |
| **6) Disposal costs** | 1 | | | |
| **Percentage of costs considered** | | **3%** | **15%** | **100%** |

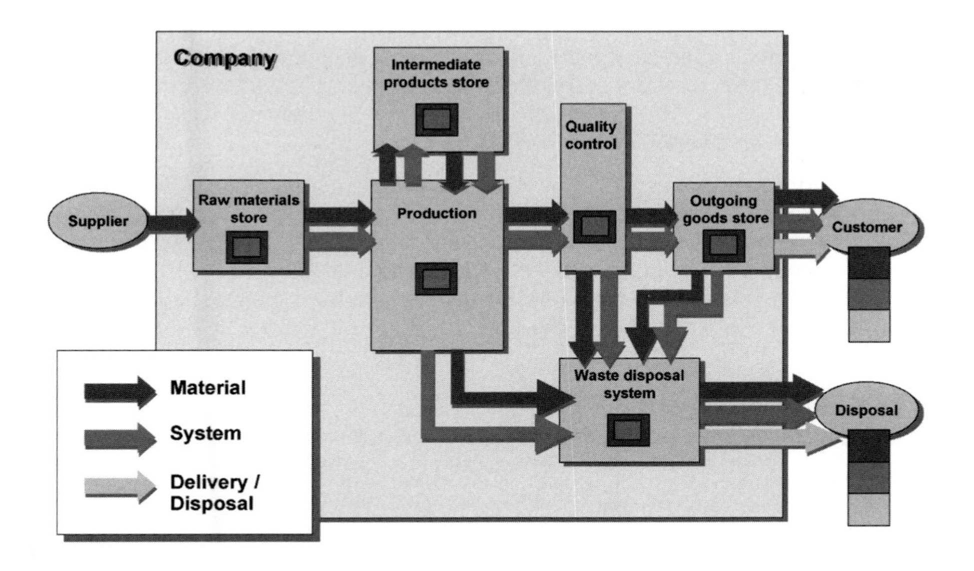


Figure 5.4. The basic idea of flow cost accounting.

To calculate the **materials values and costs**, one needs to know in physical, quantitative terms how large the materials flows and stocks are. Usually, the existing materials management systems and production planning systems provide (at least for the product materials) a comprehensive database, which has merely to be adapted and expanded. Materials costs, often also termed 'materials consumption', can then be identified by working out which materials flows are cost-relevant.

The possibility of reporting materials values and costs for materials flows and materials inventories separately throughout the company is also known as '*materials value orientation*'.[8] Materials value orientation is the core of flow cost accounting: it will show where the hotspots are.

For the purposes of assigning the **system values and costs**,[9] materials movements have to be treated as cost drivers. Irrespective of whether it concerns raw materials, intermediate or semi-finished goods, or materials losses, each in-house materials flow can be seen as a cost carrier for apportioning system costs systematically according to cause. This category encompasses costs that are incurred in the company for the purpose of keeping the materials throughput going, especially personnel costs and depreciation.

System costs are associated with the outgoing materials flows first (e.g. from the 'production' cost centre) and then passed on as system values to the flows and inventories.

Flows leaving the company need to be assigned their specific **delivery and/or disposal costs** involving payments to external companies (transport) and governmental bodies (fees).

It is most important to record and manage separately the afore-mentioned three categories of values and costs. This applies both to the materials flows and the materials inventories.[10] Practical experience shows that this can bring about truly pioneering changes in a company's perceptions, and in how it makes decisions and then acts upon them.

## 5.2. Methods of flow cost accounting

This section presents a number of methods of flow cost accounting. It discusses the necessary information base, different accounting procedures and how these can be reported.

### 5.2.1. *Methods – an overview*

The **information base** needed for flow cost accounting consists of the materials flow model and a defined database. The *materials flow model* maps the structure of the materials flow system. The *database* contains the data which is needed to quantify the materials flow model. The database refers to both materials flows and inventories, and also includes other relevant system data. It is used as the basis for calculating the quantities, values, and costs assignable to the flow model.

---

[8] In the US such approaches are termed 'Materials only costing' (MOC), see e.g. Coopers and Lybrand (1997), Lucent Technologies (1998).

[9] 'System costs' by definition are those costs that are incurred in the course of in-house handling of the materials flows (e.g. personnel costs, depreciation). System costs are incurred by the company in its efforts to ensure that materials movements can be made in the desired form. System costs being apportioned to materials flows are defined as 'system values'.

[10] In traditional cost accounting, soon after the first processing stage when the intermediate product is calculated, materials costs and system costs are already aggregated together. It thus very soon becomes impossible to list costs and values separately according to the three afore-mentioned categories, for either materials movements or inventories.

The **accounting elements** can be divided into materials flow accounting and system cost accounting. The purpose of *materials flow accounting* (see also Strobel and Wagner, 1999) is to check whether the database is in order and to assign data to the flow model, involving different kinds of calculations. Materials flow accounting consists of the following individual elements:

- materials flow quantity assessment;
- materials flow valuation; and
- materials flow costing.

*System cost accounting* is based on materials flow accounting and used in a multi-stage procedure to allot system costs to the materials flow model. System cost accounting has the following individual elements: the definition, allocation and apportionment of system costs (cf. esp. Hessisches Ministerium für Wirtschaft, 1999).

The **results and report forms** define the way in which the data in flow cost accounting is edited and made available to staff. The '*materials flow model with data*' is the most important result form. It shows much more information than previous accounting approaches. The '*flow cost matrix*' shows the outgoing materials flows such as products and packaging, and allocates materials losses according to cost emergence to the following categories: materials costs, system costs, and delivery and disposal costs. In addition, the '*flow cost report*' makes it possible to compile tables on an ad hoc basis in order to assist specific decision-making processes.

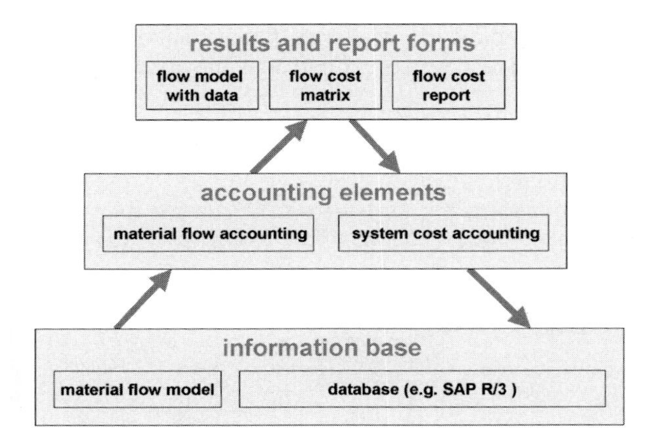


Figure 5.5. The components of flow cost accounting.

The subsequent elements of flow cost accounting consist of providing the information base, the different elements of accounting and the various results and report forms. However, when new information needs arise, there may be pressures to change the information base and to adopt new modes of accounting.

## 5.2.2. *The information base in flow cost accounting*

### *Materials flow model*

For the purposes of describing a company in terms of a materials flow system, we use the **materials flow model** (cf. also LfU, 1999). Using this model, it is possible to map out all the operation-specific materials flow structures. This helps to raise the level of transparency of all intra-company interrelationships. An organization diagram along these lines can establish a uniform overview, providing a basis for communications between various key players and decision makers.

When setting up the model, the first step is to define the **system boundaries** of the materials flow system and thus of the flow model. The system should encompass all those company units that have an impact on the materials flow. It is recommended that the boundaries should coincide with an individual production site, but in principle multi-site system boundaries can also be selected. Besides the system boundaries, other basic elements of a materials flow model are the quantity centres and the materials flows themselves (cf. Krcmar et al., 2000).

**Quantity centres** are physical units where materials are processed or stored. Typical examples of internal quantity centres are a company's storage areas, production areas, and filter systems. External quantity centres (e.g. supplier, customer, final disposal agent) form the flow-relevant outside world for the company, i.e. the source and destination points of the company's input and output flows when these cross the system boundaries. External quantity centres should, therefore, always be incorporated in the materials flow model. What materials are stored, as well as where they are stored and in what quantities, can be determined only within the framework of materials flow accounting.

**Materials flows** represent the possible materials movements from one quantity centre to another. For the purpose of flow cost accounting, these movements can be represented by one materials flow without much detail. Typical materials flows run, for example, from the raw materials store to production, or from effluent treatment to the drainage system. In this case also it is materials flow accounting that has to generate the correct quantities.

Quantity centres and materials flows should be recognisable in the data by means of code numbers. By so doing, the data used in materials flow accounting will have a direct reference to the flow model.

As well as the use of graphs, it is preferable to document the results of materials flow modelling in tabular form. These '*materials flow tables*' and '*quantity centre tables*' may

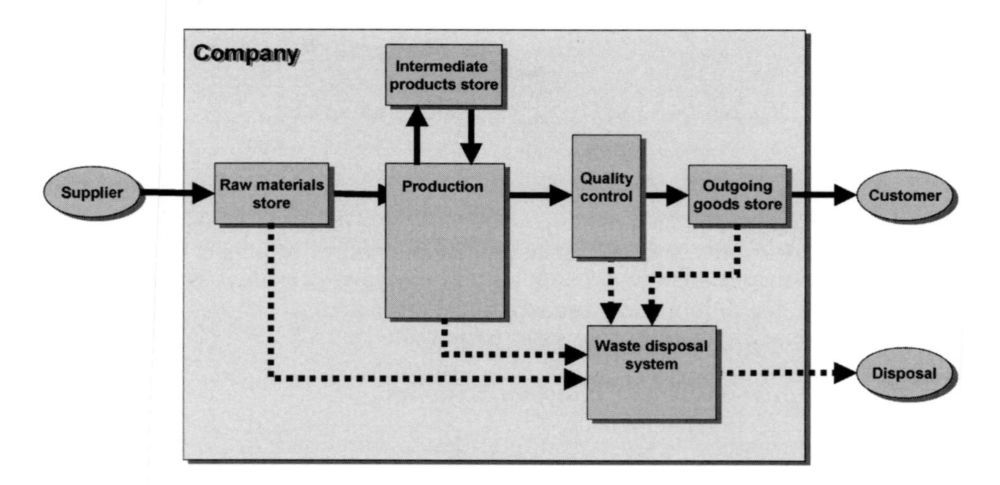


Figure 5.6. Example of a materials flow model.

also contain further additional relevant information such as areas of responsibility, competences, and computer systems.

### *Database requirements*

#### Database for materials flow accounting

A comprehensive database of high quality is a prerequisite for materials flow accounting. The database required for materials flow accounting must contain not only prices and the net bill of materials (BOM), but also quantities for each period. Quantitative data refers both to materials movements and to materials inventories. Usually, access to existing quantitative data is no problem. Materials management and production planning and control systems often contain the majority of the quantity data needed.

The quantity data describing **materials movements** can be used to allocate the appropriate quantity data to materials flows in the materials flow model. For materials movement quantities, the standard ERP[11] systems (available from software houses like SAP, BAAN, SSA, etc.) provide the following record types:

- storage postings;
- production order data;
- materials quantity postings per cost centre; and
- process and machine data.

The **materials inventories** can be allocated as start-of-period and end-of-period inventories of the quantity centres in the materials flow model. The standard ERP systems refer to inventory quantities per storage location. It is useful, therefore, to associate storage locations clearly with quantity centres. Inventory data for each storage location is recorded as a result of stock-taking, either based on a complete count or on sample-counting.

#### Database for system cost accounting

The database required for system cost accounting consists of system costs and system data. The records for system costs comprise information on type of costs, e.g. personnel costs or depreciation, and where the costs were incurred. The system data provides quantitative information on, for example, surface areas and number of plant units, etc., on the basis of which system costs can be correctly allocated to the materials flows.

## 5.2.3. *Accounting elements in flow cost accounting*

### *Materials flow accounting*

Materials flow accounting is used to quantify the materials flow system on a per-period basis, thus also permitting an ex-ante perspective. Materials flow accounting comprises the following inter-linked computing elements (see also Strobel, M. and Wagner, F., 1999): materials flow quantity assessment, materials flow valuation, and materials flow costing.

#### Materials flow quantity assessment

Materials flow quantity assessment is at the heart of materials flow accounting. It determines for each material and each individual movement (i.e. per material number, article number, part code number, etc.) the quantity of physical units (e.g. pieces, kilograms,

---

[11] ERP (abbr.) = Enterprise Resource Planning.

meters) per period. This data may either be derived from available process documentation or may be downloaded from an existing database, or may specifically result from materials flow quantity assessment.

The second function of materials flow quantity assessment is to determine per quantity centre the start-of-period and end-of-period inventories. It is typical of materials flow quantity assessment to view materials not only per processing stage (i.e. as either intermediate product, finished product, or unsorted waste) but, parallel to this view, also to show all materials in terms of **incoming materials**. Incoming materials are all those materials that enter the company from an outside source, and are tracked through all the materials flows and all the quantity centres, right up to the point where they finally leave the materials flow system. Materials may be aggregated only if they are recorded in consistent units of measurement. Based on these incoming materials (and taking account of the laws of thermo-dynamics),[12] it is now possible to calculate possible materials losses by comparing inventory figures with recorded arrivals/departures figures. Assuming that the counts done at the quantity centres are accurate, this differential calculation should produce the result 'zero'.

In practice, however, the differential result often deviates substantially from zero. Such deviations indicate either inconsistencies in the database, or materials movements which the database does not reflect. The latter usually are materials losses at quantity centres that can be related to specific materials flows in the materials flow model. Therefore, it relatively easy to reveal specific materials losses in a systematic way.

### Materials flow valuation

Materials flow valuation is based on materials flow quantity assessment. The materials value is produced, for both materials flows and materials inventories, with the following simple formula: materials **value = quantity × input price**, where the 'input price' is the price for the external purchase of the materials.

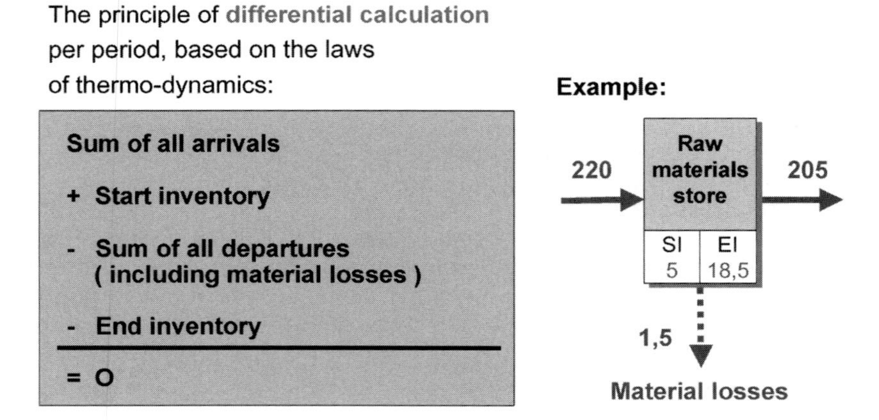


Figure 5.7. Calculating the in-house flow quantities.

[12] The first law of thermo-dynamics states that, in a closed system, materials and energy may be transformed but never generated or destroyed.

Materials flow valuation envisages a fixed input price per material used over a given period. This price may be the standard price taken from cost accounting, the start-of-period price, the end-of-period price, or the average price. This fixed input price is then used to value the materials at all the quantity centres and in all the materials flows. This applies irrespective of the actual processing stage or condition of the materials.

The same differential calculation as used in materials flow quantity assessment can also be used in materials flow valuation. The difference does not confuse quantities and prices: it is clearly a quantity difference since there is a fixed input price. Thus, using this fixed input price, it is possible to aggregate in a consistent manner the values of materials flows, inventories, and differences.

### Materials flow costing

Materials flow costing is based on materials flow valuation. However, it includes only those materials movements that are driving costs, which have to be defined. The present total costing procedure and turnover costs procedure already adopt such definitions, but in principle other definitions can also be used. The definitions indicate thresholds from where costs materialise.

According to the **total costs procedure** (or 'total cost type of accounting'), costs are understood, by definition, to be those variables caused by all materials movements in production during a certain period. The term 'materials consumption'[13] is sometimes also used in this context. The threshold in this case runs through the middle of the materials flow model.

According to the **turnover costs procedure** (or 'cost of sales type of accounting'), costs do not arise until the materials are sold in product form and leave the company. The threshold in this case runs along the system boundaries in the materials flow model and refers to all outgoing product flows.

Materials flow costing is capable of listing all the materials costs per period for the threshold in question, but fails to produce the beginning-to-end transparency that materials flow valuation can give.

### *System cost accounting*

In materials flow accounting the quantities, values, and costs of materials flowing through a company are recorded in detail and both the flow-related threshold(-cuts) where materials costs are incurred, and the outgoing delivery and final disposal costs, are defined. To enhance flow transparency even further, it is now necessary to define those costs that are incurred in the course of in-house handling of the materials flows. System costs are incurred by the company in its efforts to ensure that materials movements can be made in the desired form. The objective of system cost accounting is to define these other materials-flow-related costs (or, for short, the system costs) and to apportion these to the materials flow model systematically and on a strictly cause-oriented basis (see Hessisches Ministerium für Wirtschaft, 1999).

---

[13] In thermo-dynamics terms, this 'consumption' concept would seem somehow inappropriate. It often leads to the false association that materials somehow actually disappear, but literally this is impossible. The true picture is that one part of the materials goes into a product and the other part becomes a materials loss.

### System cost delimiting

The first step in system cost accounting is to clearly delimit the system costs. For this purpose the company's prime costs must be divided between materials costs, system costs, delivery and disposal costs, and 'non-flow costs'. The objective of system cost accounting is to allocate appropriately all the system costs to the materials flows in the flow model. Stipulating the non-flow costs determines which costs are not to be allocated to the materials flow model. Costs that are undoubtedly non-flow in nature include, for instance, the costs of sponsorship activities or of external auditors. However, it is often not possible to distinguish clearly between costs which should, and costs which should not, be allocated to the flow model. Decision aids here might be the degree to which the cost in question can be influenced, and the degree to which it varies with the materials flow. System costs always include those costs that:

- Occur inside the system boundaries;
- Are incurred in planning, implementing, and maintaining materials throughput; and
- Cannot be considered materials costs or outgoing delivery and final disposal costs.

System costs can be clearly distinguished from the latter because they are usually not connected with any payments to third parties (i.e. external agents). Costs for waste removal performed by external firms or municipal waste disposal fees are not system costs; these are already covered in the category of outgoing delivery and final disposal costs. If, by way of contrast, waste transport were performed using vehicles from the company's own fleet, the costs would have to be classified as system costs.

The exact delimitation of system costs also depends essentially on the areas to which they are to be allocated. This is the next step.

### System cost allocation

The second step in system cost accounting is to allocate system costs to the areas in the flow model where they are incurred. Depending on the objective of system cost accounting, there may be pronounced differences in the amount of system costs to be considered and in the way they are apportioned.

In simplified terms, system costs can be apportioned on any of three levels:

- quantity centres;
- cost centres; and
- processes.

The allocation of system costs to **quantity centres** is the least complicated form of system cost accounting. It considers only those costs incurred at quantity centres that directly participate in the materials flow. The next step then becomes a straightforward process of assigning the quantity centre costs to the materials flows leaving the quantity centre. Typical quantity centre costs are personnel costs for production and warehouse staff, and depreciation of a production plant and other equipment that actually comes into contact with the materials flow.

The allocation of system costs to **cost centres** located directly in the materials flow considers not only the direct quantity centre costs but also the cost centre's overheads, such as the employment costs of the head of the cost centre or its computer equipment. However, cost centres usually cover a number of quantity centres, and thus permit aggregated statements only. If the costs of the cost centres are listed as system costs, the system costs will be higher than with the above variant based purely on quantity centre costs. Consequently it will hardly be possible to allocate costs to the materials flows on a cause-oriented basis. Cost-centre-related system costs can also be characterized as arising largely through information and control activities with a more or less direct contact with the materials flow.

The allocation of system costs to **processes** is the most comprehensive variant – and also by far the most work-intensive. It usually requires its own independent activity-based cost accounting system. Having specified which processes are relevant for the company (e.g. 'procurement'), costs are recorded per process on this basis. This approach leads to very high system costs and reduces the non-flow costs to virtually zero. However, in the subsequent apportionment step, where the process-specific system costs have to be allocated, there are some very blurred edges.

#### System cost apportionment

The objective of this step is to apportion system costs, systematically and as far as possible cause-related, to intra-company materials flows. As system costs have to be derived from existing cost centre information that has been defined in terms of organizational units rather than of flows, they have a limited potential to generate flow-specific information. By means of apportionment, these rather static system costs are transformed into **materials-flow-specific system values**. These can be differentiated according to types of materials, and then tracked throughout the entire company. This makes transparent, for example, what percentage of system costs ends up as waste without creating added value.

Here, different allocation methods can be used, but they always follow the same principles – irrespective of whether this is done on quantity centre level, cost centre level, or process level. It is crucial that apportionment is performed separately for each unit (quantity centres, cost centres, or processes). Since quantity centre costs are also part of cost centre costs or of process costs, to mix them would risk certain data being included more than once ('double-counting').

First, one should focus on those materials flows that exceed certain previously determined limits. Then it should be examined whether these materials flows have a major impact on the operational costs of the unit in question, which will not always be the case.

The system costs must be allocated in full to the relevant flows, which can be done proportionately (e.g. pieces or weight in kilograms) or according to a key to be specially defined for this purpose.

### 5.2.4. *Results and report forms in flow cost accounting*

Materials flow accounting and system cost accounting produce an enormous database. The materials flow and inventory values, and the system costs and values, are acquired as records in database form (e.g. in MS Access). However, given the sheer volume of this data, these tables are too large and complicated to be used to support directly decision-making procedures. In flow cost accounting, therefore, the database must first be evaluated and compressed, either in a form which opens up a total overview of flow costs and values in the materials flow system, or which gives a number of excerpts. These excerpts can, for instance, summarize the figures of cost centres.

Whereas the '*materials flow model with data*' and the '*flow cost matrix*' provide a total overview, the '*flow cost report*' (e.g. materials value report) gives particular excerpts of the materials flow system (see Wagner and Strobel, 1999, p. 57 et seq.).

*Materials flow model with data*

To visualise the results of materials flow accounting and system cost accounting, data is given for certain periods and entered into the materials flow model. This *materials flow model with data* has the highest information content, if compared with conventional accounting approaches. It also, in our experience, provides management and staff alike with the clearest possible way of presenting and reporting results, because the materials flow structure followed is most familiar to them.

In many cases, the original materials flow model was used to map out a highly complex materials flow structure. Especially with differentiated materials flow models in which materials movement between two quantity centres is shown broken down into materials types with directional flows and arrows, it is recommended for the sake of a clearer total overview to compress the model in a meaningful way, i.e. according to objectives.

A materials flow model with data referring to materials inventory values and flowed materials values usually already allows meaningful inferences to be made. It is usually also possible, so long as this does not impair clarity, to add the system costs and values and the delivery and disposal costs. However, the latter should not impair a clear presentation of figures.

*Flow cost matrix*

The *flow cost matrix* represents in tabular form **flow cost accounting data in simplified and standardized form at a defined threshold-cut in the flow model**. Its structure remains constant even if the materials flow structure is modified. This makes it a particularly useful and meaningful form in which to present and report results. The *flow cost matrix* can be used to describe the development of a particular company site over several years, or to compare different sites within the company more reliably and exactly. It can also be used to compare reliably different companies in the same industry in terms of their respective flow cost structures.

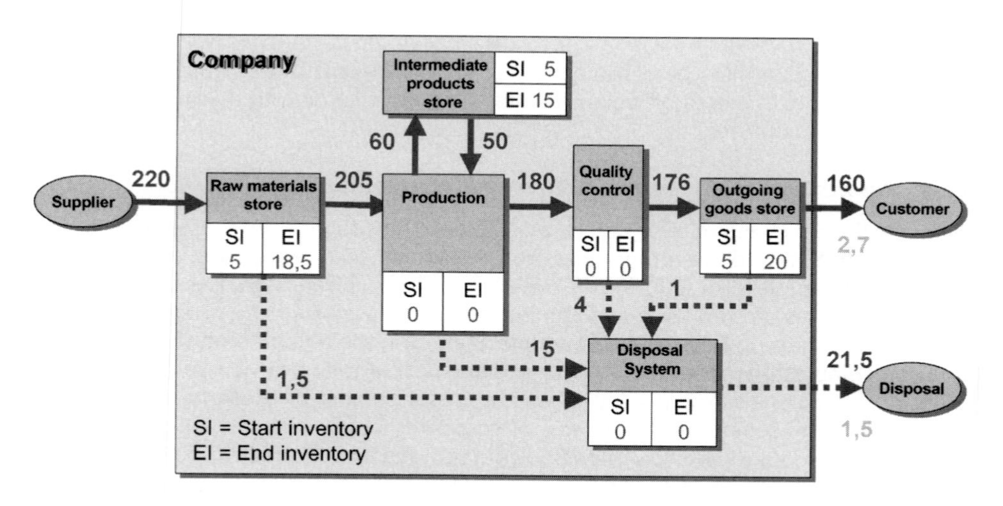


Figure 5.8. Materials flow model with materials flow values and delivery and disposal costs (in millions of US $).

Irrespective of the company's specific materials flow structure, the **structure of the flow cost matrix** is always as follows:

The matrix contains columns headed with the three flow cost categories 'Materials costs', 'System costs', 'Delivery and disposal costs'. In the rows below, the objects under consideration are broken down according to outgoing materials flows, i.e. according to those materials flows which pass from inside to outside the system boundaries for the object under consideration. The top-level distinction to be made in these rows (a basic idea of flow cost accounting), is into the two flows:

* products; and
* residual waste.

The **product flow** can subsequently be broken down into the product's components. Typically these include 'raw materials', 'packaging' or 'other trade goods'. These components can, depending on the objectives, be broken down even further, for example into raw materials according to the main types of incoming materials (e.g. metals, plastics).

The **residual waste** should be subdivided first of all into the categories 'solid waste', 'effluent', and 'exhaust'. It is definitely recommended that these too be broken down further according to some additional classification (e.g. raw materials, packaging, outside purchased parts) and possibly even further, as already shown, according to individual types of materials. As an alternative to classification by medium, residual waste can also be differentiated according to the reason for its becoming waste (e.g. rejects, cut-offs, cutting mistakes, chippings, the destruction of obsolete or perished goods, repackaging, etc.).

In compiling this data it must first be decided where in the flow model the decisive threshold-cuts for determining the flow costs are to be set (cf. subsection Materials flow costing). To concentrate on flow costs which arise after production corresponds to the **total costs procedure**, whereas concentrating on flow costs leaving the company corresponds to the **turnover costs procedure**.

The *flow cost matrix* in the following example is oriented towards the turnover costs procedure.

**Typical flow costs structure** *(example: pharmaceuticals industry)*

| Production costs ( in US $ m. ) | Material costs | System costs | [illegible] | Total |
|---|---|---|---|---|
| **Product** | 120 | 25 | 0.2 | **145.2** |
| **Packaging** | 40 | 25 | 2.5 | **67.5** |
| **Material losses** | 21.5 | 6.4 | 1.5 | **29.4** |
| **Total** | **181.5** | **56.9** | **3.9** | **242.3** |

Material costs account for a considerable percentage ! *(here: 75 % of production costs)*

A considerable share of costs is caused by material losses ! *(here: > 10 % of production costs)*

Figure 5.9. Illustration of a flow cost matric (simplified).

It should be pointed out once again at this juncture that the total costs of the aggregated flow 'residual waste' contain more than just the costs which are incurred purely for final disposal. These costs and the costs for end-of-pipe technology (i.e. the system costs incurred by cost centres downstream from production) are, as already explained in detail in Section 4.1.3, relatively small anyway. A really decisive costs block here (besides the **dominant block of materials costs involved in residual waste**) is above all the share in the system costs incurred by production. As a way to enhance transparency further it may thus in some cases be advisable to subdivide the column 'system costs' to correspond to the three in-house flow phases, into 'system costs before production', 'system costs during production', and 'system costs after production' (= end-of-pipe costs).

### *Flow cost report*

The *flow cost report* offers the possibility to select **specific excerpts** and compile these in user-defined tables for the purposes of analysing further the flow costs thus presented in focussed form. The *flow cost report* represents the minimal solution in a continuous results and reporting system. It is used in particular by staff responsible for individual quantity centres or cost centres to provide themselves with important information for monitoring success.

A variant of the *flow cost report* which has previously been used very often in practice is the '*materials value report*'. This report variant can be compiled both per **quantity centre** and per **quantity centres group** (e.g. for all the quantity centres within a cost centre or within a production area).

The **materials value report** (or in short: '*materials report*') first lists in compressed form all materials values per period entering a quantity centre and then leaving it again. Outgoing materials values are divided into two categories: firstly, the materials value output by the quantity centre in the desired form as (intermediate) products, and, secondly, the compressed value of materials leaving the quantity centre in the form of materials losses. The *materials value report* of course also lists the materials value of any inventory change within the period, calculated as the difference in value between the start-of-period inventory and the end-of-period inventory. The *materials value report* can also be differentiated for individual materials classes (cf. Strobel and Wagner, F., 1999, p. 28).

The *flow cost* report should also always contain **relevant indicators** (e.g. here, for the purposes of assessing materials efficiency). Flow cost accounting supplies comprehensive data which can be used in a wide variety of ways to form such indicators. These indicators are particularly suitable for evaluating once-only separate accounting procedures. They permit an initial assessment as to the relevance of flow cost accounting. For continuous flow cost accounting, however, time-series are preferred.

# 6. Resource-Efficiency Accounting[1]

*Thomas Orbach and Christa Liedtke*
*Project co-ordinator, resp. senior fellow, both working with the Wuppertal Institute For Climate, Environment And Energy (Wuppertal, Germany), Division For Materials Flows And Structural Chance.*
*E-mail: thomas.orbach@wupperinst.org; christa.liedtke@wupperinst.org*

## 6.1. Introduction

In response to criticism of existing eco-management accounting systems, the Sustainable Enterprise Program of the Wuppertal Institute has developed the concept of Resource-Efficiency Accounting (REA).

In this concept, not only materials are considered to be a resource, but also money: both have to be used efficiently in order to guarantee the sustainable success of a company (Orbach et al., 1998). The core thesis of Resource-Efficiency Accounting is:

> *Only if economic and ecological aspects are considered simultaneously and life-cycle-wide, can all cost reduction potentials of a company be explored. These cost reductions should be ecologically sound.*

## 6.2. Conceptual background of Resource-Efficiency Accounting (REA)

REA has both an economic and an ecological component; these are combined within the framework of Resource-Efficiency-Portfolios. This means that REA does not express environmental effects in monetary units. The economic component relies on data derived from the company's cost accounting system. As there are no standards for cost accounting, REA can be integrated into different systems (Absorption Costing, Direct Costing, Activity Based Costing). The aim is to reveal hidden costs, i.e. costs that have not been allocated to the cost centre or to the product that caused these costs. The transparency of company materials flows within REA can be used to modify existing cost allocation procedures.

The ecological component of REA is based on the MIPS concept, which has been developed by Friedrich Schmidt-Bleek at the Wuppertal Institute. The MIPS concept has adopted the internationally agreed assumption that life-cycle-wide inputs of primary materials can be used to indicate the general environmental impact potential of products and services. Increasing materials input will generally indicate a rising pressure on the environment, and a decreasing input will reflect a decreasing pressure (Schmidt-Bleek, 1994).

The MIPS concept has been developed in face of methodological and practical problems with output-oriented assessment approaches such as LCA, *Ökobilanz* (eco-balance) and *Produktlinienanalyse* (product-line analysis) (Schmidt-Bleek, 1994):

- It is doubtful whether it will ever be possible to evaluate or even to know all possible impacts of human action. Even now man cannot fully control nature or even predict the behaviour of eco-systems.

---

[1] This chapter derives from an Environment & Economy research programme on eco-management accounting sponsored by the Netherlands Organization of Scientific Research (NWO).



*M. Bennett et al. (eds.), Environmental Management Accounting: Informational and Institutional Developments,* 83–90.
© 2002 *Kluwer Academic Publishers. Printed in the Netherlands.*

- Output-oriented approaches are at odds with the precautionary principle. In the past it happened that certain harmful impacts were discovered only long after the materials involved had been discharged into the ecosystem, causing serious and often irreversible damage (e.g. the CFC problem). At best, output-oriented methods can only bring to light damage which has already been inflicted on the eco-system.
- Output-orientation favours 'end-of-pipe technologies'. These technologies have been the main weapon in the fight against toxic substances and emissions over the last twenty years. However, a high price had to be paid for filters, catalysts, purification plants, etc. Nowadays there is a common understanding about the need for prevention, or at least for integrated technologies. This is particularly relevant to reducing $CO_2$ emissions and the related reduction in the use of fossil fuels.

In recent years it has become clear that an adequate ecological policy must go beyond attempting only to combat harmful substances, to a multi-issue approach in which dematerialisation plays a major role. To lead the economy of the industrialised countries to 'a sustainable path', the enormous quantities of materials that the industrialised countries use as economic inputs have to be reduced by a factor of 10 (Schmidt-Bleek, 1998). This aim can be achieved only by means of an input-oriented ecological policy and its related instruments.

Schmidt-Bleek suggested comparing products, services and infrastructure on the basis of their materials input (related to whole life cycles) per service-unit, using a measure called **M**aterials **I**ntensity **p**er **S**ervice-Unit (MIPS). The service-unit as the functional unit makes it possible to compare different products that deliver the same kind of service (in the following, the expression 'product' includes professional services and infrastructure as well).

Materials intensity is calculated within the framework of **M**aterials **I**ntensity **A**nalysis (MAIA), in which all inputs of raw materials and resources related to a product under study are included. Using MAIA, raw materials are defined to comprise all materials (including energy carriers) moved by man from natural sites by technological means. The total materials input includes all productive life-cycle stages (such as raw materials extraction, pre-production, transport, etc.), the use of the product, and recycling/waste disposal. The measure is a mass unit, in kilos or tonnes. The total materials input minus the proper weight of the product represents the *'ecological rucksack'*, which means the total mass of all materials taken from or moved into the environment in order to produce the product without, however, themselves becoming part of the product itself. Relating the materials input to a service unit or to a specific quantity generates the measure of materials intensity. The materials intensity of 1 tonne of steel, for instance, amounts to 7 tonnes of inputs per tonne of product output (t/t). Hence this results in an ecological rucksack of 6 t/t.

The MIPS approach has several advantages in terms of materials inputs across the whole life-cycle (Hinterberger and Welfens, 1996):

- It is assumed that environmental impacts cannot be assessed for every aspect and in every detail, since eco-systems are too complex for this. It is however still possible to quantify general tendencies. MIPS can serve as a screening tool within the framework of an ecological assessment.
- Input data either already exists in companies, or is easier to record than is output data. Thus, the MIPS concept is easy to implement in companies. For special purposes, it is also possible to generate output indicators.
- The common basis for the assessment is the mass unit (kg or t), which avoids the methodological problems of LCA during the evaluation and assessment procedure. It

is not currently possible to aggregate the different categories of LCA on a scientific basis.

- The results of MIPS are reproducible and show general tendencies. The MIPS concept relies on a strict methodology, which leaves very little room for subjective interpretations.
- The MIPS concept considers the whole life cycle of products when expressing the ecological rucksack.
- MIPS is a suitable measure to indicate a company's environmental performance, as it is simple, comprehensible and insightful. The ecological impact potential of a product could be given, for example, by labelling it with an ecological rucksack.

The Division of Materials Flows and Structural Change of the Wuppertal Institute has calculated the materials intensity of the most important materials and of a lot of products. The results are published on the Internet (http://www.wupperinst.org).

## 6.3. Use at company level: Company input-output analysis

A first step in REA aims to record company-wide materials inputs and outputs (a top-down approach). During this step, the company itself is considered as a black box: all materials and energy flows are recorded and listed on a company input/output balance sheet (see Table 6.1) without assessing them (Liedtke et al., 1995).

The four main categories – input, output, stock and inventory – are divided into subcategories so as to record the materials and energy flows as exactly as possible. All data is recorded in a common weight unit (e.g. kg or t). Most of the data necessary normally already exists in conventional company information systems, such as the information system of the purchasing department or the cost accounting system. As materials and energy

Table 6.1. Structure of a company input/output balance sheet (Liedtke et al., 1995).

| *I* | *Input* | *O* | *Output* |
|---|---|---|---|
| I.1 | Raw materials | O.1 | Products |
| I.2 | Energy | O.2 | Energy |
| I.3 | Water | O.3 | Waste water |
| I.4 | Air | O.4 | Vitiated air |
| I.5 | Products | O.5 | Solid waste |
| I.6 | Merchandise | O.6 | Merchandise |
| I.7 | Communication | O.7 | Communication |
| I.8 | Services | O.8 | Services |
| I.9 | Transports | O.9 | Noise |
| *L* | *Stock* | *B* | *Inventory* |
| L.1 | Raw materials | B.1 | Land areas |
| L.2 | Energy | B.2 | Structure |
| L.3 | Water | B.3 | Plant and equipment |
| L.4 | Products | B.4 | Vehicle fleet |
| L.5 | Merchandise | | |
| L.6 | Communication | | |

flows are considered to be cost drivers there, consumption is measured in monetary values; however, consumption expressed in weight units can also be derived from the available data. Nonetheless, practical experience shows that not all data necessary can be obtained in this way – Inventory data in particular are frequently not available, meaning that this has to be measured or calculated.

The company input/output balance sheet lists all materials and energy flows which enter or leave the company. This information can be used for internal and external communications, or for the participation in EMAS that demands the recording of environmental data at company level. It gives an overview of the actual company situation and it is suitable to set up environmental management targets as well as cost targets. It is possible, for example, to define targets for reductions of the company's materials flows. Actual performance in this area is monitored by a regular input/output analysis of the company. Furthermore, it is possible to link materials flow data with economic indicators in order to measure the company's resource productivity.

With the aid of regular input/output analysis, the success of the company's environmental protection efforts can be substantiated by means of transparent and reproducible indicators. The long-term objective should be to record environmental data continuously and to have a powerful information system in place that regularly updates the company input/output analysis. The input/output analysis can also be used for the comparison of different products, companies or sectors.

### 6.4. Use at process level: the process analysis

In the second step, company-wide materials and energy flows are linked to the company's production processes in order to know at what points in the company materials and energy are consumed. For this purpose, the company's activities are modelled into a flow diagram, which may be based on existing materials flow diagrams, and which represents the production processes with their mutual dependencies. For each process, an input/output balance is set up which contains internal inputs from upstream company processes, external inputs from outside the company, internal outputs into downstream company processes (main products of the process) and external outputs (by-products, emissions and waste).

To avoid double-counting, the materials intensity of a process is calculated on the basis of external inputs, since internal inputs (coming from company processes) have already been taken into account upstream. In order to allow a life-cycle-wide perspective, the external inputs are assessed by their ecological rucksacks. The corresponding financial-economic data is derived from the cost accounting system. The process analysis serves as a screening in order to identify 'economic and ecological cost drivers' within a company. The result is the so-called Resource-Efficiency Portfolio at the process level, representing the cost categories (in $) and materials intensity (MI) for each process (see Figure 6.1). All processes are classified according to their relevance (high and low), where criterion to determine the boundary between high and low will differ from one company to another (e.g. the average of all processes).

Based on the classification of the different processes, specific strategies for action can be derived. Those processes which are classified 'high/high' are very important for both the economic and ecological success of the company (processes 4, 6, and 7 in Figure 6.1). They should be given high priority as the highest potentials for reduction are likely to be found there (Gotsche, 1995). Table 6.2 indicates possible strategies for action for the different sections of the Resource-Efficiency Portfolio.

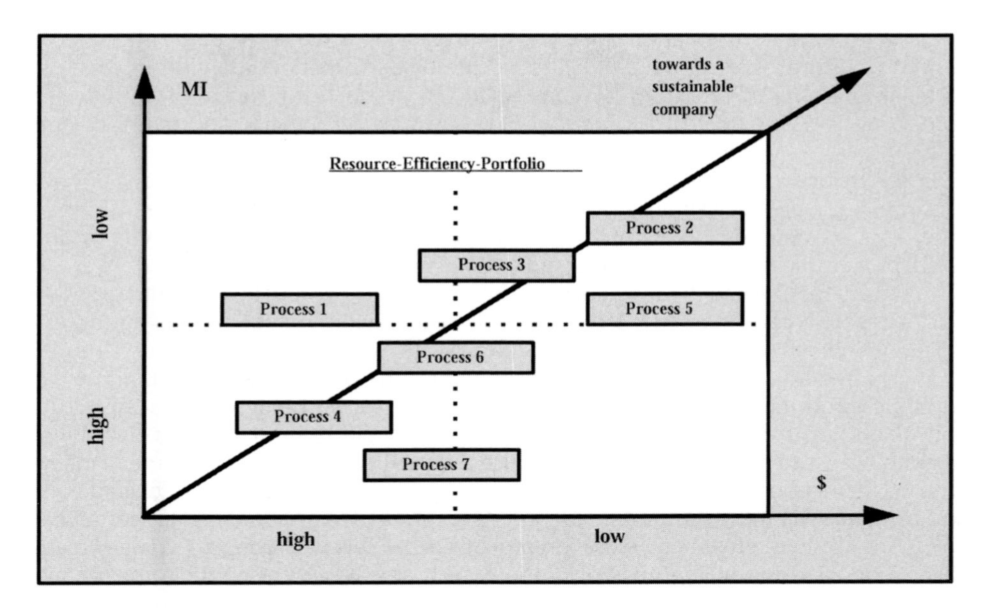


Figure 6.1. Resource-Efficiency-Portfolio at process level, considering costs ($) and materials intensity (MI) of processes.

Table 6.2. Strategies for action corresponding to the sections of the Resource-Efficiency Portfolio.

| | *Costs high* | *Costs low* |
|---|---|---|
| MI low | Selection of some processes. Review of several cost elements | Low economic and ecological relevance, urgent action not necessity |
| MI high | High economic and ecological relevance: Should be part of an environmental program, systematic review, exploration of reduction and substitution | Selection of some processes. Review of several inputs |

## 6.5. Use at product level: Mass accounting

Resource-Efficiency Portfolio at process-level aims to generate relevant information to support the optimisation of processes. This operational level is common for technical (or more general operating) staff, who facilitate the implementation of improvements at this level. Nevertheless, a powerful environmental management and eco-management system should provide information at the product-level, too. As the scope at this level is much broader, the whole life cycle of a product can be considered, so that the REA methodology can serve both process-oriented and product-oriented environmental management.

Economic information at product-level can often be taken from the cost accounting system, for example in the form of the cost of the final product or its contribution margin.

Detailed ecological information does not normally exist at product-level and has to be derived from process-level data with the aid of company mass accounting in which the materials intensity of processes is allocated to the products which are produced. Mass accounting is based on cost accounting and uses a similar structure of materials inputs and costs.

In the following, a mass accounting procedure will be demonstrated by analogy with Absorption Costing (see Figure 6.2), though mass accounting can easily be transferred to other cost accounting systems such as Direct Costing or Activity Based Costing in order to adapt it to the specific needs of companies.

Company mass accounting is subdivided into three steps, similar to Absorption Costing: mass category accounting, mass centre accounting and unit of output accounting as shown in Figure 6.2.

Mass category accounting systematically classifies the company's materials and energy flows including their ecological rucksacks. Direct masses can be directly allocated to the unit of output (the produced product) – as for example tyres in car manufacturing – whereas overhead masses are used to produce different units of output. Overhead masses, such as the materials inputs used in administrative activities, have to be allocated to the unit of output via mass centre accounting. Mass centres are all company devices (departments, installations, etc.) that are used to produce or sell several products. Overhead masses have to be allocated to the units of output which are causally responsible for the mass consumption. As the final step, the unit of output accounting accumulates all materials flows, with their ecological rucksacks, that occur up to the point when the product leaves the company.

In order to consider the whole life cycle of a product, the use and the deposit/recycling phase have also to be taken into account. Companies will increasingly find that their customers and client appreciate products that are is less materials and energy consuming in use. The European Commission, for example, has introduced a classification system which informs the client about the energy consumption of household goods. Materials and cost information for use and disposal/recycling can be obtained by calculating, measuring, and estimation, and should be included in the mass accounting procedure.

Resource-Efficiency-Portfolio at product-level can be used again to develop strategies for action. But unlike the process level, where costs and the added value are the matters of principal interest, an economic/ecological analysis of products has also to consider other economic figures such as the contribution and profit margins of products. Thus, in the company's decision-making procedure, different Resource-Efficiency-Portfolios will be set

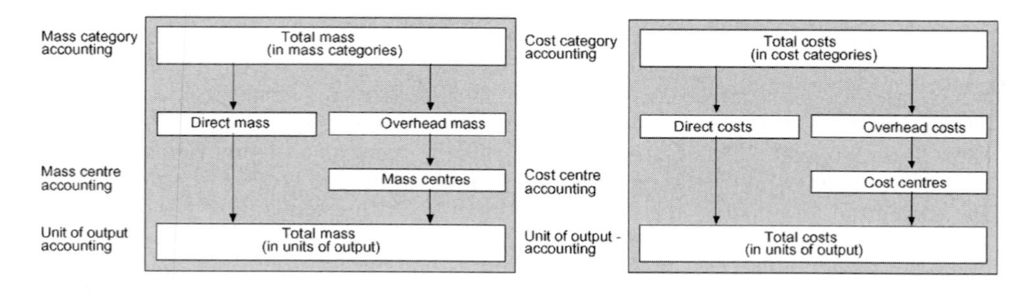


Figure 6.2. Mass accounting by analogy with absorption costing (Preimesberger, n.y.).

up representing various economic figures combined with the corresponding materials intensity.

This data becomes input to a Resource-Efficiency Portfolio at product level. Similar to the Portfolio at process level, all products of a company are classified according to their materials intensity and their costs in the categories (or other economic figures), high and low, in order to explore 'economically and ecologically cost driving products' (see product 4 in Figure 6.3).

It might be useful to distinguish between the production phase and the use-deposit/recycling phase. Products may be classified into:

- Production-intensive (mono-functional products for one single use such as packaging materials), producing little or no impact during use, or
- Use-intensive, where materials and energy consumption during use plays the major role (e.g. washing machines).

Mass accounting which includes use and deposit/recycling enables companies to improve the whole life cycle of their products including the optimisation of materials composition and design (Schmidt-Bleek; Tischner, 1995).

## 6.6. Conclusion

Environmental protection and environmental management are increasingly important for companies. This is due to several reasons such as the increasing costs of environmental protection measures, rising pressure of the markets, anticipation of future changes of economic conditions, legal compliance, and the increasing demands of stakeholders. The concepts of eco-management which have been presented in this paper try to link envi-

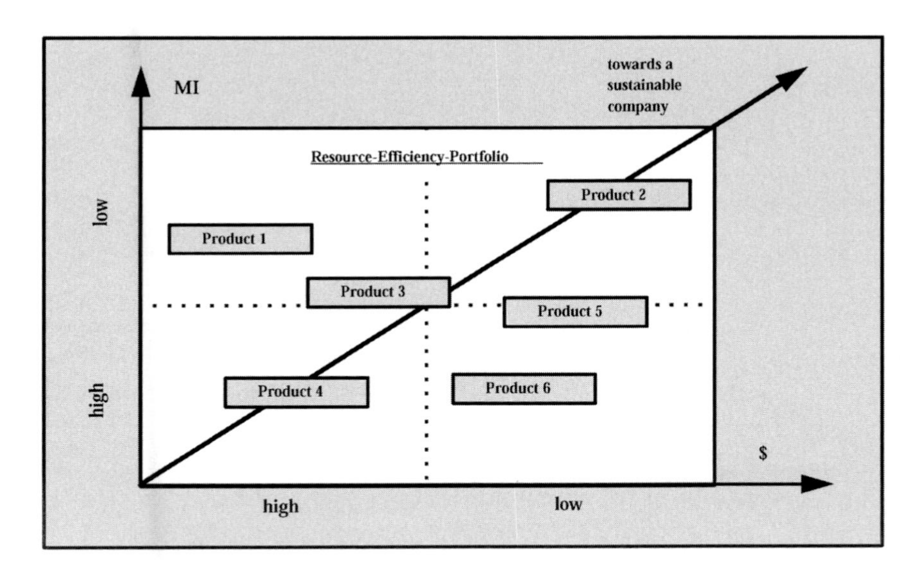


Figure 6.3. Resource-Efficiency-Portfolio at product-level, considering the costs (in $) and materials intensity (MI) of products.

ronmental aspects with the economic features of a company. The existing concepts differ regarding their scope, and generate different information.

New findings in environmental science and practical experience have shown that environmental protection can be efficient only if the company's materials flows are transparent and well known. This is why materials flows are the basis of the concept of Resource-Efficiency Accounting, which represents a pragmatic first step towards the integration of materials intensities into the decision-making process of companies. By referring to inputs, the implementation of REA is quite easy, since to a large extent most companies already have input data at their disposal.

The methodology of REA is flexible enough to be able to be integrated into different cost accounting systems. REA serves both process-oriented and product-oriented environmental management, which is an important feature since, to be effective, environmental management has to consider information at both the operational level and the product-level. The process-level is important in order to detect the potential for optimisation in production, whereas the consideration of the product-level allows a life-cycle-wide perspective. The REA methodology is flexible and supportive to decision-making, as the Resource-Efficiency-Portfolios can be set up at different company levels and based on different economic indicators. This enables the decision-maker to obtain tailor-made information. Thus, REA can support all decisions that are of economic and ecological importance.

It can be stated that a suitable eco-management accounting system will allow the exploration of cost reductions that improve the environmental performance of the company at the same time. It becomes possible to consider simultaneously economic and ecological aspects in order to guarantee the sustainable success of the company.

# PART II

INFORMATION SYSTEMS

# 7. Efficient Eco-Management Using ECO-Integral – How to Save Costs and Natural Resources at the Same Time

*Wolfgang Scheide and Georg Dold*
*Green-it, Konstanz, Germany; E-mail: scheide@green-it.de*

*Stefan Enzler*
*Institut für Management und Umwelt, Augsburg, Germany; E-mail: enzler@imu-augsburg.de*

## 7.1. Introduction

Industrial eco-management is currently undergoing a profound change. Only a few years ago, environmental protection consisted mainly of the sorting and disposing of wastes and the cleaning of emissions and wastewater. This so-called 'end-of-the-pipe' approach led to the frequent statement that '*environmental protection is expensive*'. The introduction of eco-management systems in accordance with the European Environmental Management and Auditing Scheme (EMAS) and ISO 14001 was a first step in overcoming this merely legal and technical focus, and resulted in an improved 'start-of-pipe' integration of environmental control into organizational processes.

The next step will be the efficient, economic and ecological design of materials and energy flows as the central focus of eco-management. However, this can be achieved only if environmental management is largely integrated into the decision-making processes of organizational units such as purchasing, product development, production, cost accounting and marketing. As the experiences of pilot companies clearly show, eco-management in its updated, modernized form communicates the message: '*environmental management cannot only reduce costs, but it can also save natural resources*'.

This organizational approach is based on a high degree of transparency with regard to the type, quantity and cost of materials and energy flows. The proven and effective tools of industrial eco-management (such as eco-balancing and benchmarking, and environmental cost accounting) can achieve maximum impact only where an efficient supply of information is ensured. Information systems therefore play a key role in eco-management as the critical link between materials and energy flows and decision-making processes (see Figure 7.1).

## 7.2. Methodology

The goal of the ECO-Integral project is, first and foremost, the development of a reference model that can be implemented and used across different industries. A reference model describes economic facts in a formal way – thus it is defined as a formal, data-processing-oriented and generally applicable description of an enterprise or a specific field. In the case of ECO-Integral, all relevant environmental management processes will be modeled based on an information-processing-oriented viewpoint. As result, the reference model contains the information requirements of environmental management tasks in a form which meets distinct operational requirements. The outcome is applicable as a guide-



*M. Bennett et al. (eds.), Environmental Management Accounting: Informational and Institutional Developments*, 93–111.
© 2002 *Kluwer Academic Publishers. Printed in the Netherlands.*

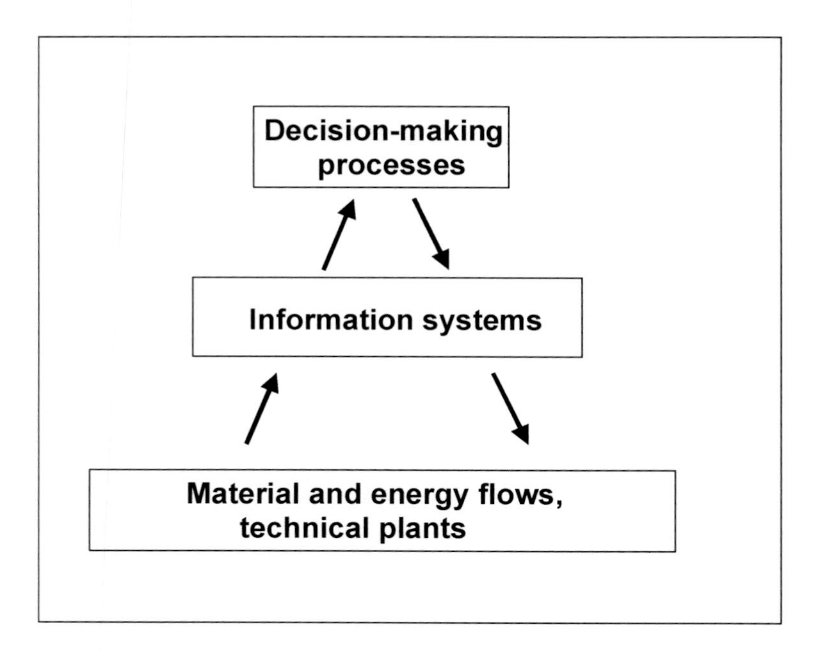


Figure 7.1. Integration of the information system into environmental management [source: own diagram].

line on how to use and how to customize standard software systems (ERP-systems) for the purpose of environmental management. It can be used by enterprises or by software developers.

The special features of ECO-Integral are:

1. It is capable of being integrated into major business applications (primarily in the areas of materials control, production planning and control as well as cost accounting).
2. The integration of the main eco-management functions into one single program.

### 7.2.1. *Project phases, project participants and work status*

ECO-Integral is carried out in three phases (see Figure 7.2):

In *phase I*, the reference model ECO-Integral was developed as Release 0.5 (based on theoretical considerations and empirical values) and subsequently evaluated at three separate industrial sites. Release 1.0 was then developed based on these evaluations.[1]

The practical implementation of the reference model in connection with individual projects run by pilot companies is currently being implemented in *phase II (called ECO-Rapid)*.[2] The project results will be summarized and published as a guideline for small and

[1] The reference model is published in Krcmar et al. (2000).

[2] ECO-Rapid started in September 1999 and will conclude in February 2002 with the publication of a guide for the implementation of the system in small and medium-sized companies. ECO-Rapid is co-financed by the Deutsche Bundesstiftung Umwelt, Osnabrück. For further information: http://www.eco-rapid.de.

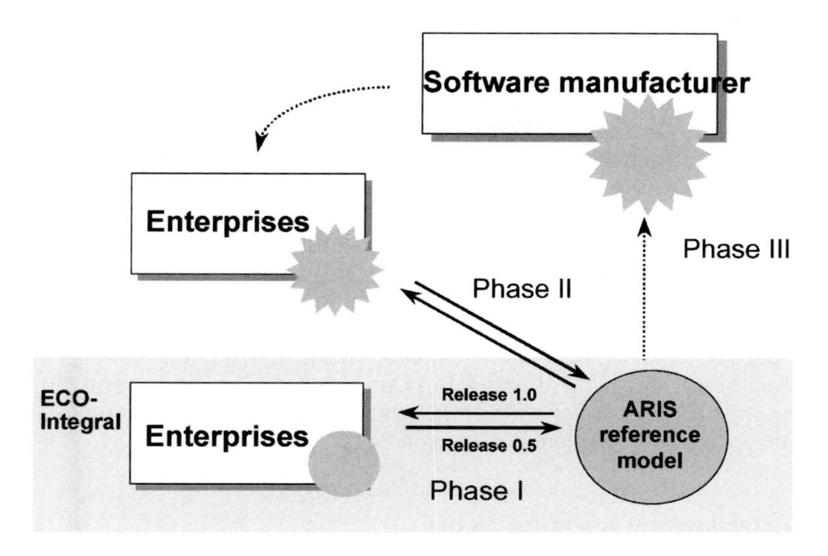


Figure 7.2. ECO-Integral project procedure [source: own diagram].

medium-size enterprises (SME's). This approach is intended to support SME's in using their information systems for systematic materials flow management at relatively little expense.[3]

Based on the experiences gained from the three pilot projects, in *phase III* the reference model will be implemented as an extension of standard software products in the area of ERP (Enterprise Resource Planning).

The project results are developed through interdisciplinary teamwork: the main contractor of the project (which is funded by the *Deutsche Bundesstiftung Umwelt*, a foundation of the German federal government for environmental issues) is the chair for business informatics at Hohenheim University (Prof. Dr. Helmut Krcmar). The cooperation partners in phase I are the Institute for Management and Environment (*Institut für Management und Umwelt*), Kienbaum Consulting, Wuppertal Institute and Dr. Werner Wohlfarth Consulting. This approach ensures the integration of EDP knowledge with expertise in the areas of business, environmental politics and environmental law.

### 7.2.2. *Methodology of reference modelling*

The description of the reference model[4] is based on the ARIS architecture (Scheer, 1995) of integrated information systems. Using ARIS, an enterprise can be described on two dimensions for software purposes.

In the first dimension, the views data, functions, control and organization are viewed with regard to their impact on an information system. Depending on the information techniques used, in the second dimension a distinction is drawn between the *description levels* technical concept, EDP concept and implementation.

Process models are the starting point for the development of an integrated informa-

---

[3] For first results from phase II (project ECO-Rapid), see Scheide et al. (2001).

[4] For further details on the topic of reference modelling, see Rosemann and Schütte (1997).

tion system in ARIS. The core elements of a process model are specific *events*, e.g. the receipt of a customer order, which trigger the execution of certain functions. Each operational function in turn generates new events or statuses. As a general rule, inter-related data is required to execute functions.

ARIS is an open modelling standard that largely supports today's generally accepted modelling techniques. The integration of the four views facilitates a comprehensive description of all vital factors within a business information system. Using the three different description levels of the program, the most important system development steps are defined more precisely.

The formal, EDP-oriented and ARIS-based description of a specialized field requires the clear definition and distinction of all terms used. A number of relatively fuzzy definitions for terms used on a daily basis became apparent in the field of 'eco-management', which is the subject of our modelling efforts in this case. The term 'eco-controlling', for instance, has various meanings. As a first step, it was therefore necessary to determine the fundamental functions of environmental management.

## 7.3. The ECO-Integral reference model

The ECO-Integral reference model consists of a joint database and a number of eco-management tools and methods to access this data base. First of all, the functions available in ECO-Integral will be outlined in the following chapter.

### 7.3.1. *Function Tree*

The function tree describes the ECO-Integral functions in form of a hierarchy. The following Figure 7.3 shows the *first* level of the function tree.

ECO-Integral covers the generic functions 'Supporting decision-making processes', 'Using tools' and 'Making data base available'.

The partial function tree 'Using tools' contains the standard methodology of materials flow management (see Figure 7.4). An additional function tree is available for each method or type of method; however, we will not go into any further detail about these at this point.

The partial function tree titled 'Making data base available' comprises all functions necessary for supplying the operational materials and energy data (flows and inventories) for the methods used. An important element in this context is the methods for capturing and computing data. The administration of master data is another important function. The

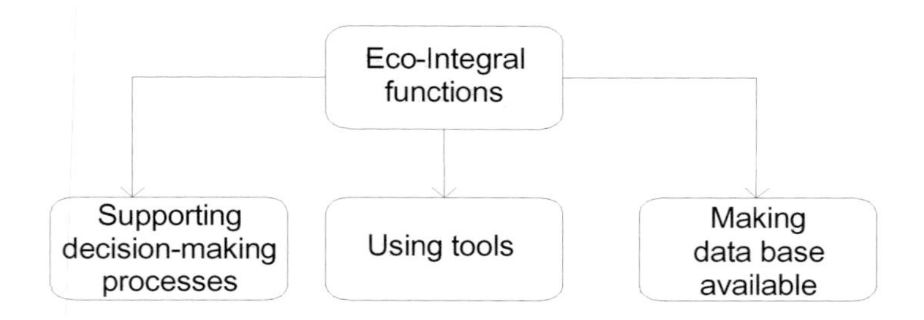


Figure 7.3. ECO-Integral function tree (1st level) [source: Krcmar et al., 2000].

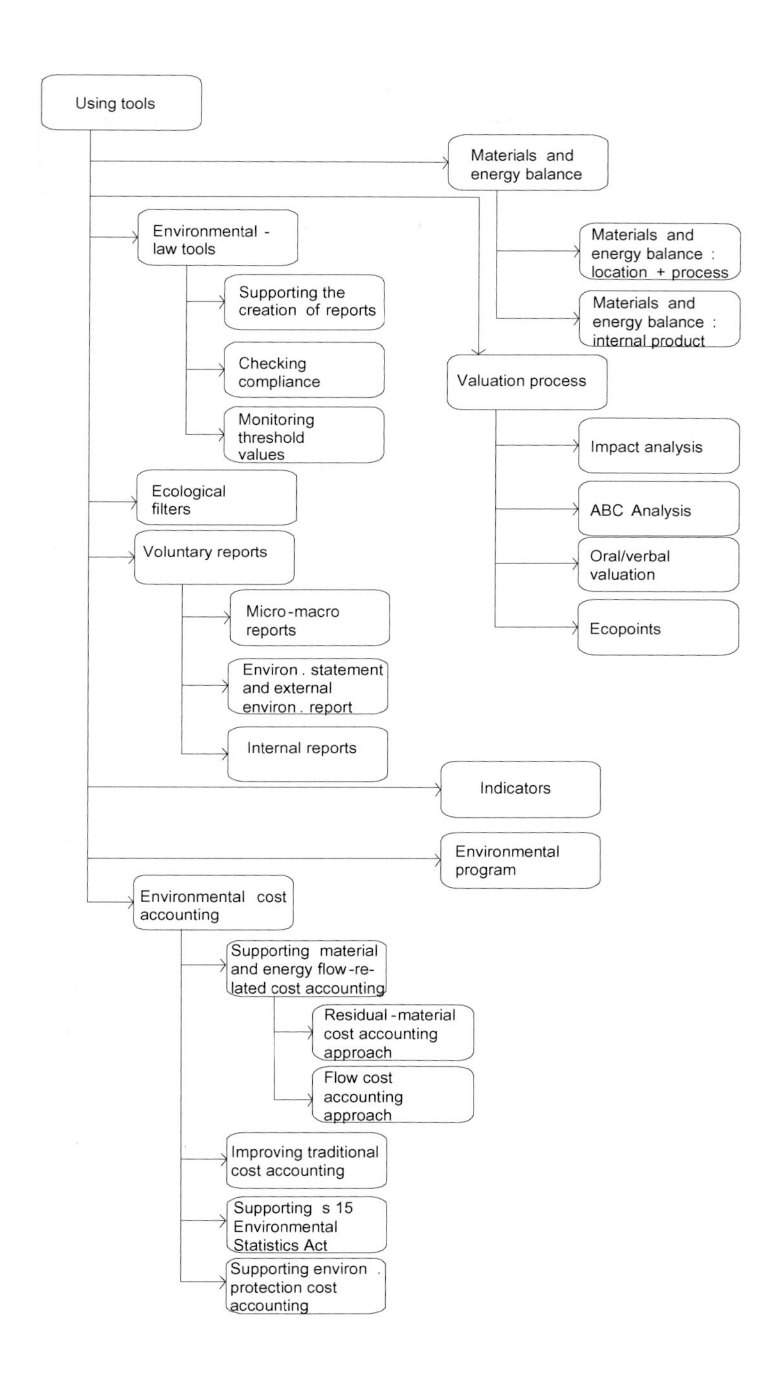


Figure 7.4. 'Using tools' partial function tree [source: Krcmar et al., 2000].

most critical master data are materials and substances as well as elements for the mapping of business materials flows (materials flow channels, transformation points, measuring points, computing rules, etc. (see Figure 7.5).

A brief outline of the most important elements of the reference model follows below. We will, however, dispense with a description of any specific function trees, data models and processes and in this respect refer you to the detailed explanations of Krcmar et al. (2000). The eco-balancing data model is shown as an example in the relevant section.

### 7.3.2. *Data base*

The database evaluated by the tools outlined below must contain all relevant materials and energy flows in a uniform format. In order to achieve the desired degree of completeness, even the flow of energy (electricity, heat, water) and residual materials (waste, waste water and exhaust air) must be captured and stored, although this approach is contrary to most existing practices. For reasons of cost-efficiency, a significant part of these additionally stored flows must be calculated instead of being captured on site. The reference model provides comprehensive computing functions for this purpose.

One of these functions is the calculation based on other flows captured and on parts lists data (analogous to the wide-spread retrograde withdrawal). For this purpose, relevant unit-variable quantities of energy or residual materials are included in the parts list as negative positions and additionally identified.

In order to store any flows in a uniform format, all materials and energies must be identified in a global materials structure and all sources and withdrawals of materials and energy flows must be identified in a global quantity-centre structure (similar to the cost-centre structure).

### 7.3.3. *Environmental cost accounting*

The environmental cost accounting tool supports the preparation of reports in accordance with section 15 of the German Environmental Statistics Act (*UStatG*)[5] as well as flow cost accounting and the allocation of 'end-of-pipe' costs in line with the causation principle. Flow cost accounting shows the entire costs of waste, waste water and exhaust air (usually 5–15% of the overall costs), thereby paving the way for significant cost reduction and resource-saving measures (Strobel and Enzler, 2001).

Based on the improved ECO-Integral quantity transparency, the cost elements are allocated to the individual residual materials flows either wholly or in part. The required allocations are derived either from the initial entries of volume, costs and services, from the permanently stored allocations, or from the current key volumes.

Based on these allocations, the relevant costs can be assigned to cost centres and cost units in accordance with the causation principle. These costs can then be flexibly evaluated by means of an auxiliary calculation (e.g. broken down into materials used, production processes, products causing costs/pollution, periods of time) and summarized as indicators for environmental cost management.

---

[5] According to § 15, enterprises have to survey their expenditures for environmental protection and their investments in environmental technology.

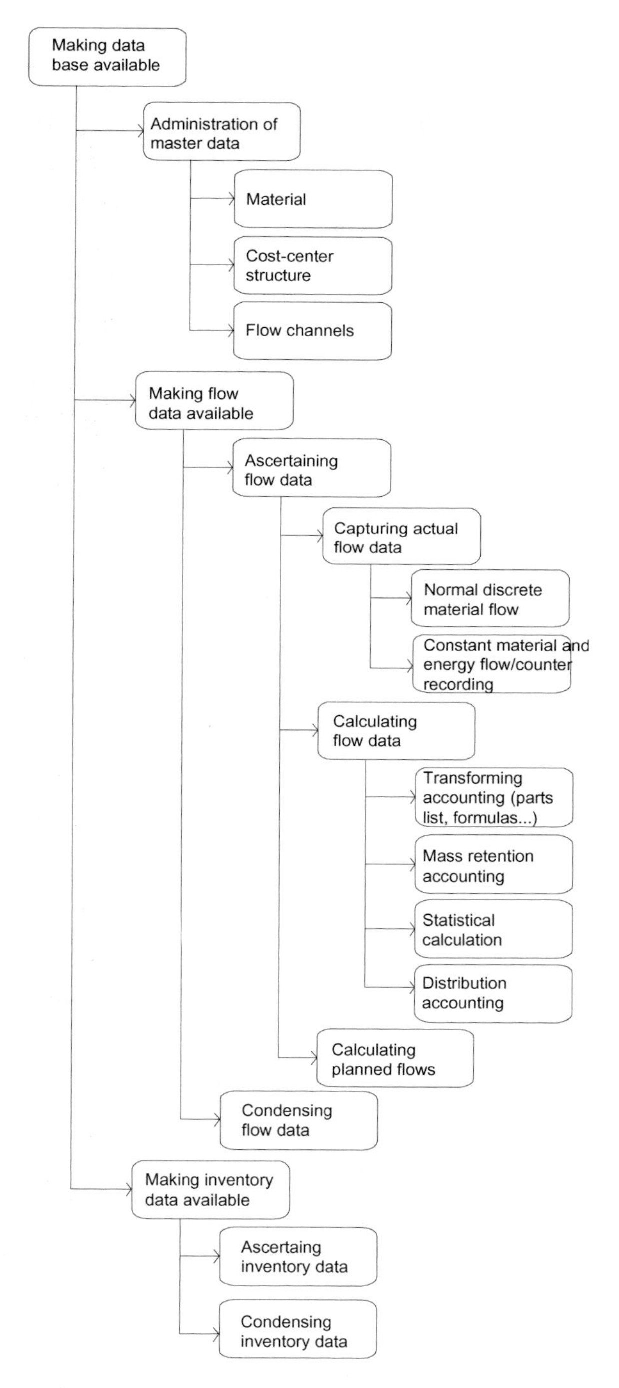


Figure 7.5. Partial function tree 'Making data base available' [source: Krcmar et al., 2000].

### 7.3.4. *Environmental Law*

Due to the restrictions of both environmental law and permits, industrial companies create an increasingly high number of reports on their materials and energy flows. In addition, these companies are required to monitor and observe both limit and threshold values. For this purpose, large quantities of data must be captured, a very time-consuming process. In addition, the quality of the data captured is often relatively poor. Hence, the ECO-Integral reference model is designed in such a way as to

- ensure the constant supply of DP-based data required for the preparation of materials and energy flow reports in the areas of waste, waste water, exhaust air and hazardous substances for government authorities; and
- support the automatic monitoring of compliance with threshold values set by environmental legislation and/or permits.

The aim is to reduce significantly the cost and time needed for the preparation of legally required reports, while at the same time increasing the quality of data for compliance control by industrial eco-management, as well as the ability to defend against liability claims and the planning and control functions of the authorities. The following data is usually required for these purpose:

- Aggregate materials flows for a specific period of time (e.g. waste balances);
- Materials inventory at any given time (maximum quantities of any hazardous substances stored);
- Place of origin and/or destination of specific materials.

With the ECO-Integral database, the fundamental data required must be entered only once. Following entry, this data can then be compiled automatically for different purposes. The requirements of the joint database were included in the data model following evaluation of important reporting requirements (approx. 100).

### 7.3.5. *Environmental Statistics/Micro-Macro-Link*

The parameters set by environmental politics as well as control activities require sound knowledge of all relevant environmental data. In its expert opinion of 14.3.1996 titled 'On the implementation of sustainable development' ('*Zur Umsetzung einer dauerhaft-umweltgerechten Entwicklung*'), the German Council of Environmental Experts (*Rat der Sachverständigen für Umweltfragen*) pointed out that the need for up-to-date environmental data had significantly increased both nationally and internationally. The amendment of the Environmental Statistics Act of 21.9.1994 and the further requirements stipulated by international standardization bodies (macro level) are designed to reflect this increased need for data.

For companies (micro level), this means additional expenditure for data capture and processing. A large portion of data refers to information on materials and energy flows. Within the reference model, the relevant reporting structures are set out in such a way as to enable companies to base their reporting on existing data. Thus, for relatively little additional expense, the reporting requirements can be met, while at the same time improving data quality.

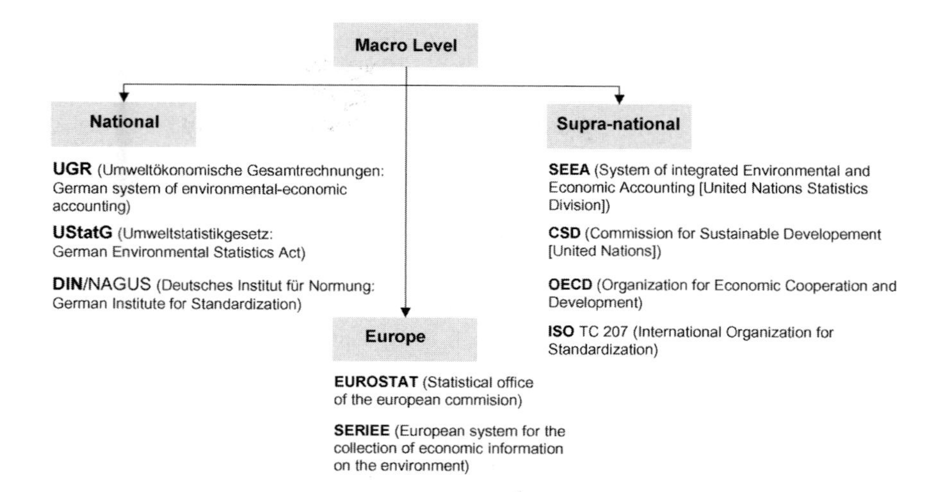


Figure 7.6. Information requirements on the macro level [source: own diagram].

### 7.3.6. *Eco-Management in accordance with EMAS and ISO 14001*

Eco-management systems, in accordance with both the pan-European Environmental Management and Audit Scheme (EMAS) and the worldwide ISO 14001 standard, focus on the improvement of industrial environmental protection. The starting point for targeted improvements is the knowledge of materials and energy flows within the company and the associated opportunities for controlled changes. The ECO-Integral reference model therefore supports the following tools which relate to materials and energy flows within companies:

* Environmental benchmarking and indicators.
* Environmental program administration.
* Set-actual comparison.
* Reporting (environmental impact statement/environmental report).

The tool 'environmental benchmarking and indicators' identifies, in a summarized form, the environmental impact of a company and facilitates the quantification of its environmental goals. Using the 'environmental program administration' tool, the environmental goals can be specified through measures, deadlines, responsibilities and additional reference values, resulting in the efficient implementation of these goals. Using the set-actual comparison, companies are able to assess how well they are doing against the set values. This enables effective progress monitoring. If and when required, the environmental goals can be adapted by means of the environmental program administration tool. Thus, the environmental performance of a company can be measured and continuously improved. In addition, ECO-Integral supports the preparation of the materials and energy-flow-relevant data section of environmental impact statements and/or environmental reports, thereby facilitating communication with third parties.

### 7.3.7. *Eco-Balancing*

The term 'eco-balancing' is now well established, despite the fact that its meaning differs within the academic fields concerned and that the term is used differently again in practice. Both simple operational input/output analyses and complex product life cycle analyses are equally termed 'eco-balance'. To avoid any confusion, the reference model contains three separate tools, which together form the generic eco-balancing tool:

- Materials and energy balance for sites and processes, which render a structured comparison of materials and energy flows of the balance object (i.e. enterprise, site, process).
- Internal materials and energy balances for products, which relate to the materials and energy flows caused by a specific product within the company.
- Impact analyses and evaluation techniques for the analysis and evaluation of aggregate materials and energy flows.

However, life-cycle analyses for products (LCA) are not supported, as this would require a separate model and therefore go beyond the scope of this project.[6]

First of all, the scope and period of the balance are determined in order to create the materials and energy balances. Then, all flow data are automatically aggregated from the joint database in accordance with a preset and self-defined balance structure. A number of different processes are available for the assessment of the impact of these materials and energy flows.

### 7.3.8. *Technical implementation of the model*

Implementation is based on the creation of a comprehensive and consistent database within the company's existing ERP software packages. Wherever possible, the enhanced database will be evaluated *within* the existing systems as a system extension. Where the integration of evaluation measures is too complex and/or costly, or where the company's existing software environment is too heterogeneous, implementing a data warehouse solution will standardize the relevant evaluation processes.

The so-called 'Data-Warehouse for the Environment' maps the ECO-Integral data model in an autonomous database. At regular intervals, the data is extracted from the previously adapted current application systems (PPC, materials management, etc) to the database and then stored. In this way, both time-related evaluation and ad-hoc links can be realized without any problems whatsoever (see Figure 7.7). In this case, data evaluation does not form an integral part of the existing software, but is carried out by regular data exchange. Data warehouses are used increasingly in the corporate sector where large data volumes are required for time-related evaluations. It is planned to realize such a data warehouse in phase II (see above).

## 7.4. Implementation of ECO-Integral – an example

The ECO-Integral reference model has been set up and put to the test in pilot projects with the companies Herlitz PBS AG (Berlin), Novartis Pharma GmbH (Wehr, Nuremberg),

---

[1] It is, however, conceivable and useful to facilitate the preparation of cross-company product balances by integrating defined interfaces between the DP systems of different enterprises. With regard to EDP support for product-related eco-balancing, see Dold (1997).

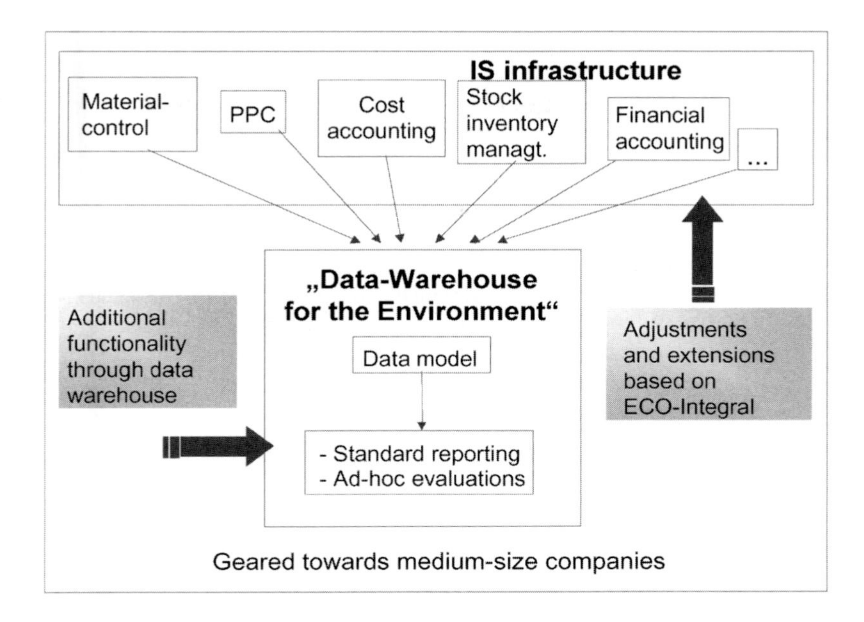


Figure 7.7. Data-Warehouse for the Environment [source: own diagram]

and Festo KG (Esslingen). In these various companies, it has been possible, thanks to the reference model's modular structure, to test successfully different component combinations within the framework of our general research project.

ECO-Integral has also in the meantime been introduced in various applications companies whose needs are mostly for individual components of the reference model. In the Merckle/Ratiopharm group (Ulm), for example, the emphasis is on flow cost accounting. At Lucent Technologies (Augsburg), combinations of several ECO-Integral components are needed. The spectrum here ranges from an integrated database – by way of the materials and energy flow model – right through e.g. to ecological auditing, environmental indicators, flow cost accounting, the requirements of environmental legislation, and support for the company's in-house environmental management system in line with ISO 14001.

In the following sections – taking the example of Lucent Technologies (Augsburg) – we describe the procedure for introducing individual components from the ECO-Integral range.

### *7.4.1. Initial situation*

Products and services in the telecommunications sector represent the Lucent Technologies concern worldwide. The corporation employs altogether some 120,000 staff and in 1997 achieved an annual turnover amounting to US$ 26,360 million and a net profit in the same year of US$ 1.507 million.

At its works in Augsburg, the 'Augsburger Kabelwerk', founded about three years ago, the Lucent Technologies Corporation manufactures high-grade optic-fibre cable. This product, given the nature of the optic fibre it uses, involves high raw materials costs.

Initially the still relatively young Augsburg works did not have an adequate in-house information system. In the absence of an information system which was capable of covering satisfactorily the particular requirements and special needs of the cable industry, most companies in this sector still work with solutions developed in-house. In Augsburg a software solution called 'ProPcs' is used. The point of departure for our project was thus characterized by an unreliable data inventory, with redundant and incorrect data, resulting from mainly manual data acquisition. This was further aggravated by certain sequences which had evolved during the corporation's expansion stage but which had not been coordinated and standardized throughout its activities worldwide.

In this general situation an instruction was issued corporation-wide to implement the 'materials only costing' concept (MOC), in which the costs of materials have to be recorded as a separate item right through the company's activities end-to-end; (see Section 4.5: 'Adapting the financial accounting system').

### 7.4.2. *Project objectives and project sequence*

The project objectives are principally to remedy the deficits described in Section: 6.4.1: 'Initial situation', and in particular:

- to improve materials efficiency with a view to reducing costs and relieving stress on the environment;
- to simplify the organizational sequences (process reliability);
- to reduce throughput times;
- to raise the standard of information regarding materials quantities;
- to enhance the quality of financial data;
- to implement the 'materials only costing' (MOC) concept; and
- to support the environmental management instruments (e.g. in-house ecological auditing, environmental indicators, etc.).

The project was implemented mainly by the company's in-house working group, with representatives from all departments and with external support provided by the 'Institut für Management und Umwelt' (IMU – Institute for Management and the Environment).

The *project sequence* can be divided up into the following phases:

1. bringing transparency to the flow of materials by setting up a materials flow image;
2. weighting the relevant materials flows;
3. devising a posting and accounting concept for the in-house information system;
4. adapting the company's financial accounting system; and
5. creating evaluation criteria for the purposes of environmental management.

### 7.4.3. *Setting up a materials flow image*

A basic prerequisite for efficient environmental management is the transparency of materials flows within the company. Most companies are unable to trace materials flows end-to-end through all value-adding stages in-house. Exact knowledge regarding materials is usually limited to incoming materials and outgoing products, and perhaps the various waste aggregates. Detailed data which pinpoints which materials, in which quantities, and at what cost, go not into the finished product but end up as solid wastes, effluent, or exhaust requiring final disposal, is available in only very few enterprises or organizations. If materials flows are not transparent, significant economic potential for reducing costs and ecological potential for relieving stress on the environment go unnoticed.

Mapping out a materials flow image brings the necessary transparency to materials flows and at the same time provides an overview of materials which is uniform for the whole corporation across the boundaries of its various departments and divisions.

In this materials flow image the individual materials flows are shown as numbered and color-coded arrows. The boxes stand for the various storage locations and transition points (e.g. individual production points, sorting areas, etc.). The coloured marking helps to distinguish between desirable and undesirable materials flows (e.g. red for waste flows).

This flow image summarizes the most important materials flows in terms of materials groups, and thus enables weighting according to relevance for the purposes of subsequently mapping these flows in the information system.

### 7.4.4. *Posting and accounting logic*

Having thus achieved materials flow transparency, the next step is to adapt the posting and accounting logic for materials flows in the company's information system. This will reflect an aggregated version of the materials flow image. Depending on the company's requirements, data referring to its individual departments and divisions can be either grouped together or detailed separately.

#### *There are three steps to modelling the posting and accounting logic*

The *first step* in developing the posting and accounting logic is to map out the storage locations and transition points. In all these stages in the flow, materials are stored until the moment that they are passed on to the next stage. The physical changes that individual materials undergo (e.g. in production) do not influence the way in which they are represented and traced in the information system. All stages are thus mapped out in the information system as storage locations.

The *second step* is to devise posting and accounting processes between the storage locations. This posting and accounting logic must be capable of mapping all movements of materials within the company. It should also be capable of mapping exceptional events in the company – in order to avoid so-called 'black holes' into which materials may 'disappear' without trace. These various posting steps thus represent materials flows between the said storage locations.

In the *third step*, the information system should be able to record not only individual postings, but also the types of movement involved. It should also be able, on the basis of these categories, to permit differentiated inquiries concerning various materials postings and to distinguish between different types of posting between the same two storage locations. This is necessary, e.g. if a particular movement or posting could be made for different *reasons* (e.g. arrival in the finished products store from production after the product has been completed, and arrival in the finished products store from production after post-processing).

#### *Integrated database for supporting environmental management*

The ECO-Integral reference model involves setting up an integrated database.

At each stage the posting and accounting logic maps out not only the desired materials throughput but also the residual substances. At Lucent Technologies these comprise mainly waste solids. Compared with the amounts involved, liquid effluent and exhaust emissions are not relevant.

Figure 7.9 shows the posting and accounting logic.

This concept also envisages, unlike other information systems, the introduction of envi-

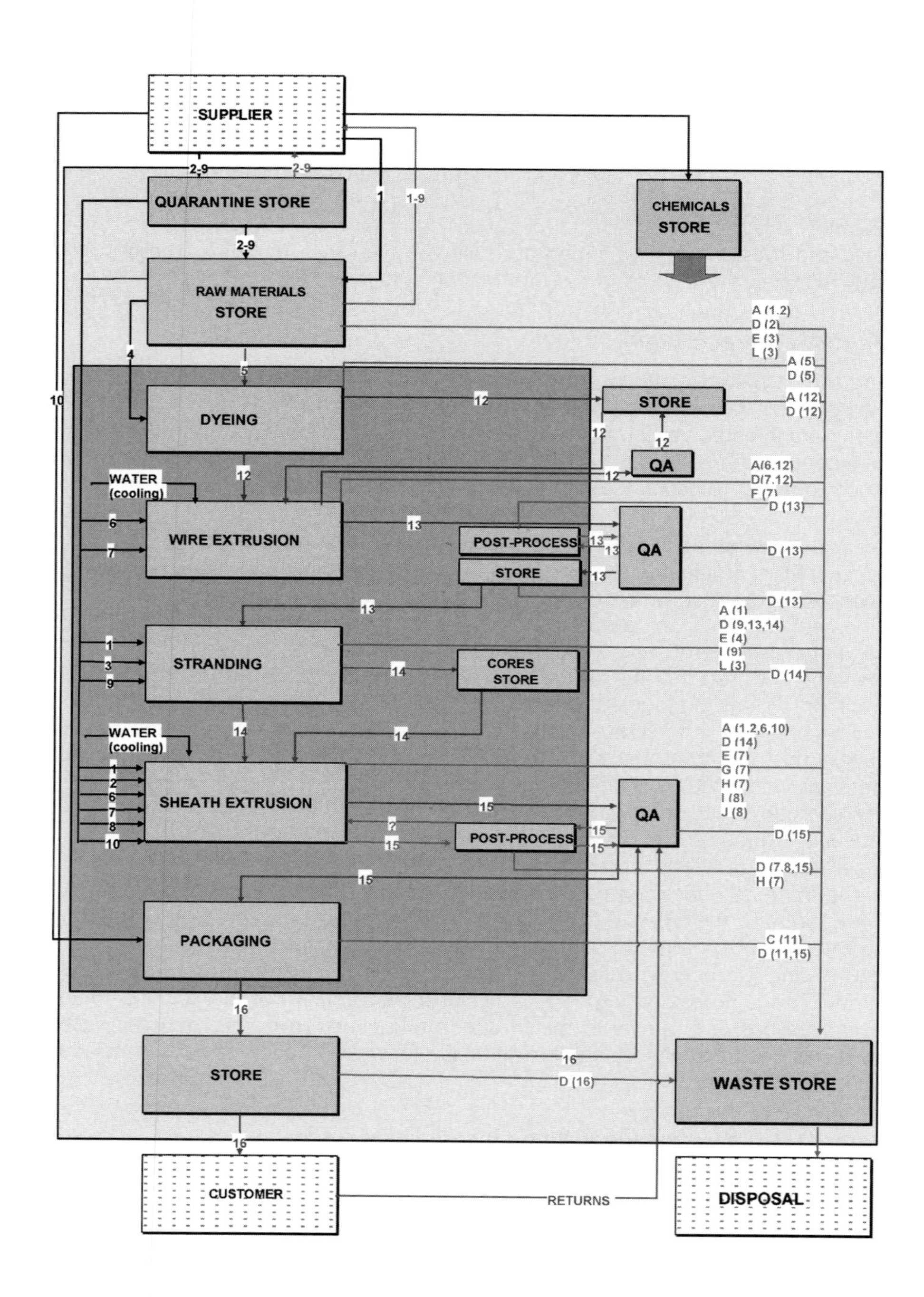


Figure 7.8. Materials flow image.

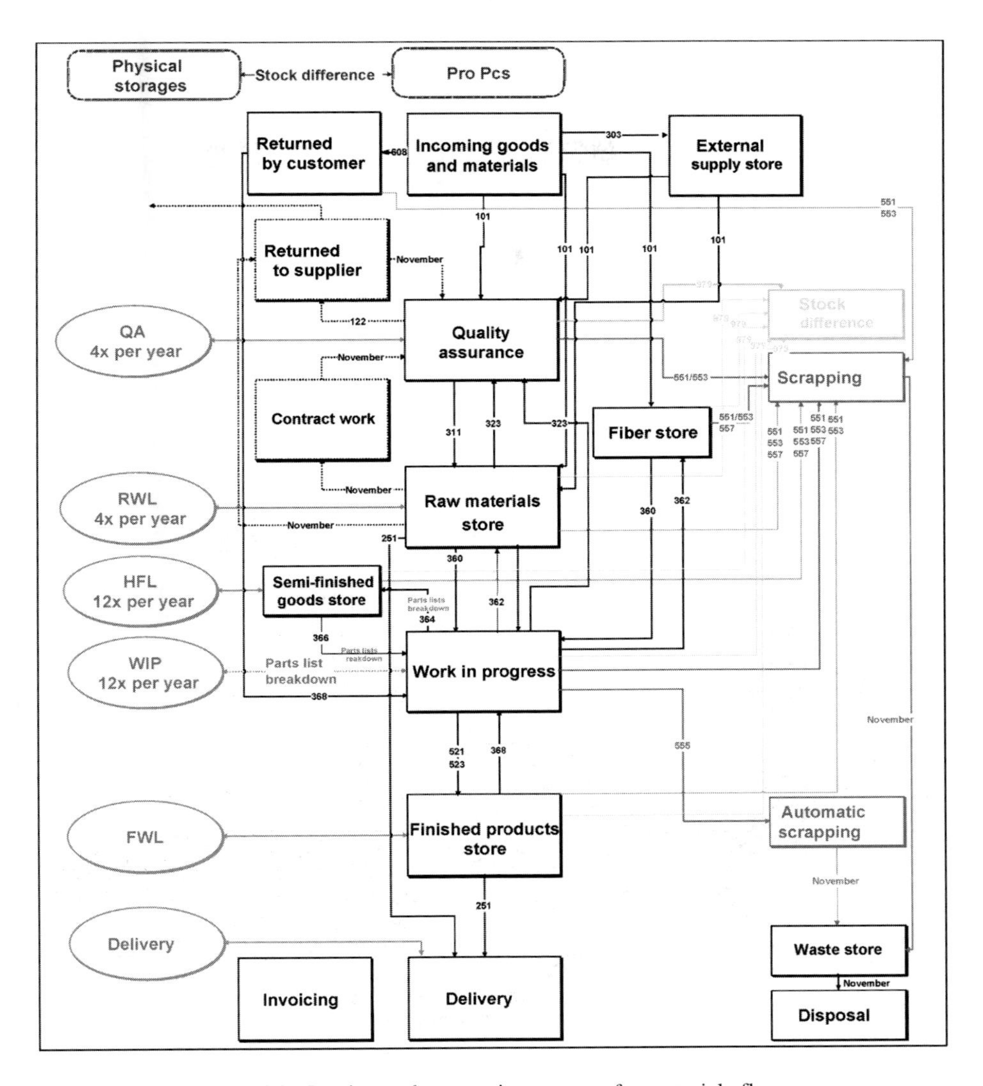


Figure 7.9. Posting and accounting concept for materials flows.

ronmentally relevant storage locations, which benefit subsequent evaluations for environmental management purposes. The inclusion of the *'Scrapping'* storage location, for example, enables staff to make an active posting for materials losses. Such materials losses (e.g. incorrect batches) can thus subsequently still be traced in terms of quantity and cost.

A materials loss which is attributable to production when manufacturing a product is listed as an extra amount in the so-called *gross parts list* – whereas the quantity of materials actually incorporated in the product is listed in the so-called *net parts list*.

The difference between the gross parts list and the net parts list for a particular product set thus reflects the planned materials loss on the articles level.

The actual materials loss incurred in the manufacture of a product can be calculated by evaluating the relevant production orders. The quantities of materials needed in the manufacture of the product per individual article are posted by the semi-finished goods store on the production order. After the finished products in question have been manufactured, an acknowledgement report is issued stating the quantity produced. By splitting the quantities produced and comparing these with the associated net parts list, the actual quantity of articles can be calculated. The difference between the materials withdrawn from stock as noted on the production order, and the subsequent net parts list split, represents the actual materials loss. To record this materials loss in terms of quantity and cost the '*Automatic scrapping*' storage location is included and posted automatically by the system.

Another important storage location is the '*Stock difference*' which automatically posts the difference between the materials quantities as physically available and those as recorded by the system. This difference can be attributed largely to materials losses which, because the quantity is difficult to estimate, cannot be actively posted by staff (e.g. unplanned cuttings on a length of cabling). To determine the exact composition of the length of cable in terms of its component materials and their respective weights would take too much time and effort to be justifiable. Recording these materials via this stock difference facility offers a sound approximation involving very little effort.

The '*Waste store*' storage location is also included. This enables a long-term comparison between weighed waste quantities with the materials as recorded in the information system. This can be used to determine exactly the composition of waste fractions regarding the individual materials used.

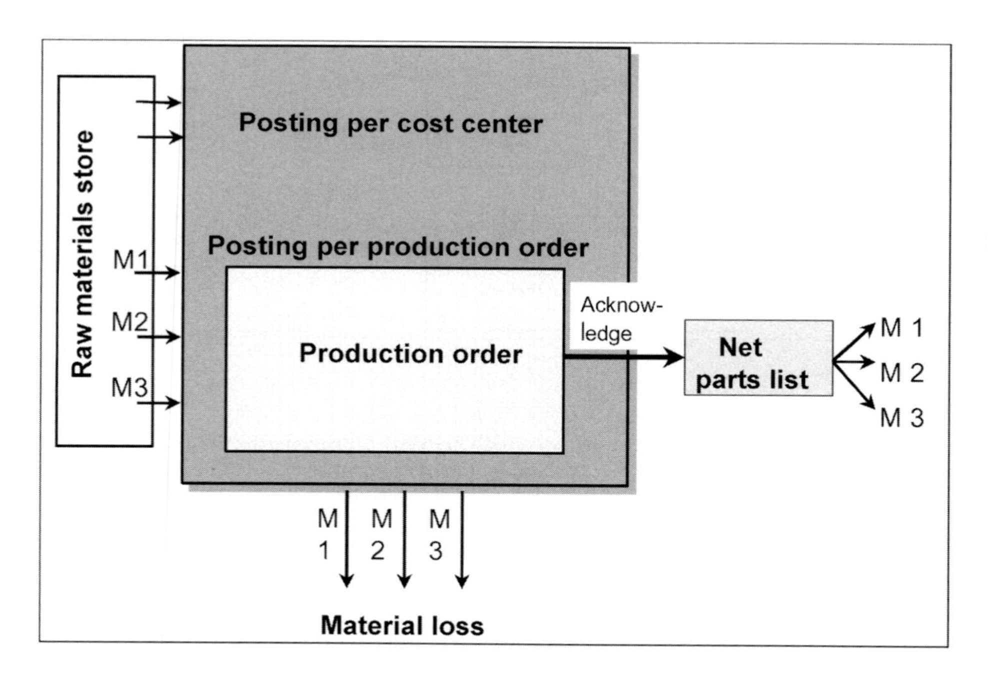


Figure 7.10. Calculating materials losses.

### 7.4.5. *Adapting the financial accounting system*

For the financial system, similarly, which is linked to the quantities system, the end-to-end transparency of materials flows, in terms of costs per individual flow, is of fundamental importance. To attain this objective a number of major corporations in the USA, all at around the same time, developed an approach known as 'materials only costing' (MOC). With MOC, only the materials value that has actually flowed in the course of the year is posted. The associated production costs are not capitalized until the end of the period.

The MOC approach has been developed for two reasons:

- In manufacturing industry, materials costs represent by far the largest block of costs.
- Personnel and plant costs which are also listed as production costs are relatively fixed and stable. The materials costs block however is large, can be influenced, and represents a potential for cutting costs (and at the same time for relieving stress on the environment), which has so far been hardly scratched.

Data in terms of quantities and costs is assigned to the flows of materials on the basis of the quantity that has actually flowed in a given period. This flow quantity is cost-evaluated using MOC only.

In most companies the evaluation criteria change in the course of the flow of materials. Thus, usually, materials going into production are evaluated according to the price of materials alone – whereas semi-finished goods and finished products leaving production are evaluated on the basis of parts lists and work schedules including materials prices and production costs.

The following *MOC example* illustrates this problem. In this example, without an end-to-end separation into materials costs and production costs, the impression arises that in the course of a year a materials value of DM 100 million goes into production, and an outgoing materials value of DM 130 million is created in the form of the product.

However, what is not clear from the data is that of the DM 100 million going into production, only DM 80 million actually goes directly into the product while DM 20 million materials value is waste and goes for disposal and the product's outgoing materials value actually contains DM 50 million production costs. The stock inventories too are subject to the same confusion between materials value alone, and capitalized production costs.

The materials loss valued in monetary terms, and thus the efficiency of in-house production, can be properly measured only if the company's financial system applies this MOC concept throughout.

### 7.4.6. *Support for environmental management*

Detailed and uniform transparency of materials flows within the company not only brings the benefits already described and the possibilities for cost saving that these reveal, but in addition the works information system developed for this purpose can also be used to support the company's environmental management system.

It is possible, without too much extra effort, by specially structuring the evaluation processes, to obtain all the data needed to compile an *in-house ecological audit*. The end-to-end transparency of materials flows makes it possible to set up an *environmental indicator system* covering the entire spectrum from company-specific right through to machine-specific indicators. An *environmental management system* certified as per *ISO 14001* is based on data on environmental objectives and environmental aspects. Environmental objectives can be formulated on the basis of the environmental indicators.

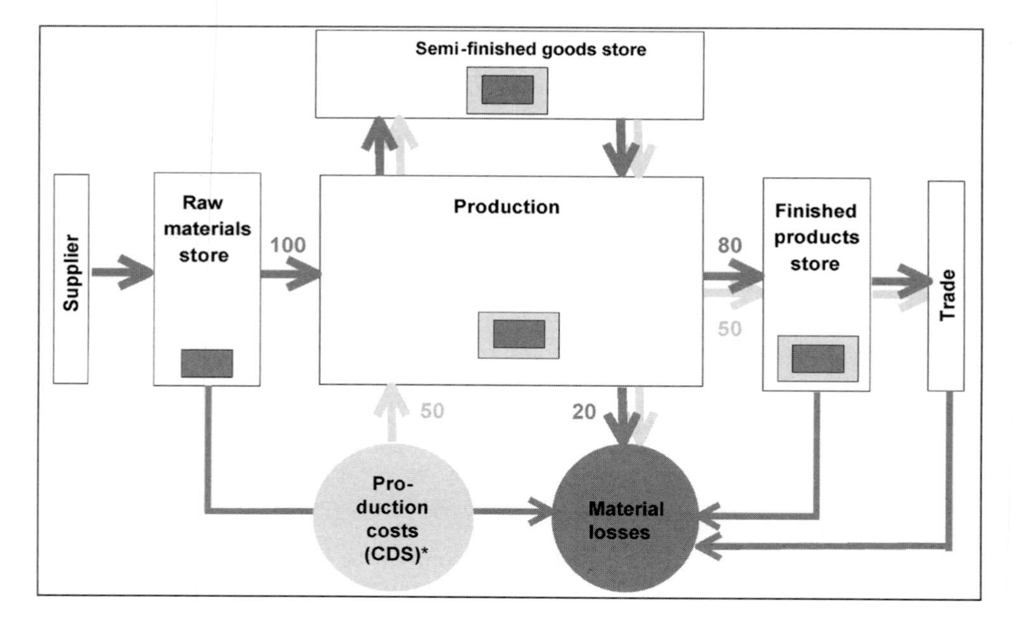


* CDS = cost distribution sheet.

Figure 7.11. Concept of 'materials only costing'.

These can, at periodic intervals, be compared with new environmental indicators, thus monitoring continuously the degree of success in attaining the objectives set. The environmental aspects can be formulated equally well via the in-house ecological audit or by appropriately structuring the environmental indicators system. Similarly, the work involved in compiling reports, required by *environmental law*, on materials flows (e.g. waste audits, etc.) can also be facilitated using the same data. Finally thanks to the uniform and end-to-end recording and administration of data referring to materials flows in terms of quantities and costs, it is possible to implement *flow cost accounting*.

## 7.5. Conclusion

The example of application demonstrates how both goals, i.e. *cost reduction* and *environmental protection*, can be achieved by consistent implementation of the 'ECO-Integral' reference model. This is made possible by the use and adaptation of the existing business information system. The objective of phase II of the ECO-Integral project (project ECO-Rapid) is to make this knowledge available to the largest possible number of companies, particularly small and medium-size companies. To this end, a guide is being planned which targets environmental and EDP experts in small and medium-size companies. This guide is designed to assist companies with planning and implementing efficient environmental and materials flows information systems without the need for outside participation.

The development and implementation of a business environmental information system based on ECO-Integral leads to the following general conclusions:

- Any such system must always be based to the greatest extent possible on the complete and differentiated acquisition of any data relating to materials and energy flows within the company.
- On this basis, it is the task of a computer-based information system to support the use of a wide array of eco-management tools, in addition to providing the database for the various decision makers.
- In ECO-Integral these instruments include, in particular, flow cost accounting and the preparation of materials-related reports and statistics, as well as eco-management information and eco-balances.
- The reference model aims primarily to provide the necessary database within existing EDP systems, as a system extension.
- In connection with the exemplary application of the system, a wide range of significant ecological and economical opportunities for intervention have already been identified and utilised in a number of different companies.
- A guide for the implementation of the system in small and medium-size companies is currently in preparation.

# 8. Materials Flow Management Based on Production Data from ERP Systems

*Gunnar Jürgens*
*Fraunhofer-IAO, Stuttgart, Germany; E-mail: Gunnar.Juergens@iao.fhg.de*

## 8.1. Introduction

Within German industry the issue of 'environmental protection' has increased in importance within recent years, as indicated by the voluntary participation in the implementation of EMAS by more than 2000 company sites. One reason for this is that a shift in the perception of industrial environmental protection has occurred over recent years. Whereas previously, environmental protection was mostly perceived as a bureaucratic burden, enterprises today take an active part in the opportunity to use environmental protection successfully in the achievement of the company's business aims through setting up environmental management systems.

This chapter focuses on materials flow management as an approach to optimise a company's economic and environmental performance. Materials flow management comprises the identification, analysis and optimisation of the procurement, usage, handling, transformation and disposal of all physical goods within a company. Physical goods in the context of materials flow management comprise not only raw materials, semi-finished and finished products, but also in particular all those materials which have not been dealt with in conventional management and information systems, such as waste, energy, commodities etc.

This chapter describes the content and potentials of materials flow management. In particular, it contains a description of the specification and introduction of an Environmental Management Information System (EMIS) which offers special support in the area of materials flow management to a machine building company.

## 8.2. Background

Although the introduction of an environmental organisation is a necessary first step, current company practice is frequently marked by a lack of transparency of the environmental relevance of the flows of materials within the company, and the costs associated with this. Within current environmental management systems, basic input-output analysis represents the central information basis through which to identify weak points and potential areas for improvement (Jürgens et al., 1997). In an input-output analysis, all incoming and outgoing materials and energy flows are summarised over an entire site and represented in a table. With this approach, approximate evaluations of the environmental relevance of a company's operations can be made. A major drawback, however, is that an input-output analysis does not support the allocation of environmental relevant materials flows to products and/or to specific processes, which means that their origin cannot be identified. In additional, a cost-effective optimisation of the company's operations cannot be achieved solely through the use of input-output tables.

To illustrate this, it is insufficient to estimate the costs for wastes based only on the expenses directly incurred in their removal. This would mean that all efforts which had



*M. Bennett et al. (eds.), Environmental Management Accounting: Informational and Institutional Developments,* 113–122.
© 2002 *Kluwer Academic Publishers. Printed in the Netherlands.*

been spent on the waste materials along their flow through the supply chain, such as in their purchase, handling, collection, separation into waste fractions, and other value-adding activities, would not be taken into consideration, although these add considerably to total costs. Specific examples in which a waste materials cost analysis has been carried out show that the real incidental costs incurred through the generation of waste materials are much larger than would be implied by a focus on their disposal costs alone would imply (Bullinger and Jürgens, 1999; Loew and Jürgens, 1999).

If a company is to improve the transparency of the environmental and cost aspects of its internal materials flows, the implementation of materials flow management is vital. This paper deals briefly with the procedure for setting up a materials flow management system, and then goes on to demonstrate how a regularly updated information set which can be used as the basis for decision support can be established within a materials flow management system through the connection of an environmental management information system to SAP/R3.

## 8.3. Materials flow management

### 8.3.1. *Potential opportunites within materials flow management*

Current approaches and concepts for production planning and control are oriented primarily towards the mapping and planning of relevant core processes, products and materials, so that auxiliary and operating materials, co-products, energy consumption, and waste materials are generally neglected. With this approach, only sub-optima will be obtained. This holds true in particular if the individual areas of responsibility within a company are organised in a decentralised manner, such as oriented to cost centres, and if those centres concentrate exclusively on cost minimisation in the company's production processes. In order to reach the aim of an holistic optimisation of the company's activities, a systematic and company-wide consideration of all materials flows is necessary.

Within the concept of materials flow management, the company's environmental performance will be measured and evaluated at regular intervals. The most important information basis for the measurement and evaluation of its environmental performance is represented by process- and product- oriented environmental effects, which are generated on the basis of materials flows. Based on this information, it is possible to identify weak points not only in individual processes, but also in entire product life-cycles or whole companies, and to initiate improvement measures such as materials substitution or operational re-arrangements. Moreover, through the direct inclusion of materials and energy costs, a new perception of a company's cost situation and development will be possible, which could not be achieved through conventional classic cost accounting (Loew and Jürgens, 1999).

The information which can be obtained through a systematic analysis of a company's materials flows are of high interest not only for environmentally-related decision processes such as within the context of an environmental management system, but more widely. If a company looks upon the costs related to waste in production only from the point of view of the disposal of that waste, it will be aware of only a limited proportion of the actual costs caused by wastage. Within materials flow management approach it is possible, however, to include also those costs which occur areas such as purchasing, materials administration, and production, until the point in production at which an undesirable by-product is dropped out of the process as 'waste'. Data taken from a medium-sized mechanical engineering company in a research project conducted by the Fraunhofer IAO show, for example, that the internal handling of packaging materials (their separation and shredding, as well as their transport to internal collection points) can on its own cause costs that are over 10 times higher than the external charges imposed on the company for the disposal of waste materials (see Figure 8.1).

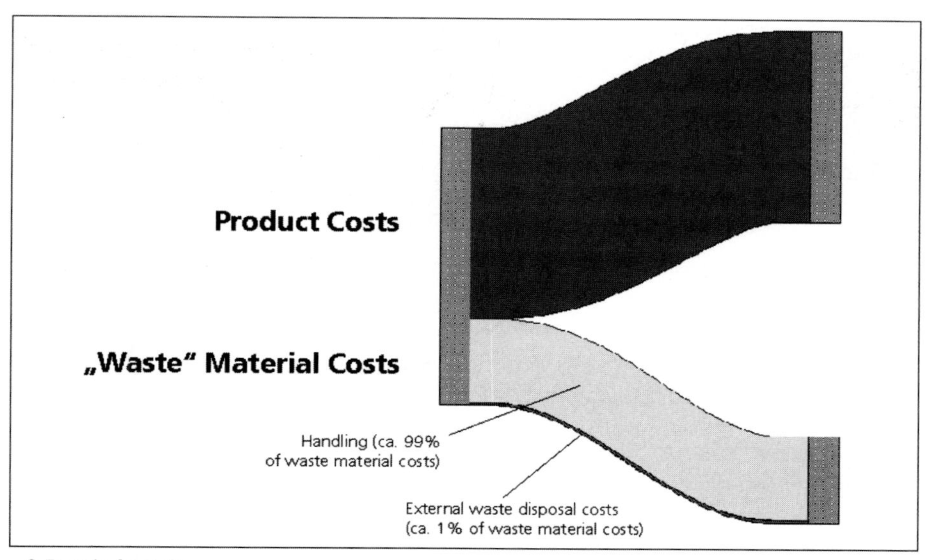


© Fraunhofer-IAO.

Figure 8.1. Internal company costs of waste handling are often higher than the charges for the disposal of waste materials by a factor of over 10.

Every decision of an employee of a company always depends on the individual information basis within the relevant business processes. Considering that, depending on the industry sector, the materials costs can constitute up to 70 % of the total costs of a company (Lied, 1999), a company-wide measurement and analysis of its materials flows is of great significance for not only the environmental department but also for other business processes.

A company-wide approach to materials flow management can also be seen as an opportunity to integrate more closely the environmental department with other departments of the company. For example, the environmental manager could take over service functions for other business processes by generating regular internal reports on materials flows in the company's production processes and the costs associated with this (see Figure 8.2).

### 8.3.2. *Use of Environmental Management Information Systems (EMIS) to support materials flow management*

The described potential opportunities which can result from a periodic analysis of materials flows in the context of a company's materials flow management, however, must be compared against the not-to-be-underestimated effort which is needed to achieve this, which is mainly caused by the following issues:

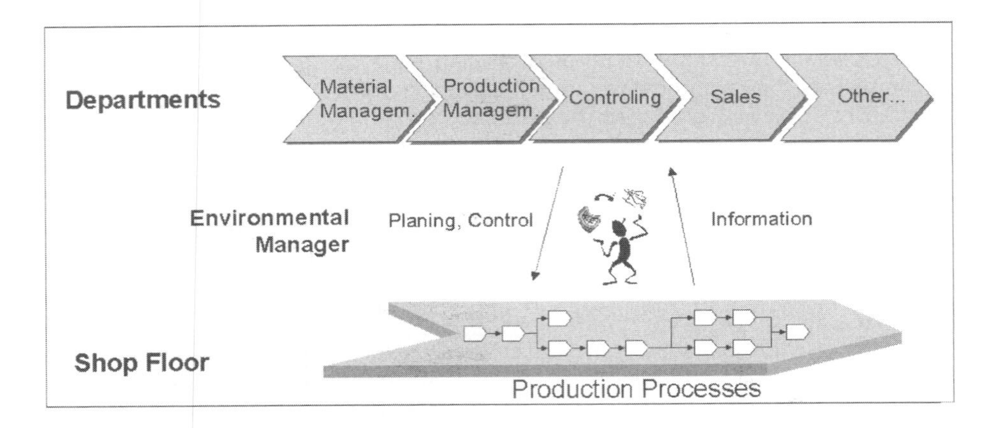


Figure 8.2. Support of business processes by an environmental manager in the context of materials flow management.

- Modelling of the company's processes in a materials flow model.
- Collection of materials flow data by means of production data, invoices, records, surveys, company interviews, etc.
- Calculation of materials flows.
- Analysis of the balance sheets, and assessment of the results by means of agreed guidelines.
- Target-group-specific preparation of the results.

In various application fields of environmental management, Environmental Management Information Systems (EMIS) are being developed (Rautenstrauch, 1999). This also applies to software systems in the field of materials flow management (Rey et al., 1998; IKARUS, 2000), which can provide significant support for the above-mentioned steps. The use of an EMIS in materials flow management systems is shown in Figure 8.3 through an example of a periodic calculation of materials flow oriented benchmarks.

The following two success factors emerge in company practise. These are seen as critical if companies, particularly small and medium-sized enterprises, are to benefit from the potential opportunities associated with the use of EMIS's (Jürgens and Steinaecker, 1999):

- The poor availability and quality of suitable data can lead to a relatively high effort being required to generate data related to materials flow, and its modelling and analysis.
- The inadequate connection of materials flow management to production planning and financial accounting can prevents the acceptance, and therefore efficient use, of materials flow analysis for an environmental- and cost-oriented optimisation of production processes.

With this background, it is important to achieve connection of EMIS to existing information systems which is extensive as possible. In data collection in particular, a connection to production planning and control (PPC) systems would be appropriate. A periodic or on-demand export of data such as bills of materials, parts lists, working plans, and manufacturing orders, would reduce the effort which might otherwise be required to establish and maintain a materials flow model in an EMIS (Mayer, 1999).

At the same time, connection to administrative accounting systems is valuable in order

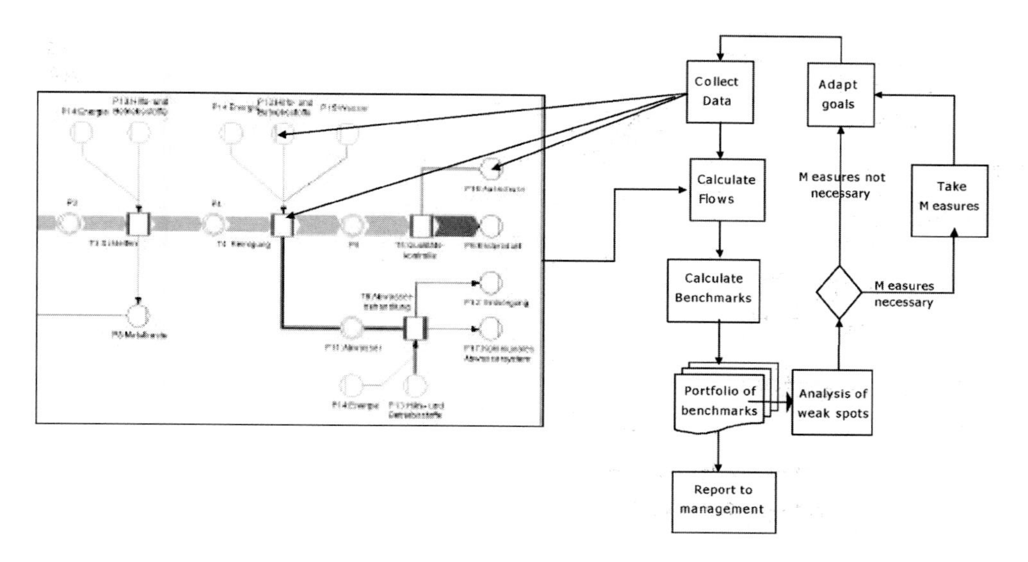


© Fraunhofer-IAO.

Figure 8.3. Support of the calculation of benchmarks by Environmental Management Information Systems (EMIS) (e.g. shown with figures generated with the EMIS Umberto 3.2).

to ensure that the results of materials flow analyses are used efficiently in all company functions. If the costs related to materials flows can be allocated directly to the cost centres and cost units which are responsible, it ensures that the company units which are affected can promote a systematic optimisation of the materials flows caused by them (Strobel et al., 1999). This is illustrated in the following section by means of an example.

### 8.3.3. *Case Study: Connection of a materials flow management system to SAP/R3*

Trumpf GmbH + Co. Machine Factory, based in Ditzingen, Germany, is well known as an innovative enterprise and a market leader in the manufacturing of metal sheet machining tools and equipment. Trumpf operates world-wide in the context of a decentralised approach to production, and the consequent organisation of the shop floor into semi-autonomous production units that this has entailed has increased the responsibility of employees and has created several opportunities for cost savings. An internal audit in 1996 showed, however, that the company could not adequately meet the high standards of performance that it had defined for itself for the environmental performance of its products and operations. In particular, significant problems were identified of a lack of transparency over the environmental impacts caused by its production processes, and a lack of support from the relevant information systems.

Trumpf has therefore has decided to strive for a holistic solution as a pilot partner in the OPUS project (organisation models and information systems for a production integrated environmental protection). This was sponsored by the German Federal Ministry for Education, Science, Research and Technology (BMBF), with the aim of developing organisational concepts at an intra-company as well as at inter-company level which support the implementation of environmental protection as an optimisation factor which is integrated within product development and manufacturing. Since the organisational concepts require adequate environmental information, the OPUS project aims also to develop information systems in order to provide an information backbone to support the processes of integrated environmental protection. The OPUS project is structured around the processes of order processing – in co-operation with several scientific institutions, methods, models and prototypes have been developed to support the business processes of product development, industrial and process engineering, production planning and control, controlling, shop floor control, and environmental management.

Within OPUS and in co-operation with the Fraunhofer Institute for Industrial Engineering (IAO), a concept for an environmental management system was developed for Trumpf which is integrated with its existing internal processes, and in particular which takes into account the distributed and decentralised nature of its organisation structure. In addition, the EMIS 'E-Bilanz'[1] of the software, which supports eco-balancing and the analysis of materials flows related to processes and sites, was further enhanced and adapted to the company's requirements.

With this enlarged information system, answers could be found to questions such as:-

- Which processes or organisation units cause which waste materials costs?
- For new production programs which are envisaged and considered, what quantities of wastes would these be likely to involve?
- Which processes are the most energy-intensive?
- How much environment-oriented costs are caused by a particular product?

Other goals within this approach were to avoid data redundancies, to reduce the effort required to maintain the EMIS, and to ensure the highest possible integration with existing organisation structures and information technology. In this context, the exchange of data between SAP R/3 and E-Bilanz played a significant role (see Figure 8.4). Through this data exchange, for example, dynamic data could be taken from SAP R/3 such as the numbers of units produced, and the machine hours worked in each workplace. These quantities could then serve as a basis for the allocation of materials flows to processes and products.

This exercise is different in a business such as Trumpf's, from a process industry. In the latter, it can be observed that there is generally a high coverage and quality of available master and dynamic data, and it is therefore often possible to calculate materials flow balances immediately on the basis of production data taken from SAP R/3. In the case of a producer of complex investment goods such as Trumpf, by contrast, the laser cutting machines which it manufactures are produced in the context of individual arrangements with the respective customer, with a high level of variety and diversity in the products which are produced. In order to calculate materials flows on the basis of the production data which was available from SAP R/3, a concept had to be developed that could provide results with sufficient precision without also requiring an intolerable effort in data collection and system maintenance. As an additional requirement, the concept was intended to be compatible also with production data exported from other PPC Systems than SAP R/3. The concept which was developed to meet these criteria will be described in the following section.

In principle the task of the concept which was developed is to allocate an overall quantity of, for example, energy consumption or waste production per site per month, amongst several workplaces or processes. The basis of this allocation should be cause-related, and the method used should be capable of reacting flexibly to changes.

---

[1] 'E-Bilanz' is a product of the software developer I-Punkt-Software, based in Homburg/Saar, Germany.

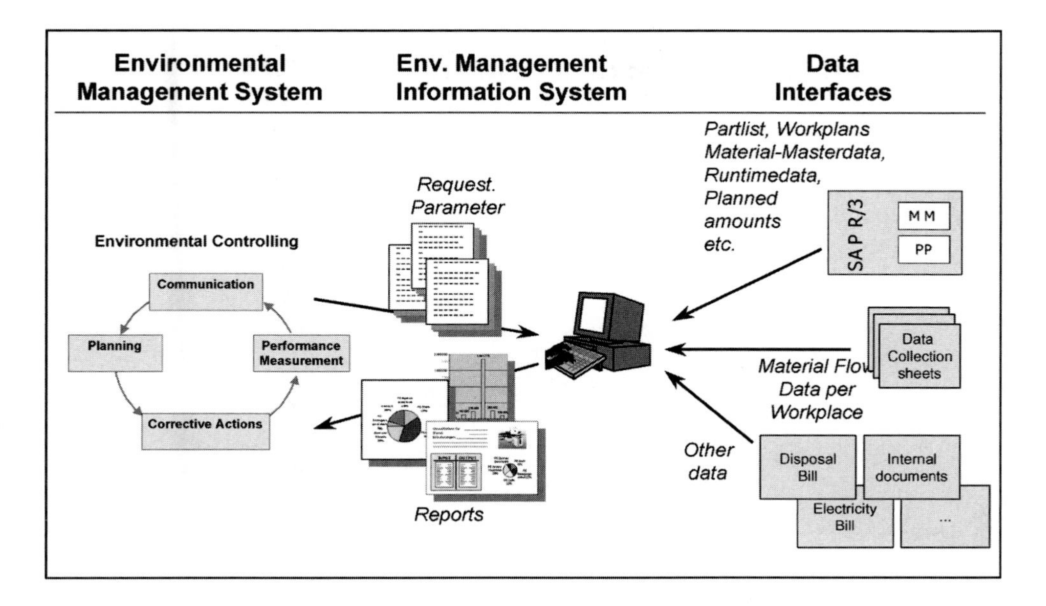


© Fraunhofer-IAO.

Figure 8.4. Use of an Environmental Management Information System (EMIS) at TRUMPF GmbH + Co. in connection to SAP R/3.

Three modes of allocation should be flexibly applicable:

- Fixed allocation:
  The same quantity of a material will be used or produced on a workplace every time period (here, monthly).
- Time-proportional allocation:
  The quantity of a material allocated to a specific workplace is dependent on the time that the workplace was in operation during the month in question.
- Quantity-proportional allocation:
  The quantity of a material allocated to a specific workplace is dependent on the number of units (output quantity) that the workplace has produced within the month in question.

For each type of materials, all modes of allocation should generally be applicable, in parallel. The above-mentioned requirements were best met by a flexible materials flow model in E-Bilanz, and a process which supports the user from the point of data entry up to the production of the requested reports.

The materials flow model within E-Bilanz is set up of balance nodes and points of measurements which are connected to each other (see Figure 8.5). For each combination of materials-type and workplace, there will be a single point of measurement. Points of measurement contain data which can be either physically measured, calculated or imported from other IT-systems, and which represent quantities of a specific type of materials which has been consumed or produced by that workplace. Points of measurement are connected

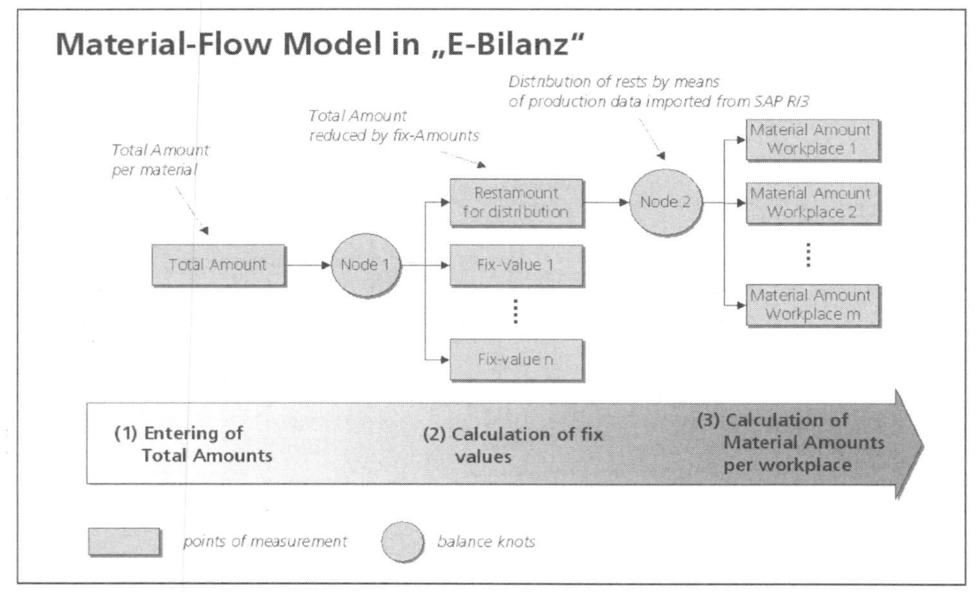


© Fraunhofer-IAO.

Figure 8.5. Materials flow model for one material type.

to each other using balance nodes, which then result in the materials flow model. A balance node can have several inputs and outputs, and describes either a physical or a logical materials flow between the points of measurement.

The user interacts operatively with the system on a regular basis (usually monthly), by executing a sequence of several steps:

1. **Input of overall consumption**
   The user enters into E-Bilanz the monthly quantities consumed or produced of each materials type, such as the consumption of energy by a process, or waste materials produced per product or per materials type. The user then starts to import SAP data, which initiates the import into E-Bilanz of ASCII or MS Excel files which are provided by the SAP system. These data files contain all output quantities of products (in their respective units of measure) which were produced by a particular workplace, and the total running times (in hours) of the machines in each workplace during the month to be analysed. These data files do not rely on specific features of SAP R/3, and can also be easily extracted from other PPC Systems as well.

2. **Allocation of fixed amounts per workplace**
   In the second step, the E-Bilanz software allocates to the appropriate workplaces the pre-defined fixed consumption amounts for respective materials types within the time period. Afterwards, the difference between the total amount entered by the user and the sum of all fixed amounts per material is calculated as a balance to be distributed (this corresponds to the content within the balance 'rest-amounts' in balance node 1 in Figure 8.5).

3. **Allocation of variable materials quantities per workplace**
   Through data import from SAP, either the numbers of products produced or the operations times are already stored for every workplace at the appropriate point of measurement (according to the selected type of allocation). On the other side of the total of these values is the input quantity in balance node 2, which has been passed on from balance node 1 through the respective point of measurement. These two values are unlikely to be identical, so the difference between them (the balance rest-amount in balance node 2) will be distributed proportionately amongst the connected workplaces. If, for example, 100 kg of waste from balance node 2 is to be allocated to two machines, A and B, which have respectively produced 20 units and 5 units of finished products, then machine A will receive 20kg plus 20/25 of the balance rest-amount in balance node 2 which is 100 – (20 + 5). Consequently machine A (B) will receive 20kg + 60kg = 80kg (5kg + 15kg = 20kg).

By this procedure the materials flow model can be calculated for each type of materials which are included in it. Using this concept, detailed and consistent balances per workplace can be generated by the relevant reports and evaluation functions.

In addition to this materials flow model, additional functions were defined to support the user in generating product-related balances. For this purpose, the bill of materials was exported from SAP and stored within E-Bilanz. The information about which share of a total amount of produced materials is, on average, produced by each workplace is captured in a second file, which is generated by SAP. The latter information is particularly relevant if a company is operating flexible and redundant process capacities, or it is considering whether to produce an input material itself or to have it produced by an external service provider (make-or-buy). The third item of information which is necessary to calculate the materials flow related to a product is taken directly from the materials flow model, which is updated monthly and describes the average amount of a certain material that a workplace produces or consumes in the production of a single unit of a product. All three information items are merged in a single query statement which produces a balance per product, containing the average amounts of materials produced or consumed during the production of this product.

With the help of this concept as it has been developed, and its calculation functions as described above, Trumpf is able not only to calculate the materials and energy flows which have been caused by past production processes, but also in its production planning to estimate the consumption of energy that would caused by the production of a certain product. The concept of materials flow management can thus be used not only in the context of accounting, but also in production planning activities.

## 8.4. Future trends

The experience gained in the context of this project at Trumpf shows that a regular assessment of materials flows within an environmental management system can lead to a substantial improvement in the information base which is necessary for the systematic optimisation of a company's environmental performance. In particular it was demonstrated that by means of the modelling capacities of an adequate EMIS, it is possible to calculate the most important materials flows of production processes on the basis of a reduced number of selected production data which are generated by PPC systems such as SAP R/3, or other systems.

Regarding the high economic and ecological potentials of the implementation of a materials flow management, the application of EMIS as a supplementary tool together with PPC systems will become an important trend in the practice of environmental management. In order to increase the benefits that can be achieved by the use of EMIS, the following aspects are of importance:

- The experience which is presently available within the field of the application of EMIS to support materials flow management should be enlarged by further pilot projects in different industry sectors.
- Research in techniques for the efficient modelling of materials flows should focus on the adaptability of concepts to different types of production processes and their type of organisation.
- Development of open interfaces between EMIS and PPC systems should be emphasized.

## 8.5. Conclusions

A pilot project with the company Trumpf GmbH & Co. analysed how to support the implementation of a company's materials flow management by the use of an Environmental Management Information System (EMIS). Thereby a concept has been developed and implemented that allows the calculation of workplace-oriented and product-oriented materials flows on the basis of production data.

The case study described in this paper focuses on the connection of an EMIS to SAP R/3. In order to be able to apply the concept to other companies, the data which is imported from SAP R/3 was reduced to a selection of production data which can also be found in most other production planning and control systems. This demonstrates that the concept can be used not only in those companies who use SAP, but also in other companies which use other information systems, especially small and medium enterprises (SME).

The major conclusions can be summarised as follows:

- The systematic improvement of a company's environmental performance requires an enhanced information basis for decision making.
- Materials flow management is an appropriate concept to generate the desired information.
- Environmental Management Information Systems (EMIS) provide a necessary and useful support for all steps of materials flow management.
- Interfaces to production planning and control systems such as SAP R/3 constitute a crucial step to make Materials Flow Management applicable.

Due to the large ecological and economic optimisation potentials that can be obtained through the regularly updated transparency of a company's materials flows, the implementation of a company's materials flow management represents an important step for the further development of environmental management systems.

# 9. 'Counting what Counts' – Raising Transparency through Environmental Management Accounting at Siemens

*Ralph Thurm*
*Member of the Corporate Office Environmental Affairs, Siemens AG, Munich, Germany; E-mail: ralph.thurm@mchp.siemens.de*

## 9.1. Introduction

Siemens is one of the world's leading companies in electrical engineering and electronics, with more than 400,000 employees working to develop and manufacture products, plan projects, create systems and installations and tailor a range of individualised services. It operates in six business segments:–

- Information and Communications;
- Industry;
- Health Care;
- Transportation;
- Energy;
- Lighting.

With research and development locations in over 30 countries and regional offices and agencies in more than 190 countries, Siemens aims to offer its customers innovative technology and comprehensive expertise in business and technical solutions. Its products and services include the installations of networks and development of products to meet future communication needs; complete solutions and service packages for manufacturing processes; increasing the efficiency of diagnosis and therapy in hospitals; enhancing the safety and environmental compatibility of rail systems and automobiles; providing the means to generate affordable power in environmentally-friendly power plants and its reliable distribution the consumer; and economical lamps and lighting systems for all kinds of applications. To deliver all these products, services and solutions, Siemens invests around 5 billion euros a year in research and development, and as a result it now holds 130,000 patent rights on 32,000 inventions in approximately 100 different countries.

Siemens with its six business fields is at the centre of the debate on how best to take action for Sustainable Development, and therefore is followed by various stakeholder groups. In its Environmental Report for 2000, Dr. Edward G. Krubasik, a member of the Siemens Managing Board with responsibility for Environmental Protection, pointed out that 'today, at the start of the 21st century, we are aware of our duty as a major international company not just to protect and sustain the environment, but also to provide the kind of innovative technologies that will benefit our world tomorrow. By making sustainability an essential element of the way we approach our business, we can enhance our customers' and our own ability to compete effectively, as well as tap into new business opportunities.'



*M. Bennett et al. (eds.), Environmental Management Accounting: Informational and Institutional Developments,* 123–135.
© 2002 *Kluwer Academic Publishers. Printed in the Netherlands.*

## 9.2. Trends and developments in the beginning of the new century

Siemens anticipates a paradigm shift towards Sustainable Development forced by demographic, social and environmental realities. By 2050 between 10 to 12 billion people will inhabit the earth, more than 90 percent of them living in developing countries. Feeding this increasing population will require a doubling of the food supply over forthcoming decades, and energy consumption will quadruple until 2050. Two-thirds of the people will live in cities, many of them in so-called 'mega-cities' like Manila, Shanghai or Mexico City, where basic needs for air, fresh water, energy, transportation, housing and jobs already exceed capacity even today.

An objective screening of the world's materials and energy flows shows that the 20% of the world population who live in the 'industrialised triangle' (North America, Europe, Japan) are responsible for 80% of the world's use of natural resources. This is a social timebomb of the first degree, considering that the already predicted migration of people from poor to rich areas of the world – mostly cities – can be expected to be reinforced by this inequality. To respond to such demographic instabilities, companies like Siemens have to adopt new ways of thinking about environmental and social issues. In a recently published brochure 'Solutions for the Cities of Tomorrow', for example, Dr. Heinrich von Pierer, President and CEO of Siemens AG, pointed out that 'the three billion people currently living in urban areas will be joined by an additional two billion within the lifetime of just one generation. It does not require much imagination to picture this development leading to a horror scenario that many experts are already warning of, with catastrophes such as the complete breakdown of mobility, housing shortages and insufficient sanitary services. Global population growth and increasing urbanisation, in particular, will therefore represent the greatest challenge of the new millennium.'

If one wishes to summarise the needed paradigm shift from the current 'greening' to the future 'sustaining' of business, the following developments can currently be observed:

- a general change of the guideline 'less than . . .' in relation to the emissions or use of materials and energy, to 'as much as . . .' – i.e. emissions or the use of materials and energy only up to an allowed sustainable maximum (that has to be clearly defined);
- from 'existing environmental damage' to 'potential environmental risks';
- from regarding emissions as environmentally harmful side-effects of production processes and the use of products, to regarding production methods and products as environmentally harmful side-effects of a lifestyle;
- from eco-efficiency to eco-effectiveness, and sufficiency;
- from a shareholder's point of view to a broader stakeholder-oriented point of view;
- from legally enforced environmental protection to market-driven and NGO-driven sustainable development.

This widespread paradigm shift recognises the absolute boundaries of the world's ecosystems, and that our conventional economic system is geared at maximising capital and work productivity whereas it is resource productivity that really matters. Other dysfunctional features are:

- the use of environmental policy as 'repair policy' rather than integrating it into all fields of policy;
- uncertainty about what may have caused certain environmental problems because of a lack of generally accepted indicators;
- uncertainty about the degree of structural economic change that is required, inclusive

of the political guidance necessary to move trade and development in the right direction.

Different movements echo a development from eco-efficiency to eco-effectiveness:

- from an end-of-pipe view to a start-of-pipe view of production, with the use of integrated technologies;
- from emission-orientated control to input-orientated control;
- from a site-orientated view to a value chain-orientated network of actors;
- from a production-orientated view to a product view;
- from an administrative environmental management function to a value-based continuous improvement of the processes and products.

These new developments help companies to ask themselves whether they are doing the right things rather than merely whether they are doing their own things right – the difference between efficiency and effectiveness. Inventing new products, services and solutions (which means the whole company) from a sustainability perspective is to be the point of departure for a sensible scenario-orientated strategic planning.

## 9.3. Environmental Mission Statement and environmental management systems

For Siemens, as a company that in many ways is connected with areas where sustainability requires concrete action, many of these changes are already approaching. Its Environmental Mission Statement therefore makes reference to the different aspects of Sustainable Development:

> *'Our knowledge and our solutions are helping to create a better world. We have a responsibility for the wider community and we are committed to environmental protection. In our global operations, featuring a great diversity of processes, products and services, our company is concerned with sustaining the natural resources essential to life. We view the economy, environmental protection and social responsibility as three key factors carrying equal weight in a liberal world market. We support the dissemination of knowledge needed for sustainable development through the transfer of knowledge in the fields of management and technology, where ever we operate as a company. For us, sustainable development in environmental protection means careful use of natural resources, which is why we assess possible environmental impacts in the early stages of product and process development. It is our aim to avoid pollution altogether or to reduce it to a minimum, above and beyond statutory requirements.'*

Environmental tasks and duties within the company are organised around a three-level model that distinguishes between general responsibility and technical responsibility. In this, members of the Siemens Managing Board, Group Executive Management, and works management are supported by Siemens' Corporate Offices and their counterparts in Groups, as well as by appointed officers at Siemens plants and locations. Every Siemens location is to have an environmental management system in place that complies with the international ISO 14001 standard, though the emphasis here is on realising tangible environmental improvements rather than merely on acquiring official certificates. Nevertheless, by the end of 1999, Siemens had reported 130 successful ISO 14001 certifications, plus more than 60 cases of validation in accordance with EMAS.

## 9.4. Environmental managerial accounting

The paradigm shift described above confronts companies with several new communication challenges. They are facing increasing concerns from various groups over their environmental and social impacts. Different stakeholder groups ask for different types of information:

- critical queries by the financial sector (banks, insurance companies and pension funds) are increasing very strongly;
- the company management itself needs information on costs, revenues and profits;
- environmental protection agencies and NGO's are concerned with environmental assets and liabilities.

Environmental managerial accounting can provide for these needs for information.

Environmental Managerial Accounting (EMA) can be defined as the generation, analysis and use of financial and related non-financial information in order to integrate corporate environmental and economic policies and build a sustainable business (Bartolomeo et al., 1999). EMA serves business managers in making good decisions in the field of capital investment, product costing and design issues. It can contribute to performance evaluations and underpin different kinds of forward-looking business strategies.

Therefore, EMA usually needs to make use of

- Cost management instruments like Activity Based Costing (ABC), Full Cost Accounting (FCA) and Life Cycle Costing (LCC); or
- Resource management instruments like Resource Efficiency Accounting (REA) and Input/Output-Analysis (IOA); or
- A mixture of both of these, by using materials and energy flow systems in connection with cost accounting systems. The result is a process-oriented approach that unites technical and financial/business orientated views that otherwise tend to develop separately (see below).

## 9.5. Raising cost transparency – an overview

To achieve continuous improvement in industrial environmental management, it is increasingly important to record a factory's usage of materials and energy and to track these during the production process and analyse how significant they are from both an environmental and economic point of view. Conventional cost accounting systems, however, typically produce inadequate data.

### *Zero-waste management*

Studies conducted in a number of industry sectors have shown that the quantity of substances and materials not turned into products – so-called 'residual materials' – can amount to as much as 30% to 50% of the total materials used (Fischer et al., 1997). Zero-waste management, therefore, takes a closer look at all the processes and quantifies the residual materials that they generate, also in terms of costs. Pilot projects have revealed that these costs can account for between 5% and 15% of an industrial company's total costs. Siemens' goal is to use zero-waste management to make the company's residual materials flows visible and transparent, all the way from the purchase of materials or energy through its transportation, storage, and handling up to final disposal, and thus to indicate potential savings.

### *Materials and energy flow management*

Furthermore, Siemens aims to enhance its materials and energy flow accounting by taking into consideration all environmentally relevant materials and energy flows associated with product manufacturing. Moreover, analysis of the environmental impact of the production processes will be refined. At the same time it aims to track relevant data for screening a product's entire lifecycle. In combination with drawing on its cost-accounting systems, it hopes to continue to make further progress in this field in the future. By using computer models to simulate trends in quantities and costs, it hopes for substantial benefits in both environmental and economic terms.

### *Internal reporting as an instrument of continuous improvement*

To lever the effectiveness and cost-efficiency of environmental protection measures throughout the company, Siemens is setting up a world-wide internal reporting system. This system has to be comprehensive, not in the least because of the large number of locations in the global marketplace; it encompasses the whole Siemens AG as well as the Regional Companies. This system builds on the existing reporting system within Siemens AG. Besides reporting on how duties are fulfilled, it also drives a process of continuous improvement. Moreover, it provides the basis for the company's external environmental reports that help to keep the public informed. These are prepared not only for the company as a whole, but increasingly for the Groups and Regional Companies as well, in order to address regional and customer-specific interests. These are supplemented with site-based environmental reports that are validated under EMAS. This new milestone in environmental reporting gives rise to the creation of a world-wide information and reporting platform that by 2003 will make available comprehensive data from those areas of the world that are most relevant to Siemens.

## 9.6. Increase in efficiency by zero-waste management

Current cost accounting systems are still based on an idealised model of output generation. The only outputs considered are the company's desired products. As a result, all inputs are taken to be transformed into them. In reality, the output of the business also includes substantial amounts of waste, unused energy, waste water and waste heat. Besides the costs of the materials themselves and the external disposal costs, these waste streams entail internal handling and other costs. 'Zero-waste management' focuses on reducing these costs that are part of overheads which eventually reduce profit margins or increase the amounts to be paid by the customers.

### 9.6.1. *Zero-waste costs*

Residual materials are undesired outputs (both materials and energy) because they are not part of the product. Zero-waste costs are all costs caused by these residual materials, minus the profits obtained from them. These costs, in fact, do not contribute to the value added. Investigations in several branches reveal an astonishingly great amount of residual materials (30-50% of the total output). Upon closer look, this is not surprising because:

- auxiliary and operating materials which are not incorporated into the product are regarded as residual materials,
- spoilage and rejects are residual materials,

- rejected and sometimes even new materials are disposed of or sold far below their original purchase cost;
- energy resources such as gas, oil, coal or wood leave the business to incinerated;
- packaging materials are not regarded as a part of a product that is valued by the customer, and therefore constitute residual waste even before its actual use.

The costs connected with the generation of residual materials are incurred:

- in procurement as external costs of raw, auxiliary and operating materials and as additional internal costs of procurement;
- in materials management as internal storage and transportation costs (sometimes causing additional costs due to a need for separate and/or special storage of environmentally relevant materials and waste);
- in production (costs for safety measures) and in logistics;
- in the collection and/or in the treatment after the production process (costs of waste collection, end-of-pipe equipment, e.g. pollution control and water protection);
- for external recovery or disposal of waste.

If the proportion of the amount of residual materials is included in a full cost accounting analysis, 5-15% of the total costs of a typical industrial business can be connected with its residual materials. These proportions depend on the degree of outsourcing; however, 5% of the total costs is a realistic minimum figure, even when the degree of outsourcing is high.

### 9.6.2. *Reasons for high zero-waste costs*

There are several reasons why a considerable and barely recorded proportion of the total costs relates to residual materials:

- costing procedures include a small fraction of zero-waste costs only. This means that the only disposal costs that are properly taken into account are those that are allocated to a particular cost unit. However, these costs are only a minor part of what is meant by zero-waste costs. Total waste costs that include the purchase value of the raw, auxiliary or operating materials over the entire chain of added value can be 5 to 20 times higher than the disposal costs, and sometimes even more.
- there is no, or only limited, explicit allocation of the zero-waste costs, as most of the time these are hidden in general costs. Therefore, the individual costs units have hardly any incentive to reduce them.
- product design, materials used and the design of production processes are the major factors that determine the amount of zero-waste costs. The managers of the individual processes do not receive support from the costing function to help to determine the scale of the zero-waste costs that they actually cause.
- almost every business function can contribute to the reduction of zero-waste costs. However, in order to achieve a total optimum, rather than being satisfied with several sub-optima, it is necessary to incorporate the zero-waste viewpoint in the company's general process management. This is an issue of process organisation that requires further decision-making. In practice, the conditions for achieving a total optimum have not yet been fulfilled.

### 9.6.3. *Benefits of zero-waste cost management*

To control the 'residual materials' process requires knowledge of the residual materials and related costs in proportion to the total materials flows running through the different production facilities. 'Transparency' is a key term here. The first application of this knowledge should be to improve materials and energy efficiency wherever this can be effectively done. Controlling of the 'residual materials' process based on such knowledge can have different impacts:

- drawing attention to the unexpectedly high amounts of zero-waste costs usually triggers initiatives to reduce them.
- zero-waste cost control requires cost structures which are informed by the link between materials and products. This calls for involving those responsible for the individual production processes in shaping the cost structures.
- insight into separate streams of residual materials enables the person responsible for a product to focus on the specific areas where most costs are being incurred.
- charting materials flows improves the accuracy of calculations, and is helpful in evaluating individual throughputs. (This is also termed 'balancing' since it measures both inputs and outputs and establishes a link between them; the balance lies in the logic that each input should be reflected in an output).
- if allocated to the appropriate cost centres, the full costs of materials often appear to be higher than the direct labour costs.
- the possibility of refunding the external and internal profits from recycling residual materials improves the calculation of cost centres. Profits from recycling are often not refunded to the cost units generating the residual material, but to cost units without any logical connection to the product production.
- zero-waste costing allows the allocation of procurement and waste management cost units according to the quantities of residual materials actually produced. This is much more accurate than traditional criteria such as (for example) the amount of floor space used.
- conventional product development processes do not usually take into account the full costs of a product over its whole life cycle. Accurate determination of product or product-group costs according to the principles of zero-waste management provides the information necessary to set sensible task priorities that take the life-cycle into account.
- zero-waste cost management provides diverse data that serves the purposes of energy management, environmental management and quality management, and supports the co-ordination of procurement and waste management, supply chain management, asset management and target costing.
- zero-waste management can be integrated into a comprehensive framework of materials and energy flow management (see next chapter).
- as an overall result, general cost units become more transparent and can be better controlled.

### 9.6.4. *Software implementation*

Zero-waste costing is not possible without a highly automated control system that is well embedded in the organisation. This control system must calculate the quantities of residual materials from existing data bases, track where residual materials are going to, and provide an IT-based analysis of zero-waste costs.

This could also be accomplished by adapting and augmenting the existing ERP (Enterprise Resource Planning) systems. Within the framework of a general strategy to integrate environmental affairs into the business – such as integration into SAP/R3 and Environment, Health & Safety (EH&S) – it has to be decided to what extent zero-waste management should be part of this integration.

## 9.7. Benefits for environmental protection by balancing materials and energy flows

Continuous improvement as required by ISO 14001 and EMAS increasingly depends on the identification and evaluation of the environmental impacts of materials and energy flows. Furthermore, the data obtained from balancing materials and energy flows can be integrated in product life cycle evaluations at higher strategic levels.

Charting materials and energy flows is also relevant from an economic point of view, because an improvement of input/output-ratios usually leads to cost savings. Therefore, there is good reason to evaluate systematically all materials and energy flows in both environmental and cost terms.

### 9.7.1. *Benefits*

The following advantages of charting materials and energy flows (the eco-balance) as an important tool of environmental management were confirmed in internal projects:

- communication and information based on materials flows: overcoming departmental thinking and a biased presentation of facts and thereby creating a common communication base by focusing on processes at a higher organisational level.
- further optimisation of materials flows instead of partial optimisations, i.e. a chain of sub-optima can be replaced by an overall optimum.
- data gaps and inconsistencies in local information systems are revealed.
- increase in transparency by visualisation of process chains, e.g. by means of Sankey diagrams that can help to identify areas for in-depth analyses and priority measures.
- accounting practices that can divide process costs into
  - input costs (especially procurement, transport, insurance, etc.)
  - internal flow costs (handling, machine allocation, air conditioning, storage, transport, etc.)
  - output costs (disposal, monitoring, transport, etc.), to be further broken down into cost classes such as materials, personnel, depreciation, energy, etc.

### 9.7.2. *Connection to SAP R/3*

To keep the costs of information at a justifiable level, the periodic updating of the materials and energy flow balances should be linked to the existing business software. Siemens is mainly working with SAP R/3, and much of the environmentally relevant data is already present there, so that this system can generate the data needed for the annual report on environmental protection or duties in waste management.

SAP R/3 cannot produce a comprehensive balance of materials and energy flows. However, with the help of a process-cost controlling tool (AUDIT(r) for R/3, certified by SAP as a complementary software product), materials and energy flows can be mapped and described on the basis of business processes. The data can be accessed by a SAP-

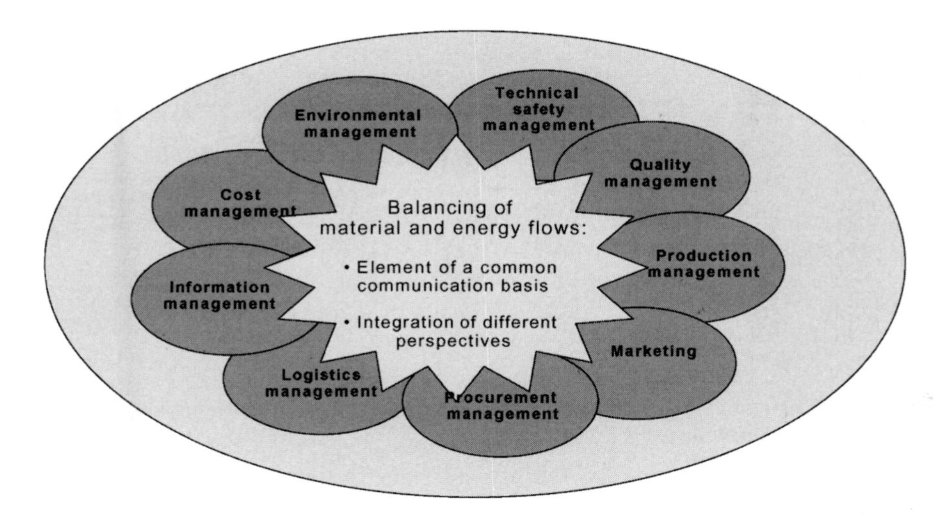


Figure 9.1. Creating common understanding with materials and energy flow balances.

certified interface by which the materials and energy flows are institutionalised and routinely updated. Depending on the IT-structure, the interface can be connected with other data bases. Weaknesses can be detected by evaluating actual values and comparing them with budgeted and normal values for work places, work plans and production contracts and by confronting them with other sources (e.g. master bills). Furthermore, controlling can optimise the efficiency of processes by simulating and evaluating materials and energy flows. This tool augments the SAP controlling model by adding a process-oriented description and evaluation of the materials and energy flows.

## 9.8. Success stories in industrial environmental protection

Thinking in terms of integrated environmental protection measures has a long tradition at Siemens. It has been over a decade since the first integrated installations were built up. The following show the effects in different ways, for example:

- environmental operating expenses and capital spending figures show that these are now significantly less than six years ago. Compared with 1993, the amount of capital expenditure has been reduced by more than half; over the same period operating expenses have come down by 37%. Capital spending on waste management has fallen by 90% compared with 1993 and operating expenses are down 35%. Spending to protect water resources is down 62% compared with 1993 levels and operating costs have dropped 51%. This is all the more significant since external charges (e.g. sewerage charges, waste disposal costs and water rates) have increased.
- cumulative cost savings: between 1993 and 1999 Siemens was able to save 30 million euros in outlaid costs on waste for disposal, and 16 million euros in the area of domestic waste. Even if these figures are balanced against the additional costs incurred through a shift away from waste disposal and toward waste recycling, the external cost saving

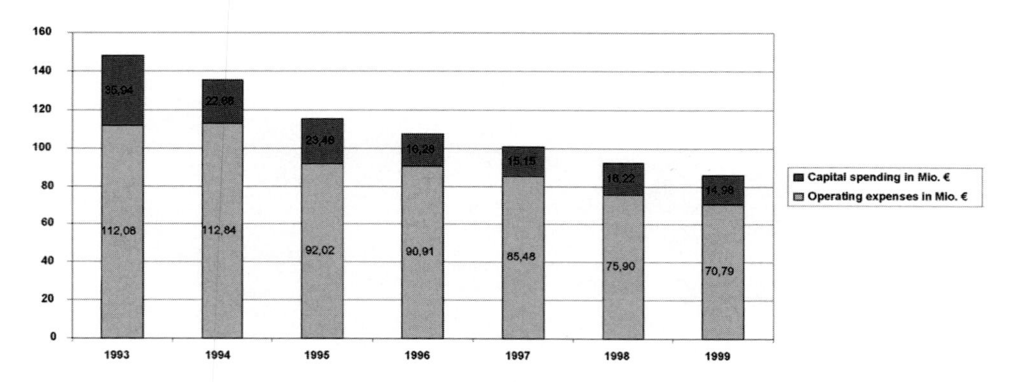


Figure 9.2. Operating expenses and capital spending for environmental protection by Siemens AG compared over time (• millions).

nevertheless runs to a net 31 million euros. Furthermore, if it is taken into account that accurate management of materials and energy streams leads to additional savings by reducing the quantities of materials and energy purchased as well as the handling overhead, it becomes clear that the real overall savings are higher than only at the end of the value chain.

- additional external cost savings also exist in the field of water and wastewater treatment: accumulated external costs from 1993 to 1999 show another 27 million euros savings.

The period between 1993 and 1999 was chosen for this analysis because it shows a good correlation between investment in integrated measures and amortisation, and moreover during this period the organisation structure of Siemens (groups, business units) was fairly stable. In 1999 a re-organisation based on the Siemens ten-point program brought many changes, so that from then on comparisons are much more difficult.

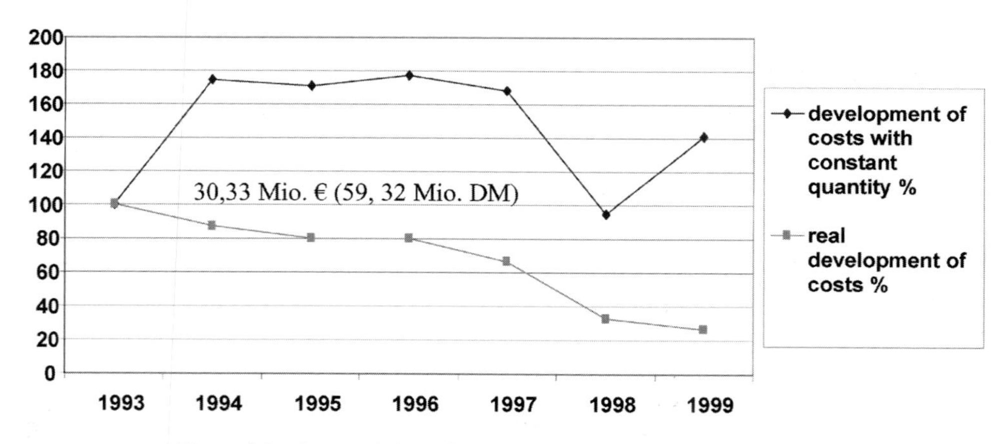


Figure 9.3. Accumulation of savings: waste for disposal 1993–1999.

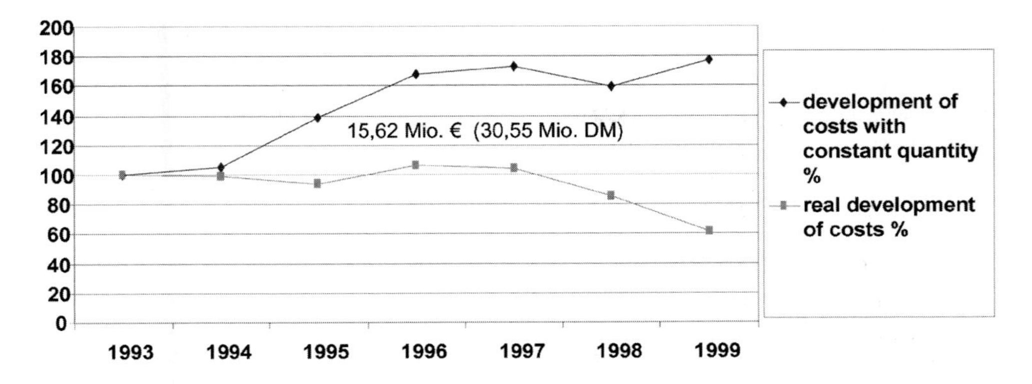


Figure 9.4. Accumulation of savings: domestic refuse 1993–1999.

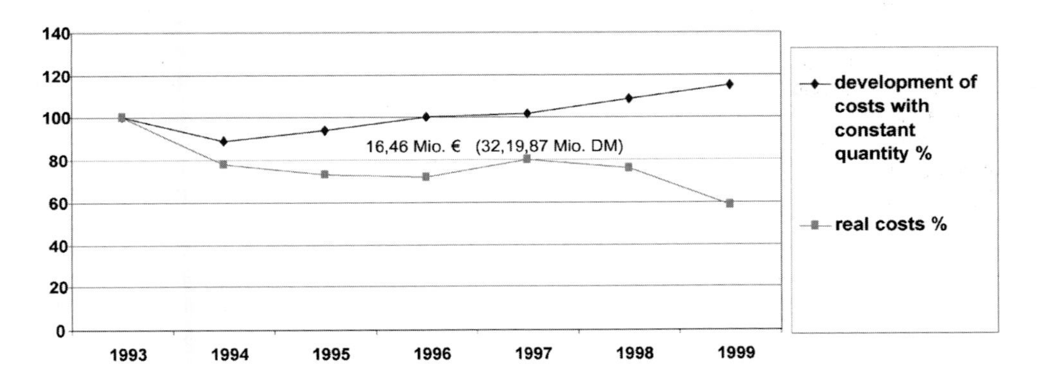


Figure 9.5. Accumulation of savings: wastewater indirect discharge 1993–1999.

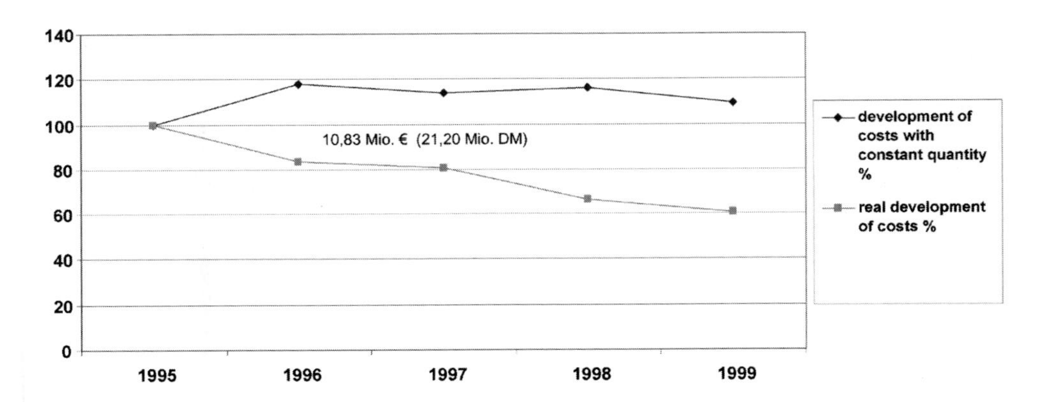


Figure 9.6. Accumulation of savings: water input 1995–1999.

## 9.9. Taking EMA one step further: Environmentally compatible product design

A very important part of the Environmental Management Systems at Siemens is the inclusion of environmental concerns into product development. The planning and development cycle covers the process from marketing to disposal. The product-related environmental organisation is managed by the product managers of the Groups' business units within the three-level-model of Siemens's Environmental Protection.

### 9.9.1. *Application of Siemens Standard SN 36350*

When a new product is about to be developed, Siemens starts with a screening, covering the whole life cycle of the product according to the Siemens standard 'Environmentally Compatible Product Design' (SN 36350). This standard complies with the internationally agreed IEC Guide 109, and application is mandatory for all new developments. The standard, therefore, contains general rules for the design of environmentally compatible products and for the integration of environmental aspects. A checklist of all steps to be taken is helpful in taking to all relevant aspects into consideration. This checklist is applicable for review steps, for life cycle screening to determine the relevant environmental processes, and for a rough evaluation of the environmental impact of a product. Unlike an LCA, this standard makes it possible to identify the phases with the highest environmental impact.

Apart from this first element, which may also be given to suppliers, the standard also includes a list of hazardous materials which are prohibited in a product (in Europe) and a list of substances to be declared and avoided. Three further parts deal with a simplified method for environmental evaluation of metals, polymers and packaging.

Many of these regulations are helpful not only to the environment but may also reduce costs. Some of these rules are known and have been tested by TQM programs. However, they now have a new focus, e.g. to reduce the numbers of types of materials and parts with a view to eventually having only pure, environmentally-friendly materials in the case of disassembly after take-back.

Effective disassembly turns screwing or gluing into obsolete technology. Reinforced plastics are also difficult to recycle, though product developers are encouraged to take a holistic view since this disadvantage can sometimes be outweighed by other factor. For example, the environmental advantages of light-weight carbon-fibre reinforced plastics in the field of transportation and energy consumption can outweigh later disadvantages in recycling.

### 9.9.2. *Preconditions and needs*

For electro-mechanical products, the application phase is generally recognised to be very critical from an environmental viewpoint. For example, up to 90% of the energy related to making or using a washing machine relates to its application so that, apart from a few of its components, it is not necessary to apply a full Life Cycle Assessment here. The manufacturer, therefore, can make a notable contribution to a better environment by improving the energy efficiency of their products. Detailed analyses like LCA make more sense at a higher complex system level which describes the application of a product such as an IT product during its application until recycling.

One constraint which has been experienced is a lack of data for which there is no immediate solution: at the moment there is not enough product information due to a missing standard or LCA data due to comparability, the age of data, too many products (with Siemens, about 1 million different products).

Recycling systems and recovery processes have to be developed and installed. Beginning with the complicated and difficult-to-organise take-back systems, an ideal recovery system has to be in place that holds over a long period of time. The possibility should exist, for example, for a customer to give back a product after 10 years. If the product is merely shredded, the application of nearly every technology or material is possible but the environmental benefit may then be very low. Therefore, to save resources, the proposed EU take-back legislation requires a high percentage of reuse and high-level recycling.

Furthermore, there are bills put forward in the EU (such as the bill concerning lead which is to be effective as from 1 January 2008) which provide for the prohibition of several hazardous substances such as Pb, Hg, Cd, Cr (VI) and brominated Biphenyls and Diphenylethers. Moreover, to improve recycling and mitigate particular certain environmental impacts after deposition, certain components demand special treatment. To be prepared for this imminent legislation, the Siemens groups are in the process of finding out where technological changes are necessary.

# 10. The Danish Environmental Management Accounting Project: An Environmental Management Accounting Framework and Possible Integration into Corporate Information Systems

*Pall M. Rikhardsson and Lars Vedsø*
*Associate Professors, The Aarhus School of Business in Denmark (Department of Accounting); E-mail: par@asb.dk and lv@asb.dk*

## 10.1. Introduction

'What are the benefits from our environmental policy?' is a question which is asked increasingly frequently in the boardrooms of Danish companies. And since Dow Jones has shown that sustainable companies may be doing better than the average company,[1] the interest in combining environmental issues and management accounting into what is often referred to as Environmental Management Accounting has intensified. Accordingly, environmental management accounting is not a question of a 'green' revolution within management accounting, but is more likely a reasoned decision combining a number of organisational objectives. Some of the key objectives often relate to statutory requirements, and to a wish for enhanced efficiency and to reduce the total costs of a company (Bennett and James, eds., 1998; Bartolomeo et al., eds., 1999; Broas et al., 1999).

Environmental management accounting (EMA) is about managing the resources used by a company for environmental management purposes, as well as about assessing costs and benefits of environmental management efforts. Mounting evidence suggests that EMA can improve the competitive position of a firm by directing the attention of managers towards potential cost reductions as well as towards possible areas for differentiation.

Recent studies seem to show an increased management focus on the value of environmental management accounting. Bartolomeo et al. (1999) in a European survey report that financial environmental performance and the future role of accounting is seen as becoming more important. Figure 10.1 below shows the percentage of companies from different industries which considered that different accounting issues would become more important for environmental management in the future.

In order to be able to assess whether corporate environmental efforts create value, it is of course necessary to measure the costs and revenues related to these efforts. But ask any CEO how much any company currently spends on environmental compliance management as well as various voluntary initiatives, and the answer will often be only a ballpark figure. The reason is that EMA has been a marginal activity in many firms – even those which have a perfect track record in managing their environmental issues – due to reasons such as the lack of a common reference framework for environmental accounting, and lack of integration into corporate information systems.[2]

Against this background PricewaterhouseCoopers initiated the Danish Environmental

---

[1] http://www.sustainability-index.com/.
[2] For a good overview, see e.g. Bennett and James, eds. (1998).



*M. Bennett et al. (eds.), Environmental Management Accounting: Informational and Institutional Developments*, 137–151.
© 2002 *Kluwer Academic Publishers. Printed in the Netherlands.*

| | *Textile finishing companies* | *Chemical companies* | *Paper & Printing companies* | *Metal plating and treatment companies* | *Electronic companies* | *Utility companies* |
|---|---|---|---|---|---|---|
| Bookkeeping | 50 | 70 | 60 | 67 | 50 | 44 |
| Budget setting | 67 | 75 | 60 | 63 | 88 | 78 |
| Budget control | 67 | 90 | 60 | 75 | 88 | 67 |
| Capital budgeting | 83 | 70 | 67 | 75 | 75 | 78 |
| Product costing | 25 | 60 | 25 | 57 | 63 | 63 |
| Financial performance measurement | 50 | 75 | 67 | 20 | 75 | 67 |
| Non-financial performance measurement | 50 | 90 | 25 | 16 | 75 | 67 |

Figure 10.1. The future importance of accounting in environmental management (see Bartolomeo et al., eds., 1999).

Management Accounting (DEMA) project in 1999 in co-operation with The Aarhus School of Business, and partly funded by the Danish Agency for Trade and Industry and the Danish Environmental Protection Agency. The main purpose of the project was to develop an environmental management accounting framework and to look at possibilities for integrating EMA into accounting information systems. The project was based on literature surveys and case studies of four Danish companies which participated in the project.

The structure of this paper is as follows. After this introduction, section 2 will present the EMA framework developed in the project and describe the deliberations on which it is based. Section 3 will discuss the potential integration of EMA into corporate accounting information systems. Section 4 will describe some of the experiences of two of the case companies, and section 5 will list some of the main conclusions of the project.

## 10.2. The framework adopted for the DEMA project

Several authors have proposed frameworks for environmental accounting.

Schaltegger et al. (1996) for example define environmental accounting as 'the subarea of accounting that deals with activities, methods and systems for recording, analysing and reporting environmentally induced financial and ecological impacts of a defined economic system (e.g. firm, plant, region, nation, etc.)'. They divide environmental accounting into three parts:

3. environmental management accounting;
4. environmental financial accounting; and
5. other types of environmental accounting such as environmental tax accounting.

This categorisation is similar to that used in traditional accounting. Using this definition, environmental accounting thus deals with financial environmental issues as well as with environmental accounting in physical units. It aims to create both financial and physical transparency regarding the results of corporate environmental management. Schaltegger et al., then propose the following framework for the classification of environmental accounting:

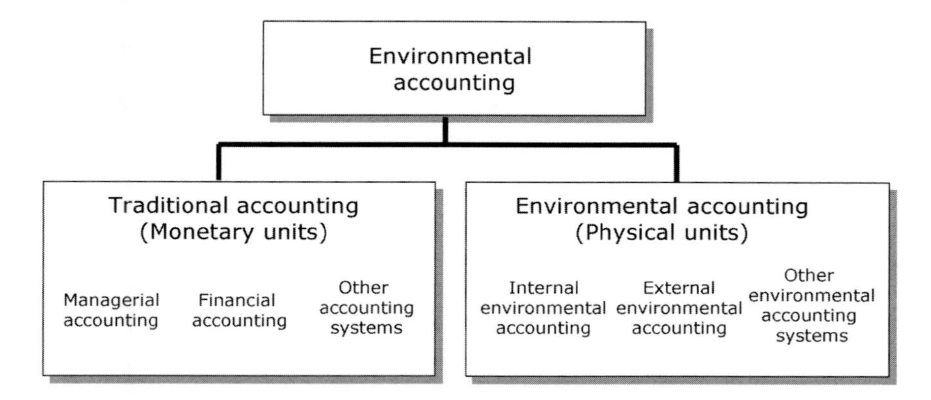


Figure 10.2. Elements of environmental accounting.[3]

The US Tellus Institute and the US Environmental Protection Agency (EPA) proposed another framework for identifying and managing the environmental expenditures incurred by companies (EPA, 1995). These are classified into:

- Conventional environment-related expenditures recorded in conventional accounting systems, such as raw materials costs, utilities and supplies.
- Indirect environment-related expenditures, which might be hidden from management because of aggregation in overhead calculations. These are, for example, expenditures associated with legal compliance (monitoring, testing, inspections, etc.), up-front expenditures incurred before operations begin (site studies, permitting, R&D, etc.), or voluntary expenditures that go beyond compliance (community relations, planning, supplier auditing, etc.).
- Contingent expenditures are those that may or may not be incurred at some point in the future. These are also called contingent environmental liabilities.
- Image and relationship expenditures affect the perceptions of management, customers, employees and other stakeholders. These are often called 'intangible' expenditures, because the benefits associated with them are often difficult to measure.
- Societal expenditures which result from environmental impacts but for which a company cannot be held legally accountable, such as degradation of ecosystems due to legal waste deposits, emissions, etc., and detrimental impacts on human health due to normal company operations.

Schaltegger et al. (2000) and EPA frameworks of environmental accounting see traditional accounting and environmental accounting as complementary. The framework used by Gray et al. (1993) on the other hand see the two as being in potential conflict. Gray et al., define environmental accounting very broadly as covering all areas of accounting

---

[3] Adopted from Schaltegger et al. (1996).

that might be affected by the business response to environmental issues. They then draw up a list of what environmental accounting is supposed to do, including:

- recognise and seek to mitigate the negative environmental effects of conventional accounting practices;
- separately identify the related costs and revenues within the conventional accounting system;
- devise new forms of financial and non-financial accounting systems, information systems and control systems to encourage more environmentally benign management decisions; and
- develop new forms of performance measurement, reporting and appraisal for both internal and external purposes.

In comparing the different frameworks Gray et al. see environmental accounting as having a much broader and important role – in both companies and society – than did Schaltegger et al., and the EPA framework. In Gray et al.'s terms, environmental accounting is just one element in the quest for sustainability, and has implicit and explicit power to change the way that managers and external stakeholders think about the environment. Schaltegger et al., and the EPA, on the other hand, see it more as a technical issue or as a tool for managers to use in the same way as they use investment appraisal methods and computers, without giving much thought for the wider implications of accounting practice.

In developing the DEMA framework and relating it to corporate information systems an important issue is the registration of financial environmental data. This includes first and foremost the identification of environmental expenditures, both capital and operational, as well as the allocation of these to products and processes (Bennett and James, eds., 1998; Holmark et al., 1995; White et al., 1991). The issue of environmental expenditures becomes important, since it influences other aspects of environmental management accounting such as product pricing, capital budgeting, performance measurement and project accounting (Bennett and James, 1994, 1996; Wolters, 1996). As yet, there are no generally accepted standards for this, and much is left to the discretion of the manager concerned.

One important issue apparent in frameworks for environmental management accounting is that many environmental expenditures are not identified as such in conventional corporate accounting systems and are thus not identifiable in later analysis and management reporting. These expenditures are collected in overhead accounts or end up in other non-environmentally related accounts. A case in point is the treatment and disposal of waste. In some of the companies participating in the DEMA project, the cost of waste treatment and disposal was mainly seen as only the payments to the waste removal contractor. However, when analysing the issue, a list of other expenditures emerge that are related to this activity but are not registered as such. Examples include the administrative time for filling out forms for waste delivery, the transport of waste, classification and reclassification, and the handling and cost of containers and other waste packaging. So all in all, payments to the waste contractor amount to only a small proportion of the total cost of waste. Giraldi (1996) reaches similar conclusions in her study of waste costs. Thus some types of environmental expenditures remain hidden in the accounting system and cannot be allocated properly.

As mentioned before, there are no generally accepted Danish or international standards for statements or reporting on a company's environmental accounting. A company that wishes to work with environmental accounting must therefore define and clarify itself the specific meaning of the concepts and how they should be applied.

Building to a large degree on the approach adopted by the EPA, the DEMA project defined environmental accounting to consist of four elements, which are: societal costs, environmental costs, investments and income. The four elements are illustrated graphically in Figure 10.3.

Societal costs are costs imposed on 'society' (globally, regionally or locally) by companies (as well as by households and public institutions) for which no 'invoice' exists that might be sent to the individual polluting entity. These costs are thus 'externalised' to society. The computation of such costs often involves a number of technical and practical difficulties, and this paper will discuss only a company's internalised environmental costs, investments and income – i.e. the financial impacts which are registered in the accounting system.

Environmental costs are in this model divided into environmental operating costs expensed in the period (period costs), and costs that cannot be expensed in the current financial period – i.e. environmental investments and commitments. If environmental costs are of such a magnitude and nature that they may be capitalised, they represent an environmental investment. Environmental investments are the capitalised expenses which are incurred in relation to compliance with environmental legislation and/or in relation to compliance with the company's environmental policy, the benefit of which is felt through several financial periods. Whereas previously many investments were in the nature of 'end-of-pipe' solutions, today's environmental investments are primarily directed at preventive measures – i.e. 'beginning-of-pipe'.

A further distinction can be made between environmental costs and investments that are *compulsory*, e.g. those that arise due to regulatory requirements, and *voluntary* costs and investments. Compulsory environmental costs are costs and investments that a company must incur to meet statutory and regulatory requirements. Voluntary environmental costs are costs and investments relating to voluntary environmental initiatives.

A company's environmental management may have positive income-generating effects. Environmental income represents the savings and/or income arising due to a company's compliance with its environmental policy and/or compliance with environmental legislation.

For environmental costs, investments and benefits, a distinction can be made between visible and hidden environmental costs and benefits. The visible environmental costs and benefits are relatively easy to define, and comprise traditional environmental costs or invest-

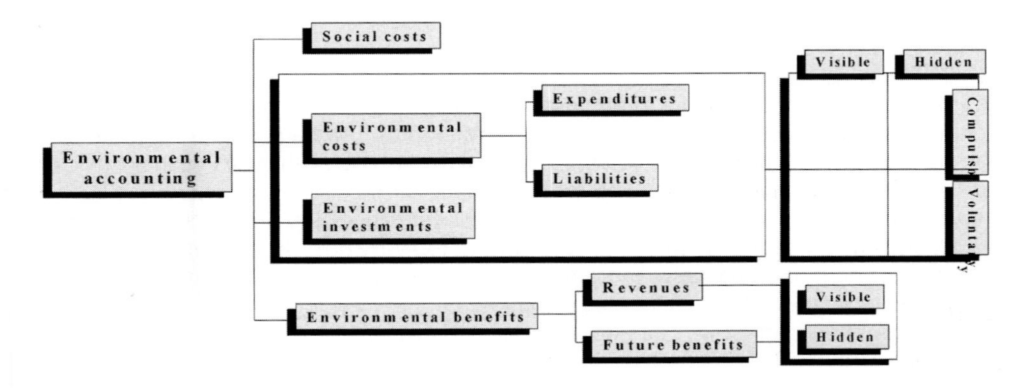


Figure 10.3. The Environmental Accounting framework of the DEMA project.

ments such as statutory investments, investments made to comply with environmental legislation, and salaries to environmental staff. Often, these costs may be deduced from a company's financial chart of accounts. Hidden environmental costs are not calculated separately but are combined in certain general items – i.e. some kind of overheads. It is therefore not always possible to identify, allocate or value these costs.

Hidden environmental costs often constitute a more significant item than may be assumed on an immediate assessment. By way of example, Siemens have declared that they can achieve a 5–15% saving on their production costs by changing the allocation and statement of a number of hidden expensed environmental costs (see chapter 9). Amoco (now BP Amoco) discovered that their total environmental costs did not constitute 3%, as originally assumed, but 22% (Ditz et al., 1995). DONG's statement of the company's environmental costs[4] would look different if the company did not include hidden environmental costs such as in the form of indirect time and materials consumption.

It is characteristic of certain types of environmental income that they may be difficult to quantify since their effect is long-term and they contain a large qualitative assessment element. This is true for example of income that is assumed to arise due to a company's improved reputation.

## 10.3. The integration of Environmental Management Accounting in corporate information systems

If environmental management accounting is to be used as a decision support tool, it is important that the registration, processing and reporting of relevant data be integrated into a company's information systems in an appropriate manner. This should be ensured, among other reasons, in order to avoid double counting, and to ensure data quality, visibility, decision relevance and reporting efficiency.

As usual, companies can choose from several strategies. Since an increasing number of companies are using standard IT systems today, we will focus on the two strategies which involve the use of standard systems to integrate environmental management accounting.

1. Integration into an existing finance system
2. Integration into a standard environmental management information system (EMIS)

It seems logical that financial environmental data should be registered, processed, filed and reported in a finance system. However, this solution involves potential advantages as well as drawbacks.

Environmental accounting is about identifying and registering financial data which must either be separated from existing entries or registered as new source data. It is therefore an advantage for financial data to be accumulated in a system that has been specifically developed to keep track of monetary data.

Finance systems often include specialised tools for processing and analysing financial data, such as product costing, scenario analyses, foreign currency conversion and elimination. Furthermore, finance systems include various control functions that are important in respect of financial data, such as reconciliation routines, period locks and access controls.

The use of a system that is already implemented also implies that the system is familiar

---

[4] See http://www.dong.dk/.

to relevant people throughout an organisation, that the technical set-up has already been performed, and that an IT support organisation has been established. Other things being equal, this means that the integration will be less expensive than if a totally new system were to be implemented.

By using the existing finance system for the registration, processing and reporting of financial environmental data – including investment requests, account coding instructions, charts of account and purpose definitions – greater visibility may be achieved than if such data were registered in another system.

There are, however, a number of potential drawbacks/barriers. One example is where the registration of data other than financial data is relevant, such as the use of volume data for environmental ratios. However, the finance system may not allow the registration of data in units other than financial units. Furthermore, there may be restrictions in terms of modifications and extensions allowed to the finance system.

If a company decides to use the existing finance system for environmental management accounting purposes, the issues mentioned below may be worth consideration.

The alternative strategy would be to integrate environmental management accounting into a standard environmental management information system. An EMIS is specially developed for registration, processing and reporting of, primarily, quantitative environmental data – typically in relation to a company's physical materials and energy flows. Today there are many EMIS's on the market, and some analysis companies project that the EMIS market will grow further in the near future.

Environmental management information systems may be described as advanced number-crunchers. Since these types of software are designed to process data about the company's physical materials and energy flows, the software is able to handle large data volumes in a number of dimensions and in many different units. Therefore, to keep track of the money unit, which is merely one among many units, is no particular challenge.

Another advantage is that all environmental data, physical as well as financial, are concentrated in a single system which typically is 'owned' by the environmental depart-

| | |
|---|---|
| **The environment as a registration dimension** | The creation of a special environmental dimension means that the finance system will have a specific dimension for environmental issues equal to other dimensions of the finance system, such as objectives, sales representatives, departments, projects, etc. |
| **The environment as an objective** | Environmental protection is defined as a separate objective equal to objectives such as production, sales and administration. |
| **Environmental accounts** | Separate environmental accounts are created in the system with separate numbering for the recording of environmentally related financial data. |
| **Marking of entries** | Use of the facilities offered by many systems of marking certain entries as environmentally related. |
| **Allocation of entries** | Allocation of environmentally related entries to other accounts is often relevant in connection with environmental management accounting. Examples include estimates of time spent on environmental purposes, and the allocation of administrative costs to various departments. |

Figure 10.4. Integration of environmental management accounting in financial systems.

ment which is responsible for the communication and use of the data. The integration between financial and physical data also means easier and more advanced analyses than if these two data types had been registered in two separate systems.

A significant drawback is the inevitable necessity of data transfer among systems. In the case of invoice-based data, the relevant invoices must either be forwarded to the environmental department for entry, or the data must be entered by the accounting function into either the environmental reporting system or the company's finance system, with subsequent transfer to the environmental reporting system.

An obvious barrier is the fact that while all companies have a finance system, many companies still do not have an environmental reporting system.

Moving environmental accounting entries to the environmental department involves a risk of the environmental accounting losing connection with the other financial management of a company. This will make the area less visible, and environmental accounting as a function will not be rooted in the company at all relevant levels.

Another barrier may be that an interface will have to be built to the finance system. Whether this actually constitutes a barrier depends, however, on the type of finance system. In modern systems, electronic data transfer does not typically present any major difficulties.

The summary below is based on a review of several standard environmental management information systems. It describes selected functionalities, which are relevant to the integration of financial environmental management accounting practices in an EMIS.[5]

## 10.4. Some company experiences

### 10.4.1. *DONG*

The oil industry and other industries operating with similar resource extraction activities have proved to be early movers when the environment emerged as an issue some decades ago. DONG A/S – an important actor in the Danish Energy sector – is no exception. The company operates in the oil and gas supply and distribution market and operates pipelines and associated logistics in the Danish oil and gas fields in the North Sea.

Today environmental management is integrated at all levels starting with the boardroom.[6] The environmental management system is a part of the Quality Department with a corporate function as well as environmental managers appointed at local levels. Company environmental policies require that decisions within DONG be made on a basis '*where all aspects are reviewed, including environment, health and safety*'. The company is a signatory to the environmental charters drawn up by ICC, OGP & EUROPIA and EUROGAS, and as an example requires its main contractors and suppliers to report environmental information which in essence has to comply with the same standards as those of DONG. DONG has for the past 4 years recorded its environmental data, for example in order to be able to publish the green accounts (compulsory public environmental reports) which are compulsory under Danish law.

Currently no generally accepted definition of environmental costs, revenues and invest-

---

[5] Carl Broas http://www.ecodat.com/, COWI, VKI http://www.mit.aar-vki.dk/www-uk/dims-uk.htm, William Hansen & Co. A/S http://www.wh.dk/, and EMISOFT http://www.emisoft.com/.

[6] DONG is an acronym for Dansk Olie og Naturgas (Eng: Danish Oil amd Natural Gas).

| | |
|---|---|
| **Data structure** | In connection with the integration of environmental management accounting, a location structure must be created, as known from finance systems, where environmental accounting entries can be linked to a geographic or organisational location. |
| **Environmental accounts** | In order to record environmental accounting entries, separate accounts must be created. Purpose, type and unit/currency are to be defined here. |
| **Data input** | Manual input views must be created for environmental accounting data. Electronic import and export of data is also possible by means of various formats, such as Excel and ASCII. |
| **Calculations/formula generators** | Calculation of environmental accounting ratios, allocation of entries based on different sharing keys, etc. All environmental reporting systems offer the option of inserting user-defined formulas drawing on registrations, constants or other ratios. |
| **Reporting** | Design of environmental accounting reports, either integrated into other standard reports or separate. Usually environmental reporting systems have an integrated report generator or they use spreadsheets as report generators. |
| **Consolidation** | Consolidation by organisation is important in relation to environmental accounting data if they are to be included in total statements at the group level. However environmental reporting systems do not generally offer the same possibilities of consolidation of data as do finance systems regarding, for example, automatic foreign currency conversion or elimination. |

Figure 10.5. Integration of environmental management accounting in EMIS.

ments exists in Denmark, which means that a host of different terms are used in companies. At DONG, environmental costs are currently defined as:

> *'operating costs (wages, costs of goods and services etc.), including investments that are expensed. Environmental costs are either independent costs primarily undertaken to achieve environmental improvements, or additional costs concerning costs that are undertaken for other reasons. A distinction is made here between enforced environmental costs, that is regulatory costs, and voluntary costs; and a distinction is further made between costs for external environment and costs for health and safety'*

This definition, however, captures many aspects of mainstream terminology in the sense that it operates with purpose-related costs. An example of costs undertaken above which would otherwise be required in a normal investment are the extra costs associated with DONG's newly built pipe-line in the western part of Denmark. In order to create a more environmentally friendly solution in connection with sand dunes, DONG incurred significant extra costs.

DONG's environmental management information system consists of 3 principal data-feeding channels, as shown in Figure 10.6 on the next page. The *Technical Settlement system* feeds the system with consumption data concerning gas quantities used for electricity, fuel gas, flare and gas supplies. From *SAP R/3*, data are retrieved from various sub-modules within R/3 including invoice-based consumption of fuel, water, electricity,

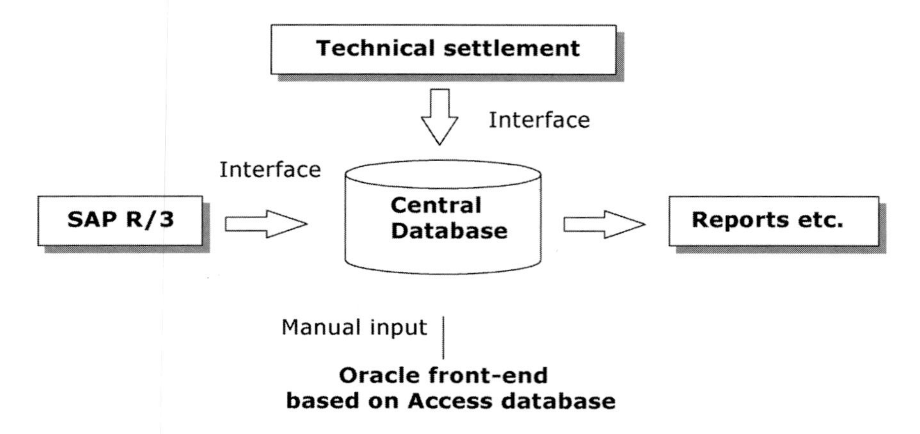


Figure 10.6. DONG's environmental management information system.

auxiliary materials as well as waste removal and waste water (quantities *and* costs generated from modules FI (Financials) and MM (Materials Management)). The 3rd source of data stems from direct recording to the main *Oracle database*. These data span non-invoiced electricity and water consumption, waste water and data not held within the SAP system, environmental data from contractors and suppliers, and calculated emissions from DONG's own electricity generation facilities, as well as calculated emissions from externally procured electricity (particulates, methane etc.).

Environmental performance reports can then be drawn from the system regarding, for example:

- district level and major project level;
- branch level (oil, gas, and major project totals);
- the corporate level;
- ad hoc reports and analysis.

Regarding the monetary side in DONG's case, i.e. environmental costs and environmental investments, these amounts are retrieved primarily for external reporting purposes and are reported from budget-responsible managers once a year. The means for this are standardised Excel spreadsheets with manual input, which are then aggregated to show costs and investments concerning gas transmission and oil transport activities. Currently there is no integration between the green chart of accounts and the financial chart of accounts.

Systems data are recorded in an environmental account structure, examples from which are shown in Figure 10.7.

### 10.4.2. *Post Danmark*

Post Danmark is the public postal service company in Denmark and, as in many other public services, deregulation efforts have spurred Post Danmark into a re-engineering process in several areas. One of these is the development of a high-profile environmental policy. The company's major environmental impacts stem from its logistical operations, mainly greenhouse gas emissions (1998 figures).

Overall responsibility for environmental affairs in Post Danmark is held within the human resources function, with decentralised responsibilities in individual postal centres.

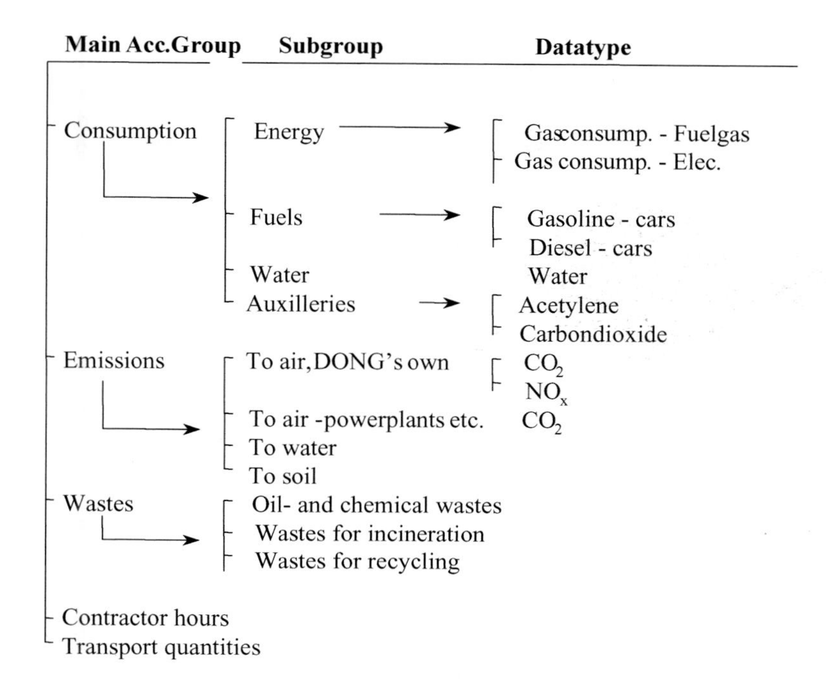


Figure 10.7. Examples from DONG's environmental chart of accounts structure.

Post Danmark's environmental management system is in principle built around the ISO 14001 model, without however aiming for a certified system. A major part of the system is then built into a Business Excellence framework with 6 different areas of efforts (Organisation, Systems etc.). Effects are then measured with regard to stated quantitative goals related to ozone depletion, greenhouse effect etc. Such a goal could for instance be a 20% decrease in $CO_2$ emissions per delivery. Environmental effects are placed under the heading of 'Effect on society'.

Utilising the Business Excellence model is a result of a stated objective of a more holistic way of managing Post Danmark.

Post Danmark's current environmental management information system utilises a number of data sources from its transport, building and production activities. An overview is given in Figure 10.9.

Environmental costs are defined in Post Denmark to include costs associated with specific consumption (i.e. gasoline, electricity, heating and water), and costs for waste disposal. Post Danmark sees green taxes as an estimate of the externalities for which the company is responsible, but does not gather additional information about non-internalised monetary effects. In 1998 the company paid DKK128 million (approx. £12 million) in green taxes, compared with a total turnover of DKK 10 billion, which is an amount considered large enough to warrant managerial actions in order to minimise it. Besides the above-mentioned costs and taxes, all the costs associated with the operation of the environmental department are regarded as environmental costs.

The collection of information regarding environmental costs is built around 6 areas of effort which are shown in the Business Excellence model in Figure 10.8. For each of these

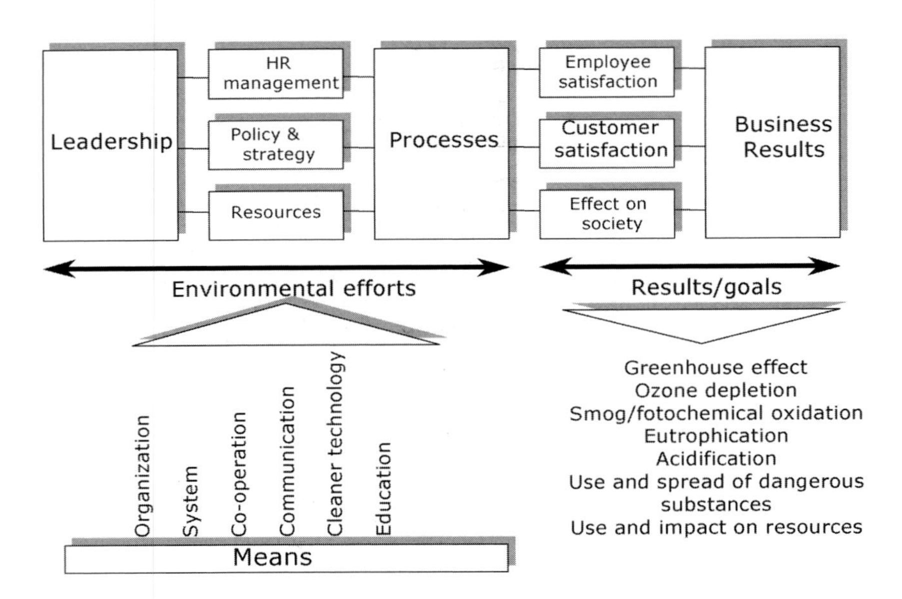


Figure 10.8. Incorporating EPIs into a Business Excellence model.

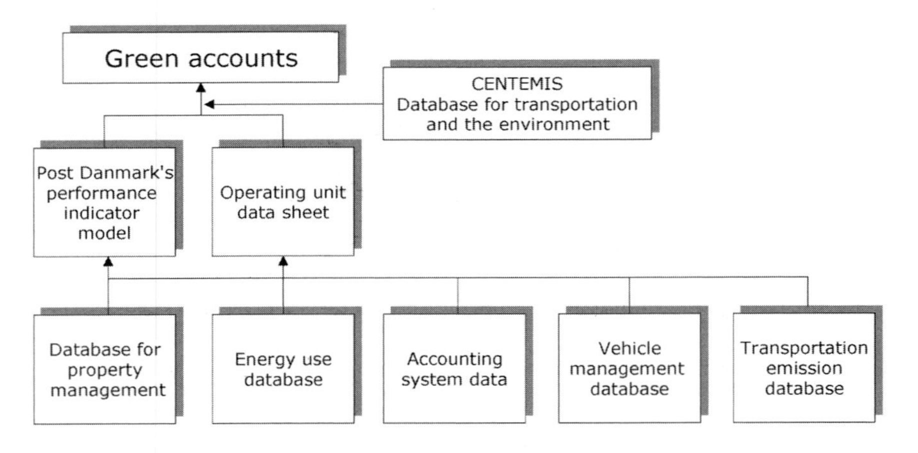


Figure 10.9. Post Danmark's environmental data sources.

areas a project code is established, under which all costs are recorded. A number of accounts in the financial system are attached to each project code, which means that Post Danmark is able to report on economic matters. For example, regarding the area of effort called 'Communication', the company records the costs that it incurs associated with the publication of its green accounts and its environmental report which involves cost items such as layout, consulting costs, printing etc. Each month, responsible managers receive a printout of costs for each of the 6 areas in the BE model.

As well as these reports, Post Danmark operates with integrated environmental economic performance measures which are based on a spreadsheet model including several

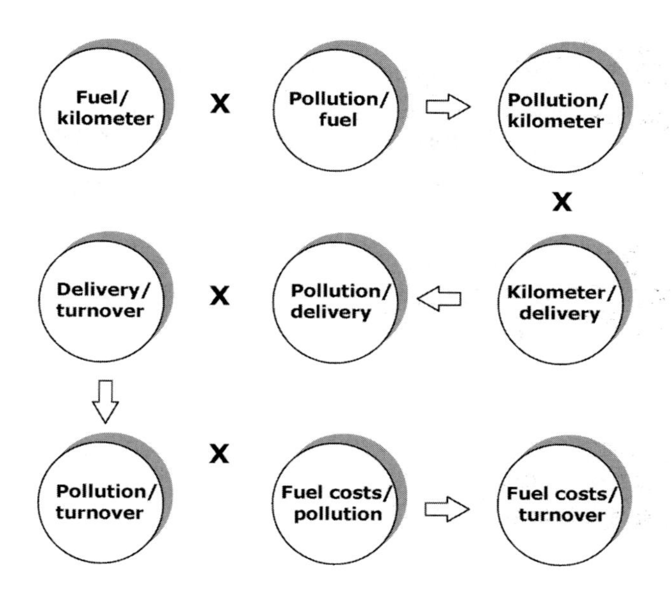


Figure 10.10. Integrated economic-environmental performance indicators – Post Danmark.

environmental performance indicators (EPIs). Such EPIs exist for Post Danmark's transport fleet and for its buildings. The figure below illustrates the integration of one type of EPI with economic considerations.

Improvements in the EPI can be achieved in the fuel component by driving more kilometres per litre or by switching to more environmentally friendly fuel types. Such improvements are partly a question of technology (emission filters in exhaust pipes, reformulated gasoline types etc.) and are partly driven by behavioural elements, since driving habits are one of the keys to reduced emissions. This means that the education of drivers will be an important parameter for future decreases in emissions. Other areas for improvements exist if Post Danmark is able to configure its logistics to align with environmental objectives by using alternative means of transportation, increasing its turnover, or altering its product mix.

The data that is input in order to integrate economics and environmental performance is taken from a diverse set of systems and files. Financial data related to electricity consumption, water, wastes etc. are all taken from the financial system. The quantity component related to wastes, water etc. is taken from an Energy System, Building Control System, or Environmental Datasheets.

Environmental investments have traditionally been evaluated on the basis of traditional economic business criteria. However, the company has just adopted a new cost-benefit model, which enlarges the range of criteria and in particular in which benefits, which are otherwise hard to assess, are modelled. This cost-benefit model covers the project, implementing and operating phases of all larger projects in Post Danmark.

| Field (or group) | Type | Subject |
|---|---|---|
| Letters, packages… | Turnover | Transports |
| Electricity concumption | Quantity, amount | Electricity consumption |
| District heating | Quantity, amount | Heating |
| Cold water | Quantity, amount | Water |
| .. | | |
| Kilometers driven | Quantity, amount | Transports |
| Fuel consumption -gas | Liters & amounts | Transports |
| .. | | |
| Calc. emissions | Quantity | Transports |
| Letters, packages.. | Quantity, amount | Transports |
| .. | | |
| Electricity consumption | Quanity, amount | Electricity |
| Household wastes | Quantity, amount | Wastes |
| Paper | Quantity, amount | Products |
| .. | | |
| Buildings, $m^2$ (owned) | Space, amount | Electricity consumption |
| Buildings $m^2$ (rented) | Space, amount | Electricity consumption |

Figure 10.11. Examples of environmental data types – Post Danmark.

## 10.5 Conclusion and possible future developments

There is a growing interest in corporate environmental accounting. This development is due to, among other things, increased competition, increasing environmentally related costs and a strategic focus on the total costs of companies. Furthermore, the Dow Jones Index shows that companies which are committed to environmental management and investments seem to offer a higher return than does the average company.

In general, companies have difficulties in defining environmental accounting. There is therefore a great need for a generally accepted standard for this area.

Because of the absence of definitions, there are wide variations on which information and types of environmental costs are included in companies' environmental management accounting. Furthermore, the documentation of a company's environmental accounting causes problems – i.e. there is a need for a systematic approach and integration into information systems.

Other barriers include:

1. Lack of internal discussion on the purposes of environmental management accounting.
2. Lack of internal discussion of opportunities and threats entailed in environmental management accounting.
3. Under-estimation of hidden environmental costs (including environmental liabilities, commitments and reputational costs) in the company, resulting in a misinterpretation of environmental accounting as insignificant.
4. Inadequate overview of the opportunities offered by the company's IT systems.

Based on the experiences of the best-practice companies involved in the project, the implementation of environmental management accounting should focus on clarifying internally the importance of the elements of environmental accounting. Such clarification should initially form the basis of data retrieval and statements in connection with internal management. This means that the focus should not be directed solely at the external financial statements when making definitions, etc.

Some environmental costs are not visible as such in the company's finance systems but may constitute a significant part of total environmental costs. It is therefore an important task to assess these costs in order to obtain an adequate picture of a company's total environmental costs. Experience shows that total environmental costs may increase by 5–10% if hidden environmental costs are included.

It is important to involve the company's IT management system. Without such integration, there is a risk that initial clarification and assessment phases will stand alone, and that practical integration into day-to-day registration and statements will not happen. Environmental management accounting can be integrated either into the existing finance system, or an environmental management information system. Irrespective of the solution selected, it is important that the data are accumulated in a system that allows extracts, analyses and reporting for decision-makers.

At this time, there are no special modules or functionalities for environmental management accounting in finance systems or environmental management information systems.

It will pay to take 'one step at a time' as regards the implementation of environmental accounting. Rather than starting an ambitious scheme, it may be an advantage to focus on a few well-defined areas such as waste water or air emissions, and to establish environmental accounting in these areas. However, the individual stages of the implementation process must be identical – i.e. clarification, risk assessments, competence assessments, design and IT support.

# 11. Life Cycle Engineering

*Horst Krasowski*
*T-Systems, Competence Centre EDM, Debis Systemhaus Industry GmbH, Germany;*
*E-mail: Horst.Krasowski@t-systems.de*

## 11.1. Introduction

This paper describes a Life Cycle Engineering approach which is able to optimise a product from an integrated technical, ecological and economic point of view. It shows the method and the importance of its integration into company business processes and IT landscapes.

Limits to the earth's ecological capacity and shortages of raw materials and energy sources will determine the limits for growth and expansion in the future (Behrendt et al., 1997). Enterprises have to face increasing costs emanating from use of resources, materials and waste, ecologically sensitive customers, and new environmental laws and regulations such as the German Kreislaufwirtschaftsgesetz (Recycling Law). The implications of this are that our industrial world is liable to significant changes, and economic growth will have to be re-directed in certain ways.

There is an increasing interest in the ecological impacts of products. Complex products such as automobiles need effective methodologies and tools to evaluate their environmental impacts without neglecting the technical and cost implications, and the consequences of developing new products and services need to be analysed. Companies need to have sound methods, and powerful tools based on them. Such methods and tools should be used from the design phase onwards, in order to generate the best possible benefits for both the company (especially cost savings) and the environment. For this purpose, Life Cycle Engineering (LCE) offers a good method; it connects different angles from which one should look at new developments and in particular, it involves technical, costing and ecological points of view. Companies using LCE are able to save money and therefore able to fulfil all demands on today's and future products and services.

## 11.2. Life Cycle Engineering (LCE)

This paper provides briefly defines LCE as it is used by Debis, then goes on to describe the LCE methods and present an integrated LCE approach.

A range of different definitions of LCE can be found in any state-of-the-art literature (e.g. Society of Automotive Engineers, 1998); however this paper will make its own definition and description of LCE as used by Debis. LCE is used here to refer to the entire product life cycle, going all the way from the design process, through product manufacturing and the use-phase, up to the end-of-life phase. The LCE target is product optimisation, taking into account economic, ecological and technical requirements.

The main focus is the product development phase, in which the analysis of the entire product life cycle should be taken into account. It is in the design stage that most of the product's eventual ecological impacts are determined, as are most (80%) of the product costs. The designers specify the kind of production processes as well as the amounts of raw materials needed. Such decisions strongly determine what is going to happen during the consumption of the product; for instance, the weight of a car which is decided during



*M. Bennett et al. (eds.), Environmental Management Accounting: Informational and Institutional Developments*, 153–157.
© 2002 *Kluwer Academic Publishers. Printed in the Netherlands.*

its design will influence its fuel consumption throughout its life. Likewise, during the end-of-life phase, the possibilities to reuse or recycle the whole or parts of the product are largely determined by the decisions made back in the design process.

Assessment of a product's life cycle is necessary, not only because of regulations that may differ from country to country, but also to find more economic and ecological benefits. For instance, it might be possible to optimise service and maintenance activities, or to decrease the costs of removing manufacturing waste.

To do so, Debis considers that three distinct methods are capable of fitting into its approach:

- Life Cycle Assessment (LCA);
- Life Cycle Costing (LCC);
- Product Structure Assessment (ProSA).

### *Life Cycle Assessment (LCA)*

The LCA method provides an overview of the ecological impacts of a product over its entire life cycle. According to the international standard ISO 14040 (ISO, 1997) this method contains 4 steps (see Figure 11.1).

### *Life Cycle Costing (LCC)*

As there are no standards available, the LCC method is more complicated to define. The literature provides several possible definitions (Zehbold, 1995; Wübbenhorst, 1984). This paper refers to the definition of Fabrycky and Mize (1991), according to which Life Cycle Costing is the economic assessment of all the money flows which are caused by the existence of a specific product. Fabrycky and Mize subdivided the 'Total System' Product Costs into:

- Research and Development Costs,
- Production and Construction Costs,

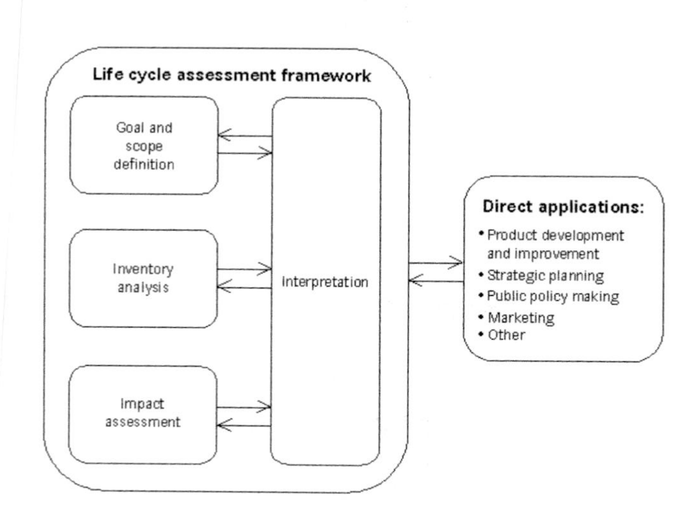


Figure 11.1. Life Cycle Assessment framework.

- Operation and Maintenance Costs and
- Retirement and Disposal Costs.

The first step is to combine and calculate all the product-related costs during the phases of production (Production Costs), use (Operation and Maintenance Costs) and end-of-life (Retirement and Disposal Costs). Given the existing LCA method, Debis combines LCC and LCA. Like material and energy flows, there are cost flows. Since its process approach is based on LCA, Debis adopted Activity Based Costing to build up its calculation schemes.

A third important method, Product Structure Analysis, is used to assess a product's features in terms of disassembly, recycling and reuse from the viewpoint of cost optimisation. Product Structure Analysis gives the inputs for calculating the end-of-life costs related to these activities.

As the application of the three afore-mentioned methods is complicated, it is necessary to make use of different supporting tools. Numerous software tools are available which support the user in a more or less user-friendly way, although almost all them are 'stand-alone' and cannot deal with more than one tool.

## 11.3. Integrated life cycle engineering method

The main requirement for an effective LCE is integration of the three LCE methods into the business processes and incorporating LCE supporting tools into the existing IT-landscape.

To be widely accepted, every new method has to be part of the business processes of a company, and this applies in particular to LCE. Therefore, the first step is that all relevant business processes should be determined. This depends on the way a company uses the whole dimension of the LCE method, and it is not necessarily appropriate for every company that the complete method should always be applied in full – management must decide in each case which of the three afore-mentioned methods will be implemented. The next step is to look at all departments and in particular business processes which should use LCE or be supported by LCE.

In accordance with the definition adopted, the product development process is seen as the most important business process. As has already been observed, this process accounts for most of the product's costs and environmental impacts. The designer determines what kind of production processes are involved and the materials inputs needed as well as the functional qualities of the product. Likewise, there are business processes which can be optimised by using LCE. These are, in particular, purchasing, product planning and assembly, service and maintenance, and product disassembly and disposal. Indirect processes such as environmental protection can also be supported with LCE.

The different processes have to be analysed in detail. For instance, one should find out what design methods have to be used, who have to be involved, and what software support is needed.

Product designers already have to meet many requirements, so to be persuaded to accept another new method requires convincing arguments. This is possible only if they see the challenges and the benefits of the method. Therefore, the designers as the main user of LCE should be consulted when decisions have to be as to how to structure and implement it.

The business process analysis shows where LCE should be applied. Later on new processes can be analysed in the same way, indicating new opportunities for using LCE.

### *LCE integrated into the IT landscape*

Connecting existing data and software systems allows an efficient use of LCE. This is especially important for LCE, since most data relevant to LCE usually already exists somewhere within the company.

The product structure was laid down first as a drawing with a CAD tool and transformed into a bill of materials, in either a CAD tool or in an Enterprise Resource Planning (ERP) software system such as SAP. The product structure together with the CAD drawing therefore form the first product-defining information during the early phase of product development. Subsequently the work schedules are to be made, and based on these, NC/CNC programs or work instructions.

The bill of materials (a list of needed materials) is the origin for almost all subsequent processes such as calculation, purchasing, product planning and sales. Like methods such as concurrent engineering or simultaneous engineering, LCE supposes a well-defined development process. Therefore, EDM/PDM (Engineering/Product Data Management) software support is a recommendable base. This enables all other relevant departments of the company to work on this data. Although this is not really new, experience from several different projects indicates that it can still be very difficult to realise in daily practice.

Different companies have different development processes and software tools. There are two main systems which can manage the relevant data: EDM/PDM systems and ERP Systems. Here, EDM/PDM systems are meant to serve as an integration platform for the systems used in the development process. This implies CAD systems, materials management systems or geometric information systems. Figure 11.2 shows the data-supplying systems and the data-demanding systems, and also the important process of product development.

Figure 11.2 gives a general outline of how integration of LCE in the IT landscape might look, but the integration is not confined to this. Before this, the different methods used

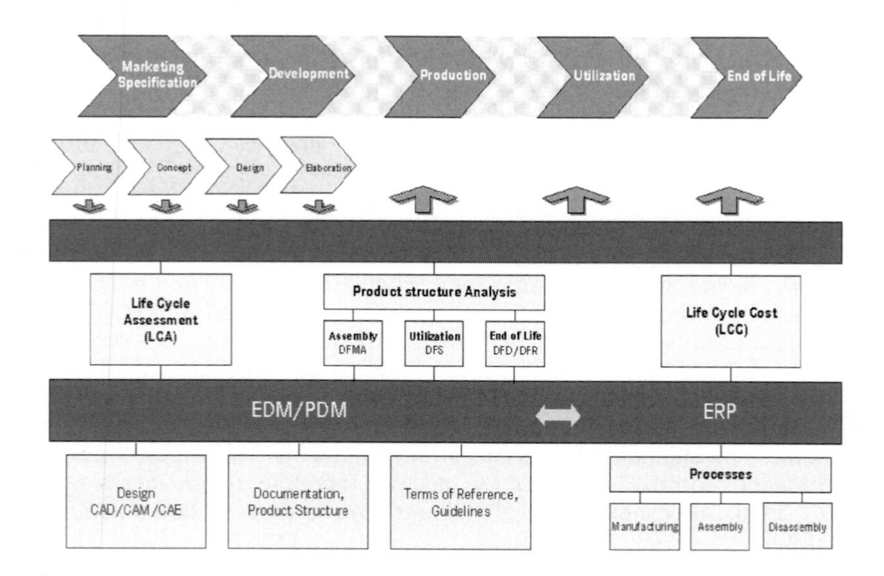


Figure 11.2. LCE integration.

for LCE have to be integrated, as has already been recognised by Debis IT Services (Krasowski and Friedrich, 1998).

To sum up, on the basis of a product structure Debis determines the optimal disassembly path and transforms it into the combined LCA and LCC software (Krasowski and Friedrich, 1998). A product can be assessed under the three mentioned methods.

*Limitation*

At present not all the data that is needed for this needed data can be obtained directly from data which has already been previously generated, so that the designers have to enter more information into the LCE tool to receive the best results. Another problem is data quality and timeliness. 'Data quality' refers to whether all data is available, applying to the right material or the weight of a material or component. If a product analysis should be done on different product versions and development phases, the timing of data availability is also important, but this can be realised with some changes in the business processes. The documentation of data, which is necessary for analysis later, should be done as early as possible. The modelling of the needed data should be as user-friendly as possible.

## 11.4. Conclusion

This paper described LCE as a method to assess and optimise a product over its life cycle by means of LCA, LCC and Product Structure Analysis. To use LCE efficiently (both cost-effectively and time-effectively) and to increase the value of a product, two aspects are particularly important. LCE has to be integrated into both the company business processes and into its IT landscape. Therefore, LCE software support is needed that enables the company to integrate LCE into its processes and to combine the three basic methods of LCE.

# PART III

# EMA POLICIES

# 12. Corporate Environmental Accounting: A Japanese Perspective

*Katsuhiko Kokubu*
*Professor of Social and Environmental Accounting, Graduate School of Business Administration, Kobe University, Kobe, Japan 657-8501; E-mail: kokubu@rokkodai.kobe-u.ac.jp*

*Tomoko Kurasaka*
*Visiting Researcher at IGES (Institute for Global Environmental Strategies), Japanese Certified Public Accountant, Kurasaka Environmental Research Institute*

## 12.1. Introduction

Environmental accounting in Japan has developed rapidly during recent years. This development was triggered by the 'Grasping Environmental Cost: A Draft Guideline for Evaluating Environmental Cost and Publicly Disclosing Environmental Accounting Information (Interim Report)' (referred to as the 'guideline draft' henceforth) which was published in March 1999 by the Environment Agency Japan (EAJ).[1] This guideline draft was the product of the environmental cost study group which was launched by MOE in 1997. Subsequently, major Japanese companies have started to introduce environmental accounting, and to disclose environmental accounting information in their environmental reports.

In July 1999 the Industrial Structure Transformation and Employment Head Office of the Japanese government, led by Mr. Obuchi (the Prime Minister at that time) as the head manager, compiled the 'Regulation Reform Proposal for Employment Creation and Enhancement of Industrial Competitiveness'. The proposal includes 'investigation of the introduction of environmental accounting'. With this, MOE proceeded more actively with preparation for setting out the 'environment accounting guideline'.[2]

MOE established two joint study groups on environmental accounting in 1999: one with the Japanese Institute of Certified Public Accountants (JICPA), the other with practical business members. In discussions in these study groups, many issues in corporate practices regarding environmental accounting and problems in accountancy were examined. MOE also launched 'Study Group for Developing A System for Environmental Accounting' in order to finalize the environmental accounting guideline draft. The final guideline was completed in March and published in May 2000; it is not legally compulsory, but is available for companies to adopt voluntarily at their discretion.

Another initiative to investigate environmental accounting was started by the Ministry of International Trade and Industry (MITI),[3] which was subsequently reorganized as the

---

[1] The Environment Agency Japan was reorganized as the Ministry of the Environment (MOE) on January 6, 2001.

[2] In the fiscal year 1999, the MOE also proposed to the Ministry of Finance a tax preferential measure based on environmental conservation cost, the calculation of which would be based on the environmental accounting guideline. However, the proposal was not approved and was considered as a more medium- to long-term objective.

[3] The authors have worked on both these projects. Both Kokubu and Kurasaka were members of the MOE committee that developed the guideline draft and the guideline. The METI committee has been chaired by Kokubu, with Kurasaka as one of the members.



*M. Bennett et al. (eds.), Environmental Management Accounting: Informational and Institutional Developments,* 161–173.
© 2002 *Kluwer Academic Publishers. Printed in the Netherlands.*

Ministry of Economy, Trade and Industry (METI) on January 6, 2001. METI established a committee for environmental accounting in August 1999 in the Japan Environmental Management Association for Industry (JEMAI). This is a three-year project. While MOE focuses on more externally oriented environmental accounting, METI exclusively targets research and development into environmental management accounting tools. In the first year, the project completed an investigation of international trends and practices including North America and Europe and a report published by JEMAI (2000). The objective in the second year is to develop some environmental management accounting tools which correspond to Japanese companies' needs.

Responding to these governmental initiatives, environmental accounting has very rapidly been diffused into Japanese companies. The latent demand for environmental accounting had already increased since 1997–1998, mainly by ISO14001-certified companies, and this trend has been accelerated by these governmental initiatives. On the other hand, some external organizations such as investment consultants for green investment started to analyze corporate environmental management practices, including accounting and disclosure, in order to evaluate environmental performance, which is also a major factor in promoting corporate environmental accounting.

This paper firstly explains the MOE guideline, then examines the current practice of environmental accounting in Japan. Finally some outstanding issues, including MOE and METI initiatives which are currently in progress, are discussed.

## 12.2. Environmental Accounting Guideline Year 2000 Version of the Ministry of Environment (MOE)

MOE published 'Developing an Environmental accounting System (2000 Report)' in May, 2000. The majority of this report consists of the 'Guideline for Introducing an Environmental Accounting System (2000 version)' (referred to as the 'guideline' henceforth). This is a final document for the guideline draft which had been published in the previous year, as mentioned above. However, MOE added such words as '2000 report' into the title of the report, because 'considering the current situation where research into environmental accounting and installation conditions is progressing steadily, we considered necessary the future reinforcement of the contents of the report as required' (MOE, 2000, p. 3). Therefore, the guideline is expected to be revised in the future as required, although the timing of this review is not at this time indicated clearly.

The key contents of the guideline can be summarized in the following three points:

- Environmental accounting system;
- Environmental conservation cost;
- Environmental conservation effects and economic effects.

### 12.2.1. *Environmental accounting system*

The guideline indicates two different functions of environmental accounting: an internal function, and an external function for communication with various stakeholders (see Figure 12.1). However, the actual contents of the guideline are considered to be oriented more to external reporting than to internal management. This is not explicitly stated by the guideline itself, but is implied in the following extract:

'This report is intended to enable the comparison of information by environmental accounting as much as possible since the report summarizes a coherent concept regarding

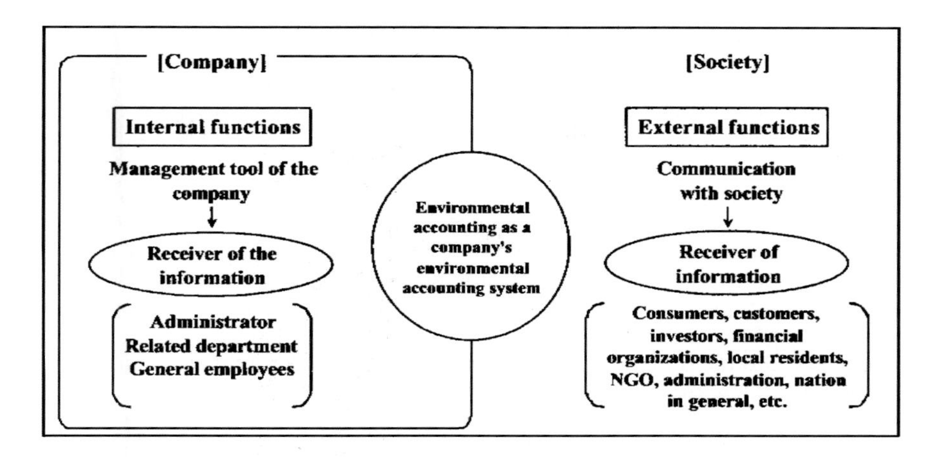


Figure 12.1. Functions of Environmental Accounting System (MOE, 2000, p. 8).

environmental accounting. Currently, only the framework of environmental accounting is incomplete and some limitation cannot be avoided due to the characteristics of the guideline that respect the independence of enterprises and the diversity of individual business categories. However, in the future, we hope to develop a system that enables the comparison of basic sections not only sequentially but also between enterprises.' (MOE, 2000, p. 5)

According to the guideline, the media to be used for the disclosure of environmental accounting information is an environmental report rather than a financial report, and this environmental accounting is supposed to be completely independent from any corporate financial accounting. This understanding is also supported by JICPA (2000).

The basic frame of an environmental accounting system is indicated by Figure 12.2. Environmental accounting is defined as a system that integrates financial performance and environmental performance. In fact these two aspects of performance are integrated by correlating the environmental conservation effects and the economic effects associated with environmental measures. At the stage of the guideline draft, environmental accounting is likely to be restricted to the calculation of environmental conservation cost; however in the guideline, the range of an environmental accounting system is expanded, in order to be a fundamental tool for environmental conservation as well as for corporate management.

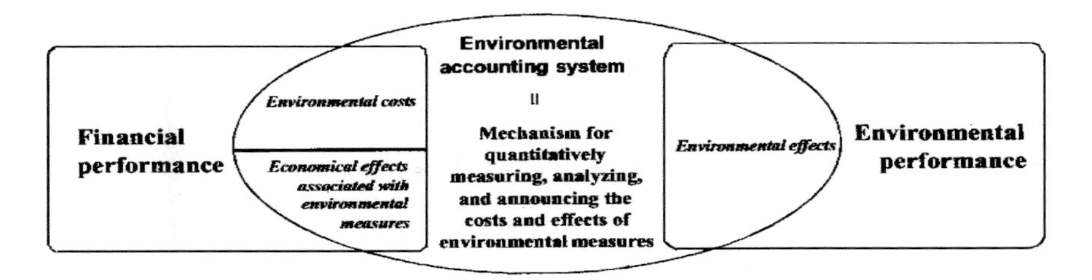


Figure 12.2. Basic Frame of Environmental Accounting System (MOE, 2000, p. 7).

### 12.2.2. *Environmental conservation cost*

The guideline expands the scope of environmental accounting, but still emphasizes the calculation of environmental conservation cost in the same way as did the guideline draft. The guideline defines environmental cost as 'investment and cost for environmental conservation'. For the definition of the investment and the cost, in principle, the definition of financial accounting is employed. The purpose of the expenditure is adopted as the criterion to identify what is environmental conservation cost or investment. If the purpose is considered to be environmental conservation, those costs and investments should be environmental. Concerning environmental conservation, three major activities, including pollution prevention, global environmental conservation, and resource circulation are indicated by the guideline.

Concerning the measurement of environmental cost, a differential calculation is recommended as a basic method when an environmental cost which has been incurred is a composite one. This method requires the exclusion of the cost which has been incurred other than for environmental conservation, from the total amount of each environmental cost item. If this method is difficult, the company can employ some simple calculations. For example they are allowed to adopt some predetermined allocation ratio such as 25%, 50% or 75% in order to distinguish the amount for environmental conservation from the amount for other purposes. This often happens when companies buy facilities that have not only an environmental protection function but also other, non-environmental, functions.

The guideline classifies environmental costs into the following six categories.

1. Environmental conservation cost for controlling the environmental impacts that are caused within a business area by production and service activities (abbreviated to *business area cost*);
2. Environmental cost for controlling environmental impacts that are caused upstream or downstream as a result of production and service activities (upstream/downstream cost);
3. Environmental cost in management activities (*management activity cost*);
4. Environmental cost in research and development activities (*research and development cost*);
5. Environmental cost in social activities (*social activity cost*);
6. Environmental costs corresponding to environmental damages (*environmental damage costs*).

The scope of the guideline is very comprehensive. However, companies do not have to calculate all cost categories in the first stage, but can choose what they consider to be the relevant cost categories for themselves. Another feature of the classification is that lifecycle thinking is introduced to the classification in distinguishing between categories (1) and (2).

### 12.2.3. *Environmental conservation effects and economic effects*

The most significant features of the guideline compared with the former guideline draft are that environmental conservation effects (benefits) and economic effects (benefits) are introduced into the environmental accounting system. This revision is to overcome the limitation of the guideline draft, which is unable to clarify how efficiently or effectively environmental conservation activities are implemented.

The effects of environmental conservation measures are classified into an environmental conservation effect that indicates the improvement in environmental performance, and an economic effect that contributes to financial performance. Basically, the former is measured in physical units and the latter in monetary units. Among these effects, the environmental conservation effect is to be checked first as a higher priority because the main purpose of environmental conservation costs should be environmental conservation rather than for economic effects.

The guideline classifies environmental conservation effects into three categories:

1. environmental conservation effects occurring within the business area;
2. environmental conservation effects occurring upstream or downstream;
3. other effects.

The guideline provides some examples of an actual index for each category. This category of environmental conservation effects is, in principle, associated with the category of the environmental conservation cost that was described before. However, since environmental conservation effects correspond to environmental conservation cost other than the cost within the business area, and the upstream/downstream cost often cannot be measured easily, these effects are summarized as 'other effects'. Measurement methods of environmental conservation effects should be standardized so that the information can be compared when the effects are reported externally. However, the guideline does not provide in detail for measurement methods.

Corporate environmental protection activities should mainly pursue the reduction of environmental impact, that is, the improvement of environmental performance. However, companies should simultaneously pursue economic benefits as well. For instance, in the introduction of an environmental management system, the main emphasis was on economic benefits such as cost saving by energy saving or waste reduction. The economic benefits specified by the guideline are classified into 'economic effects calculated based on a credible basis' and 'economic effects based on a hypothetical calculation'. Only the former is expected to be disclosed externally, and the latter is not requested to be disclosed. When the latter is reported publicly, however, the effects are to be distinguished from the 'effects based on a credible basis' and the method and basis of calculation have to be disclosed. As the 'economic effects calculated based on a credible basis', real effects such as recycling income and cost savings through energy efficiency are indicated, and the 'economic effects based on a hypothetical calculation' include the effects of the avoidance of contingent risks and profit contribution assumption effects.

### 12.2.4. *Disclosure format*

The guideline provides three types of formats as an environmental accounting statement to be disclosed.

Format A: environmental cost only;
Format B: environmental cost and environmental conservation effects;
Format C: environmental cost, environmental conservation effects and economic effects (Table 12.1).

Format C is the most comprehensive. When a company discloses environmental accounting information in their environmental reports, Format C is highly recommended if they are able to do this.

Although there are some points to be improved in the future, such as the methods of calculation of effects, the basic frame suggests a new framework of environmental accounting that integrates environmental accounting in monetary units with environmental

Table 12.1. Environmental Accounting Format C Aggregation scope: (    ) Target period: from    to    Unit (    ) yen.

| Environmental cost | | | | |
|---|---|---|---|---|
| Category | | Details of main implementation and the effects | Investment amount | Expense amount |
| (1) Environmental costs for controlling environmental impacts occurring within a business area as a result of production and service activities (business area costs) | | | | |
| Breakdown | 1) Pollution prevention cost | | | |
| | 2) Global environmental cost | | | |
| | 3) Resource circulation cost | | | |
| (2) Costs for controlling environmental impacts occurring in the upper stream or lower stream associated with production and service activities (upper/lower stream costs) | | | | |
| (3) Environmental costs in management activities (management activity costs) | | | | |
| (4) Environmental costs in research and development activities (research and development costs) | | | | |
| (5) Environmental costs in social activities (social activity costs) | | | | |
| (6) Costs corresponding to environmental damages (environmental damage costs) | | | | |

- When there are environmental costs that are not applicable to any of the categories from (1) to (6) and the costs are described as (7) other environmental costs (other costs), disclose the contents and the reason in order to clarify the scope.

| Item | Contents | Amount |
|---|---|---|
| Total amount of investments for the period | | |
| Total amount of research and development costs for the period | | |

Source: MOE(2000).

| Environmental effects | | Comparative index |
|---|---|---|
| Contents of effects | Index of environmental impact | |
| (1) Environmental effects occurring within business area (business area effects) | | |
| (2) Environmental effects occurring in the upper/lower stream (upper/lower stream effects) | | |
| (3) Other environmental effects | | |

| Economical effects associated with environmental measures | |
|---|---|
| Contents of effects | Amount |
| Revenue obtained by recycling | |
| Reduction of costs achieved by energy conservation | |
| Reduction of waste processing costs achieved by recycling | |
| | |
| | |

accounting in physical units. The environmental accounting statement such as Format C provided by the guideline must be regarded as a settlement document in an environmental report, likewise the financial statement in a financial report.

It can be said that there are two main streams of environmental accounting: the Anglo-American (Atlantic) model that is based on monetary units, and the European (the Rhine river) model that is based on physical quantity units such as those derived through an eco-balance. The MOE guideline establishes a new environmental accounting model that integrates these two types of environmental accounting.

## 12.3. Current practices of environmental accounting in Japan

The guideline played a very important role in encouraging many Japanese companies to introduce an environmental accounting system, so that the number of companies disclosing environmental accounting information in their environmental reports has been rapidly increasing. This section firstly examines current practices of Japanese corporate environmental accounting in general, and then goes on to describe and analyse Ricoh's environmental accounting as an example of best practice.

### 12.3.1. *Japanese corporate environmental accounting practices in general*

There are several investigations into environmental accounting practices by Japanese companies. Japan Accounting Association (2000) surveyed all listed companies in the first section of the Tokyo, Osaka and Nagoya Stock Exchange Market (total: 1433) as of June 2000 and found that 194 companies published an environmental report and 99 companies disclosed environmental cost information in their environmental report.

Kokubu et al. (2001) investigated all listed companies in the first section of Tokyo Stock Exchange Market, excluding the financial sector, as of 30 November 2000. The total number of companies was 1257. They found that 211 companies (16.8%) published an environmental report and 145 companies (11.5%) disclosed environmental accounting information in their annual report. If a company disclosed more than one item of environmental costs, it was included by this survey with the companies that were categorised as disclosing environmental accounting information.

The number of companies that disclose environmental accounting information is higher than in the JAA survey, because many companies published a new version of environmental report from July to September in which they included environmental accounting information. This paper explains Japanese corporate environmental accounting practices based on Kokubu et al.'s survey. Concerning some determinants of environmental report publication, please see Kokubu et al. (2001).

Kokubu et al. (2001) report that the number of companies in each industry sector which disclosed environmental accounting information in their environmental reports was as follows:

| | |
|---|---|
| Chemicals | 24 |
| Electric equipments | 22 |
| Machinery | 11 |
| Transportation equipments: | 11 |
| Electricity and gas | 10 |
| Food | 9 |
| Retailers | 7 |

| | |
|---|---|
| Glass and cement | 6 |
| Iron and steel | 6 |
| Wholesaler | 5 |
| Textiles | 5 |
| Transportation | 4 |
| Miscellaneous manufacturing | 4 |
| Pulp and paper | 4 |
| Medicines | 3 |
| Precision machine | 3 |
| Construction | 3 |
| Metal products | 3 |
| Oil and coal | 2 |
| Non-ferrous metals | 2 |
| Rubber | 1 |

Environmental accounting seems more widespread in sectors with relatively high environmental impacts such as chemicals, electric machines and electricity/gas. One reason for the high levels of disclosure of environmental accounting information by many chemical companies is that the Responsible Care principle requires them to disclose environmental cost information, although this requirement is much simpler than the MOE guideline.

Kokubu et al. (2001) also investigated whether or not those companies' environmental accounting was based on the MOE guideline, based on whether or not the companies stated either that they had based what they had done on the guideline, or that the cost category they had applied is similar to the guideline or the guideline draft. The result is as follows:

| | |
|---|---|
| Based on the MOE guideline: | 92 |
| Not based on the MOE guideline: | 53 |

It can be concluded that the influence of the MOE guideline on Japanese companies is quite strong. Most of the 53 companies which were classifying as not based on the guideline did not present any alternative methods in their place, but had been unable to apply the guidelines only due to having had insufficient time for their introduction. Therefore, the number of companies which will base their environmental accounting on the MOE guideline is expected to increase in future.

The MOE guideline requires companies to disclose environmental cost (investment or expense), environmental protection effects, and economic effects. All companies disclosing environmental accounting information include at least one environmental cost item, sometimes more. Therefore disclosures of two types of effects were analysed:

| | |
|---|---|
| Companies disclosing environmental conservation effects: | 62 |
| Companies disclosing economic effects: | 72 |
| (Companies disclosing both effects: | 58) |

While the number of companies disclosing both effects is still a minority, it is still significant bearing in mind that the guideline's explanations of those effects are far from sufficient, and can be expected to increase if the explanation is revised or other reports supplement this.

Kokubu et al.'s investigation shows that the MOE guideline is having a strong influence on Japanese environmental accounting practices. However, it was also found that there is an inadequate degree of comparability in the environmental accounting information which is disclosed in environmental reports, though to achieve a significant improvement in this would requirement first the accumulation of sufficient environmental accounting practice and knowledge.

Since the MOE guideline has strongly influenced Japanese companies, the Japanese environmental accounting practices are much more inclined towards external disclosure rather than to internal use. Toyokeizai (2001) indicates an interesting point. Toyokeizai, which is a publishing company, sent a questionnaire on environmental accounting to 1208 Japanese companies, of whom 562 responded. One question asked what had been the purpose of the introduction environmental accounting. The study found that 57.0% of the companies responding attach more importance to external disclosure, and only 17.8% companies lay the main stress on internal decision-making. Therefore, it would be of importance to develop environmental management accounting tools for Japanese companies, which is an objective of the METI project which will be discussed later.

### 12.3.2. *A case of environmental accounting in Japan: Ricoh's environmental accounting*

Ricoh is a leading manufacturer of OA facilities such as photocopy machines, printers, etc. The company is famous for its environmental protection activities – in the Nikkei Newspaper's ranking of environmentally conscious companies, Ricoh has been ranked as the top company in each of the past three years (1998–2000).

Ricoh have published an environmental report annually since 1998, with an environmental accounting statement first included in the report in 1999. Ricoh divides its environmental accounting into two parts: 'corporate environmental accounting' and 'segmental environmental accounting'. 'Corporate environmental accounting' is defined as the environmental accounting of overall corporate activities, which is made public. This accounting is intended to be useful for external disclosure as well as internal management. Ricoh's environmental report (2000) explains as follows:

> *'Corporate environmental accounting aims to identify all environmental impact reduction effects and economic benefits in each investment area. It is used as a decision-making tool in identifying and publicizing the Group's achievements, as well as in making effective environmental investment' (p. 31).*

On the other hand, 'segmental environmental accounting' is Ricoh Group's unique environmental management accounting system in order to estimate the efficiency of environmental facility investments in business sites or new projects. The company understands that segmental environmental accounting is effective in evaluating small-scale investments, such as investments in divisional facilities. In Ricoh's environmental report (2000) 'corporate environmental accounting' plays a role in disclosing overall environmental accounting information, and three cases of 'segmental environmental accounting' are descriptively explained.

The Ricoh 2000 corporate environmental accounting is indicated in Table 12.2. The statement is one of the most comprehensive among all Japanese companies in disclosing environmental accounting information. The Ricoh Group discloses not only economic benefits and environmental conservation effects, but also some eco-efficiency ratios. While Ricoh's environmental cost category is the same as the MOE guideline, the scope of its economic benefits is broader than those indicated in the guideline. The benefits are divided into three types: substantial effect, expected effect, and incidental effect. These benefits are defined by the company as follows:

Table 12.2. Corporate Environmental Accounting of the Ricoh Group for Fiscal 1999.

| Item | Costs | | Economic benefits | | | Effect on environmental conservation | | | | Environmental impact | Eco-ratio[3] | Converted Value | Conversion |
|---|---|---|---|---|---|---|---|---|---|---|---|---|---|
| | Environmental costs | Main costs | Monetary effects | Category | Item | Environmental impact reduction (t) | Reduction rate | EE value[1] | Converted quantity of reduction | Total (t) | (\100 million/t) | of reduction | coefficient[5] |
| Business area costs | ¥1,670 million | Environment-related facility depreciation and maintenance costs | ¥960 million | a | Energy savings and improved waste processing efficiency | $CO_2$ ............ 1,317 | 4.1% | 139.7 | 11,317 | $CO_2$ ....... 262,053 | 0.0105 | 262,053 | (1.0) |
| | | | ¥5,090 million | b | Contribution to value-added production | $NO_X$ ............ 2.006 | 2.9% | 0.0248 | 12.44 | $NO_X$ ........ 67.11 | 40.95 | 416.1 | (6.2) |
| | | | ¥700 million | c | Avoidance of risk in restoring polluted environment and avoidance of lawsuits | $SO_X$ ............ 7.404 | 33.7% | 0.0914 | 6.663 | $SO_X$ ........ 14.53 | 189.1 | 13.08 | (0.9) |
| Upstream/ downstream costs | ¥2,410 million | Costs for collection and reassembly for recycling used products | ¥580 million | a | Scales of recycled products, etc. | | | | | | | | |
| Managerial activity costs | ¥1,790 million | Costs for the division in charge of environmental measures; costs to establish and maintain the environmental management system | ¥200 million | b | Improved efficiency in environmental education and establishment of the environmental management system | BOD ............ 1.726 | 4.5% | 0.0213 | 0.1726 | BOD ........ 36.61 | 75.05 | 3.561 | (0.1) |
| Research and development costs | ¥660 million | Research and development costs for environmental impact reduction | ¥50 million | a | Cost reduction through eco-packaging | Final waste disposal amount ............ 3,458 | 34.6% | 42.69 | 359,632 | Final waste disposal amount ........ 6,538 | 0.4203 | 679,952 | (104.0) |
| | | | ¥2,040 million | b | Contribution to value-added research and development | | | | | | | | |
| Social activity costs | ¥390 million | Costs for preparation of environmental reports and advertisements | ¥70 million | b | Environmental advertisement, etc. | PRTR substances (178 substances, including toluene and dichloromethane) | | | 47,120 | PRTR substances (178 substances, including toluene and dichloromethane) ※See page 54. | | 250,683 | (Ricoh standards per substance) |
| Environmental restoration costs | ¥130 million | Costs for restoration of soil pollution and environment-related reconciliation | | — | None | | | | | | | | |
| Other costs | ¥50 million | Other costs for environmental conservation | | | | | | | | | | | |
| Total | ¥8,100 million | | ¥9,690 million | | | | | | 418,088 | | | 1,193,121 | |
| | | | | | | | | | 0.0516 EE index[2] | | | 230.3 Eco-index[4] | |

a= Substantial effect (actual gains from cost and energy reduction as well as sales of property, plant, and equipment) b=Expected effect(amount to which the environmental measures contributed) c=Incidental effect (amount of additional costs avoided stemming from such problems as pollution and lawsuits).
1. Eco-efficiency (EE) value (unit: ton/\100 million)= Environmental impact reduction amount/total amount of environmental costs.
2. EE index= Total converted quantity of environmental impact reduction /total environmental costs (thousands of yen).
3. Eco-ratio (unit: \100 million/ton)= Total sales profit/total environmental impact amount.
4. Eco-index= Total sales profit/total converted value of environmental impact reduction.
5. Conversion coefficient is based upon literature related to LCA impact evaluations. For final waste disposal amounts and PRTR substances, the converted coefficient is set according to Ricoh's internal standards.
Source: Ricoh(2000) pp.31-32.

- *substantial effect:* actual gains from cost and energy reduction as well as sales property, plant, and equipment
- *expected effect:* amount to which the environmental measures contributed
- *incidental effect:* amount of additional cost avoided stemming from such problems as pollution and lawsuit.

Comparing these effects with the MOE guideline, 'substantial effect' corresponds to 'economic effects calculated based on a credible basis', and 'expected effects' and 'incidental effect' are similar concepts to 'economic effects calculated on a hypothetical calculation'.

Ricoh discloses two kinds of environmental ratios (EE value and Eco-ratio) in their environmental accounting. The EE (eco-efficiency) value is calculated by dividing the environmental impact reduction amount in a given year by total amount of environmental costs for the same year. Eco-ratio is an index calculated by dividing total sales profit by the total environmental impact amount.

Ricoh both calculates these ratios separately for different environmental impact, and also integrate these different impacts into a single figure. The company used LCA and its original weighting method, and integrated six kinds of environmental impacts including $CO_2$, $NO_x$, $SO_x$, BOD, waste disposal amount and PRTR[4] substances. Then they generate two integrated indices: the EE index and the Eco-index. While the method of integration can be controversial, if the company continues to use this method these ratios should be a very important index to compare current environmental performance against the past.

Ricoh recognizes environmental accounting as one of the most important tools to promote environmental management in the company. They constructed an environmental accounting information system which is linked to the corporate accounting system. Within the company, an environmental performance index has been introduced into its corporate performance evaluation system. Ricoh employed the balanced scorecard model as a tool of performance evaluation and has added an environmental dimension to this system.

Ricoh's environmental accounting has, therefore, been constructed for an external disclosure purpose as well as for an internal management purpose, which is unusual in Japan. The method that Ricoh developed in this area seems to be useful for other companies.

## 12.4. Some outstanding issues of Environmental Accounting

### 12.4.1. *Metrics of effects and new indexes – the MOE Guideline on Environmental Performance Indicators and other efforts*

Although the guideline gives some guidance on metrics for environmental conservation effects and economic effects, it does not provide very detailed instruction because at the time when the guideline was published these two elements had been the subject of far less investigation than had the other element, environmental conservation cost.

Concerning environmental conservation effects, the MOE launched a new project and published a draft guideline 'Environmental Performance Indicators of Business Operators: Draft for Inviting Opinions' in November 2000. The MOE had invited opinions from the public until late November and published a guideline in February 2001. This guideline is expected to be used when companies consider metrics for their environmental conserva-

---

[4] The Pollutant Release and Transfer Register (PRTR) system, a remedial system for hazardous waste sites.

tion effects. Concerning the economic effects, since many examples of metrics of economic effects are found with electric/electronic companies' environmental reports, MOE launched the electric/electronic working group on environmental accounting and requested it to study this issue. The working group is going to publish a concluding report by the end of March 2001.

Although the above projects are expected to help the guideline to provide more detailed instruction on metrics of the effects, there remains a need for further study. For example, new indexes that integrate both physical and monetary information should be developed, so that the environmental accounting standard can really work as the settlement document in an environmental report.

### 12.4.2. *Environmental Management Accounting Tools – the METI project*

Another important issue for Japanese companies that introduce environmental accounting is how to integrate the guideline into corporate decision-making. When management accounting is undeveloped, financial accounting is utilized for internal management as well as for external reporting. However, since decision-making in companies has its own specific purpose such as investment decision, price setting and performance evaluation, the integrated environmental conservation cost calculation system provided by the guideline cannot simultaneously meet all such individual purposes to an adequate extent.

In order to solve this problem, it is necessary to develop various environmental management accounting tools. While in Japan environmental management accounting practices have developed only slowly, Japanese companies started to recognize the importance of those tools for internal use. The METI project which was described at the beginning of this paper targets the development of tools of environmental management accounting. In this sense, the MOE project and the METI project should be complementary to each other.

The METI project started in 1999 and has been working on a three-year research plan. In the first year, it held discussion from various perspectives including financial accounting, quality costing, life-cycle assessment and costing. It also conducted a research on related programs/tools of the world mainly in the US/Canada and Europe. The research results were published as a first year report by JEMAI, whom METI entrusted with the research.

Based on the outcome of the first year research, four working groups were established in the second year to develop tools for specific management purposes. The first (WG1) is discussing tools for environmental capital investment decision-making. WG2 is investigating tools for environmental cost management. WG3 is going to develop tools for environmental and financial performance evaluation. WG4 is examining material flow cost accounting and will pilot-test this with a Japanese company. Some of these tools will be developed in 2001 and the project will be concluded by March 2002. As mentioned above, since Japanese environmental practices are inclined towards external disclosure, the METI project should be important in developing the other aspect (internal use) of environmental accounting.

### 12.4.3. *Environmental Accounting in the Asia-Pacific region*

Environmental accounting is expected to play an important role in corporate efforts toward sustainable development. This is true not only for the Japanese business community but also for other parts of the world. Particularly in the Asia-Pacific region, where further rapid economic growth is expected, it is important to find ways to promote corporate voluntary activities including environmental accounting. Based on this understanding, the

Institute for Global Environmental Strategies (IGES) Kansai Research Center is planning to conduct a 3-year research program on business and the environment from 2001, which includes extensive studies on environmental accounting in the Asia-Pacific region.

## 12.5. Conclusions

The development of corporate environmental accounting in Japan has been accelerated by the MOE guideline. The number of companies that disclose environmental accounting information in their environmental report is increasing, with most either already following the MOE guideline or likely to follow it in future. While the guideline has some limitations, MOE have already published some new reports in order to address these, and will be publishing more within the near future.

While the emphasis of the MOE guideline is on external disclosure, environmental management accounting should be developed as well. The METI project will play an important role in developing environmental management accounting tools for Japanese companies. Japanese external environmental accounting is ahead of US and European practice, but internal environmental accounting lags far behind them. While environmental accounting has been moving into the second stage in Japan, the big issue at this stage is to develop environmental management accounting so as to catch up with the external environmental accounting system.

# 13. Environmental Accounting in Korea: Cases and Policy Recommendations

*Byung-Wook Lee, Seung-Tae Jung and Yun-Ok Chun*
*Environmental Management Center of POSCO Research Institute (POSRI),147 Samsung-dong Kangnam-gu Seoul, Korea; E-mail: bwlee@mail.posri.re.kr, iready@mail.posri.re.kr, yochun@mail.posri.re.kr*

## 13.1. Introduction

Environmental problems such as exhaustion of resources, global warming, ozone depletion, acid rain, desertification, species decimation, and marine pollution have generated substantial concern in recent years, and to address these problems many countries have established or reinforced environmental laws, provisions and international agreements. These environmental measures are sometimes closely connected with international trade. so the environment has become an important factor in international business. This context has an important effect upon corporate business activities so that the relationship between the environment and business management is of great and growing importance.

In line with this trend, rapid increases in environmental costs are now causing companies to begin to integrate environmental aspects into managerial decisions at all levels. However, the measuring and reporting of corporate environmental performance is still at an infant stage in spite of the development of a number of methodologies and practices. In this context, environmental accounting has recently been advanced as one of the most significant tools in promoting successful environmental management. This reflects the view that conventional accounting, which ignores most environmental externalities, is not appropriate for encouraging companies to manage their activities in an environmentally benign way.

Environmental degradation is almost inevitable given current accounting practice. Conversely, many companies have now come to recognize that environmental accounting can play an important role in both the prevention and restriction of negative environmental responses, and the facilitation of positive and proactive responses. Under these circumstances, environmental accounting has been introduced or implemented in many leading companies, especially in Europe, North America and Japan. Compared with these advanced companies, however, most companies in developing countries are still well behind in understanding, developing or implementing environmental accounting in their business practices.

## 13.2. Overview of environmental accounting in Korea

As a wide range of stakeholders such as shareholders, financial institutions, government, and local communities have been interested in corporate environmental performance and its disclosure, since the mid-1990s some Korean companies have begun to examine the introduction of environmental accounting.

Environmental investment and costs of pollution prevention have increased in Korea, as shown in Table 13.1. This is in line with the emergence of green consumerism, non-governmental organizations' (NGOs) environmental activities, and international trade barriers related to the environment. Some leading companies in Korea, such as POSCO,



*M. Bennett et al. (eds.), Environmental Management Accounting: Informational and Institutional Developments,* 175–186.
© 2002 *Kluwer Academic Publishers. Printed in the Netherlands.*

Table 13.1. Corporate Pollution Abatement and Control Expense in Korea (million Won).

| Field | 1993 | 1994 | 1995 | 1996 | 1997 | 1998 | 1999 |
|---|---|---|---|---|---|---|---|
| Air | 700,789 | 797,651 | 916,888 | 957,276 | 1,284,333 | 46,034 | 1,140,798 |
| Water & Soil | 684,537 | 805,863 | 1,030,374 | 1,162,034 | 1,040,543 | 18,498 | 939,515 |
| Waste | 625,837 | 744,300 | 833,827 | 1,024,743 | 1,050,808 | 901,423 | 975,759 |
| Noise & Vibration | 68,502 | 92,583 | 74,599 | 79,849 | 62,830 | 50,054 | 69,785 |
| Others | 73,643 | 115,583 | 122,550 | 117,302 | 99,666 | 84,492 | 80,002 |
| Byproduct sales in waste treatment (–) | 7,801 | 9,363 | 11,659 | 12,164 | 16,297 | 17,152 | 20,793 |
| **Sum Annual** | **2,145,507** | **2,546,617** | **2,966,579** | **3,329,040** | **3,521,883** | **2,883,349*** | **3,185,066** |
| **Growth Rate (%)** | **(12.8)** | **(18.7)** | **(16.5)** | **(12.2)** | **(5.8)** | **(–18.1)** | **(10.5)** |

* In 1998, the Korean economy went through an abrupt recession because of a monetary crisis in the region.
Source: Bank of Korea, Pollution Abatement and Control Expense in 1999, 2000.

Samsung Electronics and LG Chemicals, have begun to consider environmental costs in management decisions, because environmental costs have continually increased as a proportion of total production costs.

Furthermore, financial institutions such as banks and insurance companies have become interested in appraising corporate environmental risk and performance when they lend or invest money. These changes have pressured Korean companies into finding cost-effective ways to enhance their environmental performance. As this continues, many companies are beginning to realize the importance of proactive environmental management strategy and environmental performance reporting, but these changes are still at an early stage. Leading companies such as POSCO, Samsung, LG and Hanhwa experience many difficulties with the introduction or implementation of environmental accounting. On the other hand, many other Korean companies do not recognize the concept of environmental accounting or understand how to implement it.

Meanwhile, in order to promote environmental accounting practice in Korea and Asian developing countries, the Korean Ministry of Environment (KMOE) has introduced a special project on 'environmental accounting systems and environmental performance indicators', funded by the World Bank. In January 2000, the Korea-World Bank Environmental Cooperation Committee (KWECC) was organized to promote environmental management in Asia, and launched three related projects including 'environmental accounting and environmental performance indicators'.

Of these, the project on environmental accounting has been carried out by the POSCO Research Institute (POSRI) under the sponsorship and supervision of KWECC from March 2000 to February 2001. This project aimed to develop a useful toolkit for assessing a company's environmental costs and performance more precisely, and to suggest a comprehensive methodological framework for the introduction of environmental accounting and performance evaluation schemes at a corporate level. The project also considered a guideline for environmental accounting which can be utilized in developing countries, and recommended some policy options that can facilitate the introduction of these toolkits into business practice.

In line with the project, the Environmental Management Accounting Network – Asia Pacific (EMAN-AP) was initiated in February 2001, during the World Bank Environmental Forum held in Korea. EMAN-AP plans to link the various efforts of organizations and individuals in the region towards the development and promotion of environmental management accounting. EMAN-AP will be launched as a regional network for corporate environmental management accounting and independently operated in close relationship with EMAN-Europe and other regional networks. The Network will be run with fourteen initial member countries including Korea, Japan, the Philippines, China, Indonesia, Taiwan, Thailand, Malaysia, Singapore, Hong Kong, Vietnam, India, Australia, and New Zealand.

At the same time, KMOE is developing a scheme for companies to include environmental accounting information in their environmental reports. Through this regulatory change, KMOE is trying to encourage Korean companies to implement environmental management in the whole range of their business processes.

In 2001, the Korea Accounting Institute (KAI) also published a report on an 'Accounting Standard for Environmental Costs and Liabilities', which covers a wide range of issues on environmental financial accounting. The report aimed to provide theoretical reviews and to propose relevant ways to introduce environmental financial accounting in Korea. The report mainly covers the definition and fields of environmental accounting, the conceptual framework for environmental financial accounting, practices of environmental accounting in Korea, and a draft environmental accounting standard.

## 13.3. Cases in environmental accounting in Korea

As mentioned above, Korean companies have a growing interest in environmental accounting and a few companies have actually accumulated a little experience in it. Three case studies are presented in this paper, of POSCO, Samsung Electronics and LG Chemicals, which have had some practice with environmental accounting and have produced information on environmental costs.

### 13.3.1. *POSCO*

#### *Profile of the Company*

Founded in 1968 as a public corporation, Pohang Iron and Steel Corporation (POSCO) operates two steel works in Pohang and Kwangyang, producing hot rolled sheet, cold rolled sheet, wire rod, electrical steel, and stainless steel, and is one of the world's largest steel-makers with an annual production capacity of 28 million tons. In 1999, POSCO employed around 20,000 people and had a turnover of 10,696 billion won (US$9.5 billion).

Since commencing business, the company has recognized that environmental preservation is one of the most important aspects of doing business. It therefore enacted the 'POSCO Environmental Policy' in 1995 and adopted an environmental management system based on ISO 14001 standards in 1996. Furthermore, POSCO has recently switched its environmental policy from the conventional passive monitoring of activities to a proactive effort aimed at preventing environmental accidents and constantly enhancing environmental performance in cooperation with the local community. The company has invested nearly 10 per cent of its total investment in environmental protection for this purpose, and is gradually planning to increase the scale of its investment. As a result of its proactive effort and investment, POSCO has achieved cleanliness ratings that are four to five times higher than the level stipulated by relevant laws.

#### *Environmental accounting practices of the Company*

POSCO has produced information on environmental costs since the 1990s, but the information did not satisfy company management so in December 1999 the company launched a special project to develop its new environmental accounting scheme. For this project, a research team was organized with the staff of the company's Environment & Energy Team and experts of the Environmental Management Center in the POSCO Research Institute (POSRI). Before beginning the research in earnest, the research team established the following four stages for the work:

1. identifying environmental costs which are hidden in overhead costs;
2. allocating environmental costs to each cost center which causes the costs;
3. calculating and reporting environmental benefits and liabilities;
4. integrating information on environmental accounting in management decision-making.

However, POSCO recognized that it is difficult to calculate environmental benefits and liabilities because of the arbitrariness that is unavoidable in their calculation, and POSRI therefore decided to tackle the first and second of the four stages for the first trial. The company considers, however, that environmental benefits and liabilities will have to be calculated in the near future.

Based on the scope of this project, the company defined environmental costs as follows:

> *Environmental costs are direct or indirect costs related to the operation of environmental equipment used to remove or reduce air and water pollutants. Moreover, they also include costs for disposing of or recycling waste and for other environmental activities.*

Under the definition, the company divided its environmental costs into costs for preserving air quality and water quality, costs for disposing of and recycling wastes, and other costs. The detailed cost items are shown in Table 13.2.

Because the above-mentioned environmental costs are mostly incurred through the operation of environmental protection equipment or facilities, it is necessary to define the conceptual characteristics and scope of environmental assets before calculating environmental costs. It was, however, difficult to find any general definition or scope of environmental assets. Therefore, POSCO defined environmental assets as follows:

> *Environmental assets are all equipment and facilities operated for the prevention of environmental pollution.*

Under this definition, when certain equipment or facilities are purchased mainly for the purpose of environmental protection, the company recognizes them as environmental assets. In general, however, much of its equipment and facilities is multi-purpose or multi-functional. In such cases, it is normally very difficult to decide whether certain equipment is an environmental asset, and POSCO is no exception to this norm. To solve the issue, the company determined that when certain equipment or facilities are used for environmental protection for over 50 per cent of the time, they should be recognized as environmental assets. The judgment on whether the figure of 50 percent or more is met by any particular asset is made by the person working for environmental preservation in factories. This is a somewhat arbitrary figure, but it can be a useful method in practice.

After defining environmental assets, POSCO re-arranged the coding structure of all its assets in order to recognize in its computerized costing process environmental costs incurred in operating environmental assets. Even though it has had some difficulties in adopting a new coding system, it is a different case in POSCO because the company is in process of re-arranging its assets coding structure prior to the launch of an 'enterprise

Table 13.2. Classification of Environmental Costs in POSCO.

| *Level 1* | *Level 2* | *Level 3* |
|---|---|---|
| Air Quality Management | Depreciation Costs | |
| | Electricity Costs | |
| | Material Costs | – Costs for chemicals |
| | Repair or Maintenance Costs | – Material costs<br>– Costs for external service<br>– Labour costs |
| | Labour Costs | – Labour factory costs<br>– Labour office costs |
| | R&D Costs | |
| | Costs for Energy Substitution | |
| | Emission Charge on Air Pollution | |
| | Others | – Test or measurement fees of equipment discharging air pollutants<br>– Measurement costs of dust collectors<br>– Test costs for Tele-metering System<br>– General expenses |
| Water Quality Management | Depreciation Costs | |
| | Electricity Costs | |
| | Material Costs | – Costs for chemicals |
| | Repair or Maintenance Costs | – Material costs<br>– Costs for external service<br>– Labour costs |
| | Labour Costs | – Labour factory costs<br>– Labour office costs |
| | R&D Costs | |
| | Emission Charge on Water Pollution | |
| | Others | – Test or measurement fees of equipment discharging water pollutants<br>– Costs for preventing sea pollution<br>– Costs for external service<br>– General expenses |
| Waste Management | Transportation Costs | |
| | Incineration Costs | |
| | Reclamation Costs | |
| | Costs for By-Product Processing | |
| | Recycling Promotion Costs | |
| | Costs for Wastes Processing | |
| | Costs for Disposing Wastes on Commission | |
| | Labour Costs | – Labour factory costs<br>– Labour office costs |
| | R& D Costs | |
| | Others | – General expenses |
| Others | Education Costs | |
| | Costs for Operating EMS | – Post-audit costs<br>– Costs for publishing environmental report |
| | Costs for External Cooperation | |
| | Costs for Afforestation | |
| | Labour Costs | – Labor office costs |
| | Environm. Improvement Charges | |
| | R&D Costs | |
| | Others | – Test costs on soil pollution<br>– General expenses |

resources planning' (ERP) system in mid-2001. Further, POSCO plans to measure and allocate environmental costs more accurately through an Activity-Based Costing (ABC) method to be introduced in mid-2001.

### 13.3.2. *Samsung Electronics*

#### *Profile of the Company*

Founded in 1938, Samsung Electronics is the world-leading manufacturer of memory devices, and also leads the world semiconductor industry in development after designing a 256-megabit DRAM (dynamic random access memory), a one-gigabit DRAM, and the entire production process technology for 4-gigabit DRAM. In 1999 the company achieved net sales of US$22.8 billion with 43,000 employees.

Samsung Electronics has recently positioned itself in four main business units: Digital Media, Semiconductors, Information & Communications, and Home Appliances, producing the world's most innovative digital components with the aim that these will become generally recognized them as being the best in the world. At the same time, Samsung Electronics has tried to improve the quality of life by engaging in business activities that respect both people and nature. For the purpose, the company first announced its 'Environmental Policy' in June 1992, and declared the 'Samsung Green Management Charter' in May 1996. Now the company's philosophy focuses on minimizing the environmental impacts created by its business activities.

#### *Environmental Accounting Practices in Onyang Plant*

Onyang Plant of Samsung Electronics was established in 1990 as a Semiconductor Assembly and Testing Plant. In 1998, the plant was very interested in calculating environmental costs, but did not have a company-wide guideline for calculating environmental costs. In consequence, in 1998, the plant developed its own guideline and calculated its first specific environmental costs using this guideline.

In the company, environmental costs include the following:

- Costs related to environmental facilities, including both pollution-prevention and damage rectification facilities;
- Costs related to waste disposal; and
- Costs for improving the efficiency of pollution prevention facilities.

Under this definition, its environmental costs are divided into 4 categories: air, water, waste and others. The costs are classified between direct costs and indirect costs. The former are those which are directly traceable to each category, while the latter cannot be directly traceable to a specific category and need to be allocated. Detailed environmental costs of the plant are classified as shown in Table 13.3.

Environmental costs that are calculated are not allocated to each cost center using a sophisticated allocation basis, though the company recognizes that a sophisticated allocation basis is required in order to calculate the environmental costs of products. On the other hand, there is no specific evidence that the available information on environmental costs has in fact been used for decision-making in the company, although the information is reported to the most senior executives.

Table 13.3. Classification of Environmental Costs in Samsung Electronics.

<table>
<tr><th rowspan="2">Category</th><th colspan="2">Cost Items</th></tr>
<tr><th>Direct Cost</th><th>Indirect Costs</th></tr>
<tr><td rowspan="6">Air</td><td>Depreciation costs</td><td rowspan="13">– Indirect support costs: authority and permission, information collection, others<br>– TMS: Depreciation costs, Labour costs, Repair costs<br>– Laboratory: Labour costs, Chemical costs, Equipment depreciation costs, Repair costs, Costs for measuring pollution around plant, External test costs, U/T indirect labor costs<br>– Operating & Maintenance labor cost</td></tr>
<tr><td>Labour costs</td></tr>
<tr><td>Electricity costs</td></tr>
<tr><td>Repair costs</td></tr>
<tr><td>Material costs</td></tr>
<tr><td>Chemical costs</td></tr>
<tr><td rowspan="7">Water</td><td>Depreciation costs</td></tr>
<tr><td>Labour costs</td></tr>
<tr><td>Electricity costs</td></tr>
<tr><td>Repair costs</td></tr>
<tr><td>Material costs</td></tr>
<tr><td>Chemical costs</td></tr>
<tr><td>Costs for waste water treatment</td></tr>
<tr><td rowspan="6">Waste</td><td>Depreciation costs of weighing machine</td><td rowspan="6">– Indirect supporting costs: Authority and permission, information collection, others<br>– Indirect labour costs<br>– Lift depreciation costs</td></tr>
<tr><td>Warehouse for waste: Depreciation costs, Labour costs, Repair costs</td></tr>
<tr><td>Attached facilities depreciation costs</td></tr>
<tr><td>Waste crusher: Depreciation costs, Repair costs</td></tr>
<tr><td>Waste acid: Depreciation costs of waste acid treatment site, Labour costs, External service costs, Repair costs, Energy costs</td></tr>
<tr><td>Costs for analysis of waste acid sludge</td></tr>
<tr><td>Others</td><td colspan="2">Education costs, Association fee, External relation costs, Costs for publication, Other labor costs, General expense, External service costs for night soil treatment</td></tr>
</table>

### *13.3.3. LG Chemicals*

#### *Profile of the Company*

Founded in 1947, LG Chemicals is the largest chemical company in Korea with eight manufacturing sites in the country. Its major business fields are life science, information and electronic materials, petrochemicals, health care and household goods. In 1999 its sales were US$3,969 million and its assets totalled US$4,911 million, with around 11,000 employees.

LG Chemicals considers environmental protection as of the utmost importance in order to become an enterprise practicing environment-focused management. To realize this, in 1997 the company declared its 'Environmental Policy' and set up an 'Environmental Safety Committee'. In particular, its eight plants have had experience of practising environmental accounting. This study focuses on its Cheongju plant, which is a large facility for the production of many kinds of chemical products such as cosmetics, household goods, flooring, and information and electronic materials. Even though it is one of the biggest chemical works in Korea, it does not discharge any wastewater.

### *Environmental Accounting Practices in Cheongju Plant*

The Environment and Safety Team in LG Chemicals initiated the environmental costing project in 1996 to standardize the measurement process of environmental costs. The project focused on the classification of environmental costs, the segregation of environmental costs from non-environmental costs, and the calculation and systematic management of environmental costs.

LG Chemicals classifies its environmental costs into proactive environmental costs and ex-post environmental costs. The specific classification is shown in Table 13.4.

In Table 13.4, proactive environmental costs are those which are incurred in pollution prevention activities, and consist of costs for pollution prevention at source, pollution treatment/disposal costs, and stakeholder costs. Ex-post environmental costs are those incurred to remedy or restore the environmental damage that has already occurred. The Ex-post

Table 13.4. Classification of environmental Costs in LG Chemicals.

| Cost Items | Level 1 | Level 2 |
|---|---|---|
| Proactive Costs | Pollution Prevention Costs | R&D |
| | | Facility Replacement Costs for Clean Process |
| | | Utility Replacement Costs |
| | | EMS Costs |
| | Pollution Treatment | Acquisition & Installation of Environmental Facilities |
| | | Measurement Costs |
| | | Maintenance & Operating Costs of Environmental Facilities |
| | | Environmental Utility Costs |
| | | Treatment or Disposal Costs |
| | | Environment-related Operations & Administration Costs |
| | Stakeholder Costs | Legal Compliance Costs |
| | | Public Relation Costs |
| | | Advertising Costs |
| Ex-post Costs | Taxes & Charges | Taxes |
| | | Environmental Charges |
| | | Environmental Deposits |
| | Fines & Penalties | |
| | Compensation to Third Parties | |
| | Opportunity Costs | |

costs include fines and penalties incurred from non-compliance with environmental regulations, and compensation to third parties for loss or injury caused by environmental pollution and damage in the past.

After classifying the environmental costs, the company examined which cost accounts in the conventional accounting system match with items of environmental costs. However, the examination did not provide any objective criteria about the distinction between environmental and non-environmental costs, which makes unreliable the cost information which has been collected. For this reason, the information on environmental costs which has been generated is not currently sufficiently utilized in the company.

### 13.3.4. *Implications*

The three companies were concerned about, and introduced, environmental accounting for the following common reasons in the 1990s:

- To identify more precisely the environmental costs which are hidden in indirect costs;
- To establish and implement a comprehensive environmental management system;
- To evaluate the performance of their environmental management;
- To invest in environmental projects more efficiently; and
- To consider information on environmental costs in product price decisions.

The practice of environmental accounting in the three companies is now primarily focused on management accounting. They are measuring only environmental costs, since the measurement of environmental benefits is as yet still at an early stage. Moreover, the three companies mainly manage environmental costs which related to end-of-pipe environmental facilities and equipment, and still do not include social or global environmental costs such as ozone depletion or climate change. The three companies do not disclose information about environmental costs in their annual environmental reports. However, they are trying to produce credible information on these and following the trial, they intend to disclose environmental accounting information.

Three issues found through these case studies are summarized below:

*the need to develop a specific guideline for the calculation and allocation of environmental costs.*

The practical measurement and allocation of environmental costs is currently mainly based not on any theoretical framework or specific guideline, but on the environmental department's intuition or experience. Moreover, two of the companies (the exception being POSCO) have no specific guidelines for the allocation of environmental costs to each cost center. This is a crucial problem because incorrect cost allocation can distort corporate decision-making. Accordingly, it is necessary first to carry out a specific field survey and then the three companies can develop better guidelines for the measurement and allocation of environmental costs. It may be appropriate for ABC to be adopted, since in the process it could convert many of the overhead costs incurred in manufacturing which are related to the environment, into direct costs. Hence, the appropriate selection of environmental activities and cost drivers through ABC allows companies to trace many environmental overhead costs to cost objects, and may give the management of the company a better overview of environmental costs.

*More understanding about utilizing environmental accounting information.*

To utilize information produced about environmental costs successfully, it is necessary for a company's management to have an understanding about its general and specific uses.

*Needs close cooperation with the accounting department.*

It was found in all three cases that the information on environmental costs was produced only by the environmental departments, who have no professional knowledge of accounting practices. This is a common situation in Korean companies because accounting staff are normally not familiar with environmental accounting, since most accounting managers are conservative and reluctant to change their practices. However, to measure and allocate environmental costs effectively, it is necessary for the environmental department to cooperate closely with the accounting department. Accordingly, companies have to encourage accounting staff to participate actively in environmental accounting projects.

## 13.4. Discussion of policy options

To promote the introduction and implementation of environmental accounting in Korean companies, it is first necessary for the government to provide an environmental accounting guideline and then to stimulate various stakeholders in their demands for information derived from corporate environmental accounting systems. To this end, government needs to develop appropriate policy options for corporate environmental accounting. In this context, it is recommended that a step-wise approach be adopted as follows, in three stages:

1. Establish infrastructure by organizing a working group and benchmarking best practices of environmental accounting in advanced companies
2. Develop and provide an environmental accounting guideline and run pilot programs
3. Activate environmental accounting through environmental reporting and auditing.

### 13.4.1. *Stage 1: Establishment of infrastructure*

As an initial measure in the introduction of environmental accounting, a working group should be organized, to be composed of government officers, environmental accounting experts, and corporate accounting and environmental managers. Cooperation and common understanding between these participants are crucial factors for establishing the infrastructure for promoting environmental accounting. The main roles of the working group would be to:

- Survey international and domestic studies on environmental accounting;
- Analyze various guidelines and best practices;
- Build up a network with international expert groups such as EMAN-AP;
- Develop an environmental accounting guideline to consider country-specific business practices;
- Establish a nation-wide program to introduce and implement environmental accounting; and
- Assign roles and tasks to related government bodies such as Environment, Industry, Finance & Economy, Financial Supervisory Service, etc.

The working group would also hold seminars to disseminate international trends and the state of the art on environmental accounting, and to share its importance with corporate

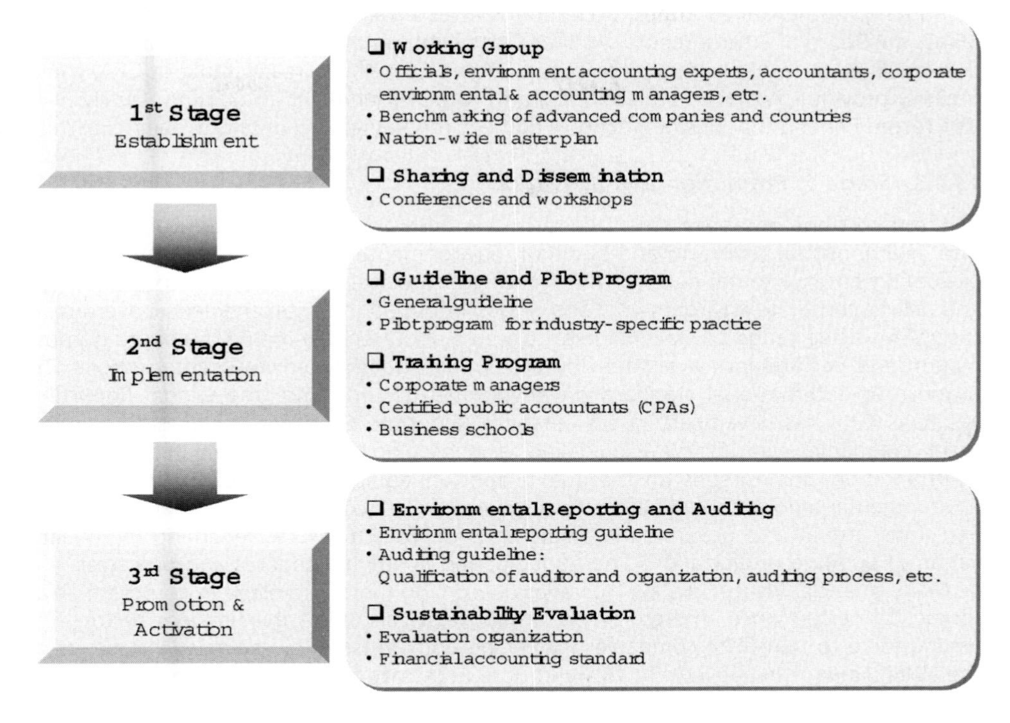


Figure 13.1. Framework of Government Policy for Environmental Accounting.

managers. Through these efforts, it may be possible to expand recognition of environmental accounting issues amongst managers and to gain acknowledgement of the importance of environmental accounting from top corporate management.

### 13.4.2. *Stage 2: Implementation*

In addition to the first stage, it is necessary that also in the second stage the government plays an important role in implementing environmental accounting in corporate practice. This would have two components. The first is to provide a country-specific guideline on environmental management accounting, which can be developed by the working group. The second is to run a pilot program to apply the guideline to several leading companies.

Based on the results of the pilot program, it would then be necessary to review and revise the guidelines. In the process of setting these it is necessary to examine and reflect upon the substance of international guidelines. The guidelines may cover the definition, scope and classification of environmental cost, and measurement methods. As these guidelines will show a general way of implementing environmental management accounting, it is necessary that more specific and sophisticated guidance for each industry sector also be developed.

In addition, the government can offer training opportunities to company staff in the practical application of environmental accounting. Certified public accountants (CPAs) also need to take part in this training program in relation to their role in environmental accounting. In the United States, accountants attend training programs managed by the BEAC (the Board of Environmental Auditor Certifications), after completing which they are then qualified to audit environmental reports. The KICPA (the Korea Institute of CPA) could similarly provide CPAs with training programs on environmental accounting. Finally, it is also recommended that business schools add environmental accounting to their curricula.

### 13.4.3. *Stage 3: Promotion and activation*

At the third stage, the government needs to establish a regulatory framework for corporate environmental reporting and auditing. Environmental reporting is a useful tool for evaluating environmental performance which can be closely related to corporate value, and deliver corporate environmental accounting information to stakeholders. Government can raise a wide range of stakeholders' concerns about environmental accounting information and performance evaluation by promoting published environmental reports. To propose an international standard on environmental reporting, the Global Reporting Initiative (GRI) has developed its Sustainability Reporting Guidelines. With some adjustment, companies can utilize this guideline for publishing their environmental reports.

In addition, some issues on the qualification of auditors and auditing processes of environmental reports should be carefully examined. To audit environmental reports fairly and transparently, the government should prepare some measures regarding the qualification of auditing organizations and auditors, and auditing standards and processes.

Many financial institutions are nowadays becoming more interested in corporate environmental performance. Therefore, the government can utilize the financial sector as a driving force to transform companies into being greener (see, for example, the efforts of the UNEP Finance Initiatives). To this end, it is necessary for the government to support the finance sector in developing useful tools for environmental risk assessment. When the finance sector actively assesses corporate environmental risks and performance, and also demands environmental accounting information, it will become common practice for companies to introduce and implement environmental accounting. At this stage, the establishment of an organization that appraises corporate sustainability in a professional way can be considered. The roles of such an organization would be:

- To rate corporate sustainability by assessing environmental, social, and economic performance and risk; and
- To provide information to financial institutions.

## 13.5. Conclusion

Even though Korean companies are still at the early stage in environmental accounting they have a great potential for introducing and implementing environmental accounting. External pressures from the government, international standards, and NGOs also play an important role for companies to increase their interest in environmental accounting.

The policy options recommended in this paper offer a possible way to apply environmental accounting to other countries as well as Korea. However, this paper does not cover the area of environmental financial accounting which is another equally important area. In the near future, therefore, it will be necessary to examine how to include environmental aspects in financial accounting standards.

# 14. Government Strategies to Promote Corporate Environmental Management Accounting

*Stefan Schaltegger and Tobias Hahn*
*University of Lueneburg, Germany; E-mail: Schaltegger@uni-lueneburg.de, Tobias-Hahn@uni-lueneburg.de*

*Roger Burritt*
*The Australian National University, Australia; E-mail: Roger.Burritt@anu.ed.au*

## 14.1. Introduction

The last decade has witnessed a considerable consolidation of the concept of corporate environmental accounting (see e.g. Gray et al., 1993; EPA, 1995a, b; Schaltegger et al., 1996; Fichter et al., 1997; Fischer et al., 1997; Tuppen, 1996; Schaltegger and Burritt, 2000). Within the framework of environmental accounting, environmental management accounting (EMA) represents internal company accounting and reporting for company-related environmental impacts (see e.g. BMU and UBA, 1996; Bennett and James, 1998; Bouma and Wolters, 1999; Bartolomeo et al., 2000; Schaltegger et al., 2000; Burritt et al., 2001). Environmental costs in companies are still frequently underestimated and remain hidden in overhead accounts. By enabling managers to track, trace, and allocate environmental costs, EMA helps to visualize the full amount of these costs as well as to calculate the complete profitability of environmental protection measures (see Schaltegger and Müller, 1997). However, many companies have not implemented EMA in spite of the various advantages flowing from its application in companies.

Government bodies at different administrative levels may find it desirable to promote the widespread application of EMA as an economically viable approach to environmental management in order to enhance corporate eco-efficiency. Within this context, the United Nations Division of Sustainable Development (UNDSD) has launched an initiative to enhance the role of governments in the promotion of EMA (UNDSD, 2000). This article investigates the different approaches that governments can adopt to stimulate the use of EMA in companies. Its investigation will be conducted by addressing the following three core questions about the relationship between governments and corporate EMA:

- *What* is environmental management accounting (EMA)?
- *Why* should governments stimulate the use of EMA in companies?
- *How* can governments best stimulate the use of EMA in companies?

The article is structured according to these three questions. After giving a common and broad definition of EMA (section 2), it outlines the motivation for government activities in the field of EMA promotion (section 3). The main focus of the article is on the discussion of different approaches of government activities for the promotion of EMA and the proposition that a strategic policy approach be adopted (section 4). Finally, section 5 points out the shortcomings and limitations of the proposed policy approach, and comments on the outlook for further and more far-reaching government activities.



*M. Bennett et al. (eds.), Environmental Management Accounting: Informational and Institutional Developments,* 187–198.
© 2002 *Kluwer Academic Publishers. Printed in the Netherlands.*

## 14.2. What Is EMA? – A common definition and framework of EMA

In the past, environmental accounting literature has defined environmental management accounting (EMA) in two fundamentally different ways. In the first approach EMA is represented by internal environmental accounting based only on the use of a monetary measure (see e.g. Schaltegger et al., 1996; Schaltegger and Burritt, 2000). The second approach accepts that EMA includes both monetary and non-monetary measures of internal accounting phenomena (see e.g. Bennett and James, 1998; IFAC, 1998; UNDSD, 2000, p. 39), thereby reflecting a more encompassing term for corporate internal environmental accounting.

Recently, a framework for a common definition of EMA has been proposed (see Schaltegger et al., 2000; Burritt et al., 2001). This definition adopts the broader view encompassed by the second approach, while at the same time maintaining the clear distinction between monetary and physical measurement tools that is proposed by the first approach. According to this definition, EMA is defined as a generic term that includes Monetary Environmental Management Accounting (MEMA) and Physical Environmental Management Accounting (PEMA), as illustrated in Figure 14.1. Thus, EMA is defined as internal company environmental accounting expressed in monetary and in physical units.

*Monetary Environmental Management Accounting (MEMA)* is the central accounting source of information for most internal management decisions, and addresses the tracking, tracing, and allocation of environmentally-induced costs and benefits. MEMA is an accounting system for the financial impacts of environmentally-induced activities. *Physical Environmental Management Accounting (PEMA)* also serves as an information tool for internal management decisions. However, in contrast with MEMA, it focuses on a company's impact on the natural environment, expressed in terms of physical units such as kilograms or joules. It addresses the classification, collection and recording of environmental impact information in physical units for internal use by management.

> *EMA can therefore be defined as including the range of accounting tools which are used for internal company purposes and which deal with environmental issues in monetary or in physical terms.*

The range of specific EMA tools can be distinguished in three ways:

- First, EMA can be distinguished into *past-* and *future-oriented* tools (see outer column on the left side of Figure 14.1).
- In addition, the accounting tools can be classified according to their perspective into those with a focus on the *short term*, and those focusing on the *long term* (see upper white row in Figure 14.1).
- Finally, from the viewpoint of internal management decisions, both past and future as well as short and long term orientated tools can be further distinguished into those used for the *generation of routine information* (accounting tools that routinely produce information for management) and those used for *ad hoc generation of information* (accounting tools that produce information for particular decisions) (see inner column at the left of Figure 14.1).

Any promotion of EMA needs to consider the different decision contexts and uses as well as the strengths and weaknesses of the various EMA tools.

| | | Environmental Management Accounting (EMA) | | | |
|---|---|---|---|---|---|
| | | Monetary Environmental Management Accounting (MEMA) | | Physical Environmental Management Accounting (PEMA) | |
| | | **Short-term focus** | **Long-term focus** | **Short-term focus** | **Long-term focus** |
| **Past Oriented** | Routinely generated information | Environmental cost accounting (e.g. variable costing, absorption costing, and activity based costing) | Environmentally induced capital expenditure and revenues | Material and energy flow accounting (short term impacts on the environment – product, site, division and company levels) | Environmental (or natural) capital impact accounting |
| **Past Oriented** | Ad hoc information | Ex post assessment of relevant environmental costing decisions | Environmental life cycle (and target) costing<br>Post investment assessment of individual projects | Ex post assessment of short term environmental impacts (e.g. of a site or product) | Life cycle inventories<br>Post investment assessment of physical environmental investment appraisal |
| **Future Oriented** | Routinely generated information | Monetary environmental operational budgeting (flows)<br>Monetary environmental capital budgeting (stocks) | Environmental long term financial planning | Physical environmental budgeting (flows and stocks) (e.g. material and energy flow activity based budgeting) | Long term physical environmental planning |
| **Future Oriented** | Ad hoc information | Relevant environmental costing (e.g. special orders, product mix with capacity constraint) | Monetary environmental project investment appraisal<br>Environmental life cycle budgeting and target pricing | Relevant environmental impacts (e.g. given short run constraints on activities) | Physical environmental investment appraisal<br>Life cycle analysis of specific project |

Figure 14.1. Framework of Environmental Management Accounting (EMA) (see Schaltegger et al., 2000; and Burritt et al., 2001).

## 14.3. Why should government promote EMA? – Motivation and relevance of government activities

The main stimulus to integrate (monetary and physical) environmental aspects into internal business decision-making and accountability processes is the growing evidence that environmental factors can have a substantial effect on the profitability and environmental performance of a business (Bartolomeo et al., 2000, p. 35). However, in spite of the fact that most of the direct benefits derived from the use of EMA by business managers remain within the company, governments may also have a considerable interest in promoting the widespread application of EMA in the private sector:

- Fundamentally, EMA serves to *integrate environmental aspects into mainstream business decision processes* (Bennett and James 1998, p. 33). By providing transparency about company-related environmental impacts on nature as well as on the financial position of a business, EMA enables managers to take environmentally and economically sound decisions.
- EMA helps governments to *achieve pollution prevention at minimal cost to government and with minimal political resistance* (UNDSD, 2000, p. 15). Once business decision-makers have full transparency about their environmental costs and benefits and the environmental impacts related to their activities, they are able to implement the full range of environmental measures which are economically favorable.
- In addition, EMA can *increase the effectiveness of new environmental policies* (UNDSD, 2000, p. 15). Companies that use environmental management accounting are in a better position to comply with environmental regulations and have greater awareness of the benefits from undertaking further reductions in environmental impacts that go beyond compliance with legislation.
- Furthermore, EMA provides an appropriate basis for *transparency and accountability of businesses, and encourages companies to act in the public interest*. Only those companies that have created the necessary internal information system are able to demonstrate responsible behaviour in relation to corporate environmental impacts and to provide high quality information to third party stakeholders.
- Finally, EMA is an economically viable approach designed to encourage greater environmental protection and to *enhance eco-efficiency of industry* (Schaltegger and Sturm, 1990, p. 274). Governments may also wish to promote the more widespread use of EMA in companies to achieve a more eco-efficient performance of the macro level economy.

These benefits illustrate that EMA serves the interests of private sector businesses and also encourages the involvement of governments. Once governments at different administrative levels have recognized that these benefits are available from the broad use of EMA throughout business, they may decide to play a more active role in its promotion. A number of activities in different countries, as well as the initiative taken by the UNDSD, have shown an increasing commitment of governments to the promotion of EMA (see e.g. UNDSD, 2000).

## 14.4. How Can Government Promote EMA? – Strategic policy and policy approaches

### 14.4.1. *Fundamental problems of conventional policy approaches*

The third core question is how governments could best promote the use of EMA in companies. Conventional policy approaches that are used to formulate and implement government programs are characterized by a number of deficits and shortcomings (see Cohen and Kamieniecki, 1991, p. 9):

- Typically, public policies are *incremental and partial* rather than *rational and comprehensive*. In the main, policy-making adopts a reactive position with the aim of moving *away from* problems rather than *towards* solutions;
- Conventional policy approaches do not take appropriate *views of different actors* into account, and instead are based on the logic of regulatory bodies. They are single view-

based. As a consequence, many policies take into account the interests of only the policy-makers and regulatory bodies;

- Consequently, there are a considerable number of *symbolic policies and regulations*. They lack a clear, problem-oriented goal. The main goal of such policies is to demonstrate activity and to substantiate and secure the power of the policy-makers rather than to solve problems, or effectively influence any target group's behaviour.
- In addition, many policies and regulations suffer from a *lack of enforcement* because of government resource restrictions. Thus, many policies are formulated by regulators but are never actually implemented by corporations. This leads to the result that even well designed policies often contribute little towards solving real world problems.
- In the majority of cases governments are not in a position to judge the extent to which their policies have led to the desired outcome. This is because very few policy programs are evaluated using an *effectiveness control* program. Consequently, government policy-makers have little feedback about the results of their efforts. This leads to a lack of control over the policy-making process.

This fundamental approach can be characterized as 'muddling through'. In order to overcome these fundamental problems and the shortcomings of this widely used conventional policy approach, there is an urgent need for a strategic policy approach to be adopted.

### 14.4.2. *Strategic policy approach*

The overall strategic goal of any policy is to influence the behaviour of a specific target group in a desired direction, i.e. to achieve compliance with well defined policy goals. A strategic approach to policy-making represents a systematic process that analyzes the different methods available for influencing corporate behaviour in direction desired by policy-makers. It takes into account the positive and negative incentives of all the groups that are involved and affected. This strategic focus helps to overcome symbolic and poorly implemented policies by adopting problem-oriented policy formulation (Cohen and Kamieniecki, 1991, p. 12 and p. 27).

A strategic approach to policy-making is defined as a multi-step process designed to formulate, implement and control policies in an effective way, i.e. a way in which goals are achieved. In principle this approach is generically applicable to any policy area, although in this article it is applied to the promotion of corporate EMA. The strategic policy approach proposed consists of four major steps as illustrated in Figure 14.2.

#### *Definition of the Desired Outcome*

First, the desired outcome of a policy has to be defined. In this fundamental step, the target groups and the desired behaviour they are required to comply with have to be defined. This is a critically important step as it serves to outline the goals of the government activity. In the case of the promotion of corporate EMA, the desired outcome of any government initiative is a sophisticated and more widespread use of EMA by business. An exact and in-depth definition of good EMA practice can be taken from the literature (see e.g. BMU and UBA, 1996; Bennett and James, 1998; Schaltegger and Müller, 1997; Bouma and Wolters, 1999).

#### *Choice of Links and Partners*

The choice of links and partners highlights the fact that the effectiveness of government programs and initiatives depends, to a considerable extent, on the actors and links involved

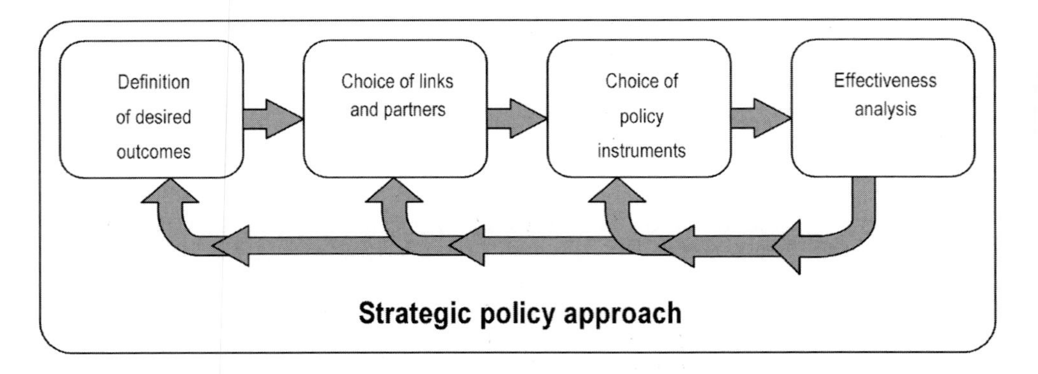


Figure 14.2. Model of the strategic policy approach.

(see Cohen and Kamieniecki 1991, p. 13 and p. 28; Burritt and Welch, 1997, p. 537 and Burritt, 2001). In this next step suitable pathways are selected through which the policy can be launched. In this context a *pathway or link* is defined as a path through which government can promote the use of EMA in companies. It refers to the relationship between different actors and decision-makers regarding EMA. The purpose of the choice of suitable links is to reveal those paths, leverage points, and stakeholders that appear to have the most potential for government promotion of EMA use from a structural viewpoint (for an in-depth analysis of EMA links see Schaltegger et al., 2001).

A fundamental point is that a distinction can be made between *direct* and *indirect* links between governments and potential users of EMA in companies. *Direct links* address the unmediated relation between governments, willing to promote corporate EMA, and managers. Direct links can be broken down to the level of specific MEMA and PEMA tools. The bi-focal relationship between government policy-makers and business managers is crucial for assessing the suitability of direct links as paths for the launching of policies. Thus, the different decision-making contexts and incentives of these two groups have to be taken into account.

On the other hand, *indirect links* engage intermediate elements and stakeholders in order to establish indirect ways of influencing corporate EMA use. Such intermediate elements include a range of corporate and national accounting systems (e.g. conventional financial accounting and reporting, and national economic accounting) as well as different management systems (such as financial management systems, management control systems, environmental management systems, etc.). This fundamental approach is outlined in Figure 14.3. For each intermediate element, specific stakeholder groups are identified. These intermediate stakeholders can be targeted by governments through any indirect influence that will encourage the use of EMA in companies. Because different accounting or management systems act as intermediate elements in indirect links designed to provide specific information to different groups of stakeholders in order to serve their interests, it is very important to take into account and to analyze carefully the viewpoints of these intermediate stakeholders, separate from the views of management as the main user of EMA and government agencies as promoters of EMA. Intermediate stakeholders that are both capable of and interested in contributing to the promotion of corporate EMA use can be considered as partners for government programs.

In order to assess the suitability of the different direct and indirect links for promoting

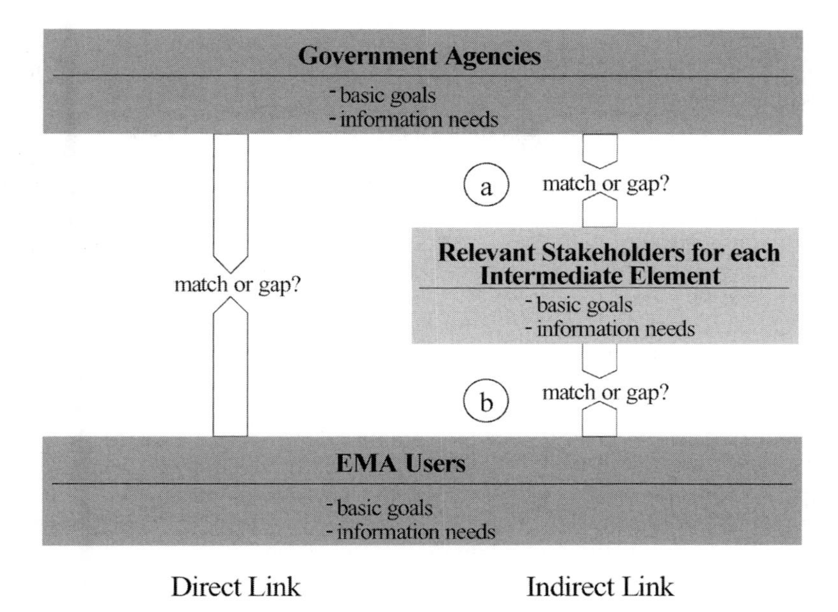


Figure 14.3. Direct and indirect links, and the integration of stakeholders (Schaltegger et al., 2001, p. 4).

corporate EMA, a set of analytical criteria is proposed. As argued above, links are strongly characterized by the groups of stakeholders involved. Therefore, the match or gap between the basic goals and information needs of these stakeholders represents the core analytical factor behind direct and indirect links. Consequently, all the stakeholders involved in any of the links being examined as a suitable path to promote EMA should be characterized according to their main general and EMA-specific interests, goals, and information needs.

For the analysis of direct links, the *interests, goals, and information needs* of different government agencies (as potential policy-makers) and of different corporate management departments (as potential users of EMA) are compared directly. For the analysis of indirect links, the same approach is taken with the only difference being that two separate partial links have to be addressed. Partial link (a) addresses the relationship between government policy-makers and the different intermediate stakeholders in an intermediate element. Partial link (b), on the other hand, refers to the complementary relationship between these intermediate stakeholders and potential EMA users in companies. The degree of match between these interests and information needs provides a first strong indicator about the suitability of the different direct and indirect links.

The *anchorage* of each direct and indirect link provides a further criterion for judging the suitability of a link as a promising path down which government policies for promoting EMA could be launched. Anchorage reflects two characteristics. On the one hand, it refers to the breadth of the match between interests and goals or information needs, i.e. it answers the question of how many stakeholders within government agencies, corporations, or – for indirect links – connections with intermediate elements, demonstrate a high degree of matching. On the other hand, the anchorage of a link also reflects the background of the stakeholders making up a link, that is, whether they are all concerned with a single issue or whether concern is over a diverse range of issues. Taken together, the anchorage criterion identifies the stability of links.

The assessment of the suitability of an indirect link cannot be judged solely according to the structural relations existing between the different stakeholder groups involved. In addition, prior to investigating the match of interests, goals, and information needs of the different stakeholders, as well as the anchorage of the links, the *method proximity* of each intermediate element with EMA has to be addressed. This criterion serves to reveal how close the methods and tools associated with any intermediate element are to the different EMA tools.

Together, these three aspects serve to identify the *crucial* links and critical paths, as well as the most promising partners for governments to cooperate with in order to promote the use of EMA in companies in a successful way. An extensive analysis of EMA-links, based on this approach at a generic level, has been conducted within the scope of the UNDSD initiative on governments' role in promoting EMA (see Schaltegger et al., 2001). The results of this work indicate that the *direct* link towards monetary environmental investment appraisal, and the *indirect* link via conventional financial accounting and reporting and external physical environmental accounting and reporting, should be the most promising paths for successful promotion of EMA by governments. The partners with the highest potential for successfully stimulating corporate managers to use EMA at the generic level seem to be industry associations and professional accounting associations. However, in the setting of more specific government policy-making, other paths and partners might be found to be more attractive.

### *Choice of Policy Instruments*

Once government policy-makers have identified the structural paths on which policies should be based, as well as the partners with whom they want to cooperate, the decision on the appropriate policy instruments has to be taken. From the viewpoint of the strategic policy-making process, policy instruments represent the technical means through which the desired behaviour of the targeted groups should be achieved. Based on this function of policy instruments, two core aspects of the choice of appropriate policy devices can be identified. First and foremost, the technical functionality and ability of any policy devices to influence behaviour in the desired way has be discussed. Second, the conditions for successful deployment of the behaviour-influencing potential of each suitable policy instrument has to be addressed.

A broad range of different policy instruments is available to government policy-makers. Knowledge of differences in the degree of coercion exerted, legal implications and complexity of policy devices provides the basis upon which functional instruments are chosen for the purpose of achieving desired outcomes. The following list provides an overview of potential policy instruments (see Cohen and Kamieniecki, 1991; pp. 69–84; Pearce and Turner, 1999; Tietenberg, 1992):

- Market solutions and economic incentives;
- Self-regulation;
- Taxes, fees, and other negative incentives;
- Education and (mandatory) information disclosure;
- Voluntary agreements;
- Licensing;
- Permits;

- Standard setting;
- Penalty setting;
- Inspections, etc.

The core question to be asked when choosing the most suitable policy devices is: does the policy instrument being considered have the necessary potential to bring about the desired behaviour of the target groups? Only those instruments which are considered to have the potential to motivate companies to apply EMA should be adopted for further analysis.

Technical functionality of a chosen policy device is a necessary precondition in order to achieve the desired compliance behaviour. In order to ensure that the actual effect on the behaviour of the target groups is in line with the desired effect, the factors which can influence how well a policy instrument works have to be analyzed (Cohen and Kamieniecki, 1991, p. 50). This step requires further stakeholder analysis. To avoid the common problem that environmental policies are designed with the viewpoint of only policy-makers in mind, all stakeholder groups affected by, or involved with, the intended policy program need to be considered. The following groups may be analyzed (see also Cohen and Kamieniecki, 1991, p. 35):

- *Policy-makers:* Issuers of government programs cannot be considered as a black box. Instead, their role and decision-making context (e.g. environmental department, treasury, etc.) represent an important factor in the implementation of effective policies for the promotion of corporate EMA use.
- *Target groups:* The parties to whom government programs are addressed (the addressees) complement the various government policy-makers. The desired outcome of any policy program, as defined in the first step, addresses these groups. As the achievement of the desired outcome usually requires a change in the behaviour of the addressees they have to be carefully analyzed. Within companies, the target groups for EMA promotion are various corporate EMA users such as product management, production management, investment managers, and top management.
- *Affected groups:* This group includes any third parties (i.e. non-government and non-target groups) that are affected positively or negatively by the policy that is to be implemented, i.e. the 'winners' and 'losers' from government programs. These groups can exert considerable influence over the success and effectiveness of policy programs.
- *Cooperation groups:* The partners in government programs have already been identified through the analysis of links in the previous step of the strategic policy making process. However, even if the chosen partners reveal considerable complementarities with the overall goal of promoting corporate EMA, they still have to be examined more closely in relation to the chosen policy device.

The reactions to a policy program by the groups affected have a considerable influence on its success, effectiveness and efficiency. The desired outcome does require a behavioural change from these target groups and, in addition, all other parties involved have to act favorably towards or, as a minimum, act in a neutral way to the policy. Major reluctance or resistance to the policy, even from third party groups, can impede the successful promotion through the chosen policy device of corporate use of EMA. Policy-makers should therefore try to predict the expected responses of each group to the stimulus set out by the policy program and to understand the motivations or reasons for their responses. All identified groups have therefore to be assessed in respect of their strengths, weaknesses, resources, and motivation. The following core questions can be used to analyze the likely response of each of the groups listed above to a specific policy instrument designed to promote the wider use of EMA in companies (see Cohen and Kamieniecki, 1991, p. 35; Schaltegger, 1999):

- *How well organized is each group?*
  Only well organized groups can bundle their resources in an effective way. The degree of organization (see, e.g. Becker, 1983; Olson, 1968) represents an important indicator of the ability of a group to resist or push a policy for the promotion of EMA.
- *How powerful is each group?*
  The relative power of each group is another factor that has to be taken into account when analyzing the probable response of groups involved to any policy program. The power of a group can be derived from different sources, such as political power, economic strength, strong network connections, etc.
- *How susceptible is each group to influence?*
  The susceptibility of a group to influence addresses the willingness or openness of a group to accept new behavioural concepts. A high susceptibility to influence can be either a sign of strength, when a group is forward-looking and innovative or eager to learn, or a sign of weakness, in the case of a low potential for resistance to being influenced.
- *What motivation to promote, oppose, ignore, or comply with a policy instrument does each group have?*
  This criterion refers to the major goals of each group and their relation to the purpose of the policy. It addresses the question of what incentives exist to promote action in a way that is favorable to the policy to be implemented, or to promote action that opposes it.
- *What resources does each group have available to promote, oppose, ignore, or comply with a policy instrument?*
  The amount of resources available to each group involved is a core factor of the intensity of its expected response to a specific government policy. Even if a group has a strong motivation and preferences in favor of or against the policy, and is highly organized, a lack of resources considerably weakens its position.

Such a prospective analysis of the probable response to a planned policy by the groups involved empowers government policy-makers to choose and design adequate policy instruments that will successfully promote the use of corporate EMA. This approach fundamentally overcomes the difficulties of the conventional approach to 'muddling through', based on the singular perspective of government policy-makers and often resulting in ineffective policies.

### *Effectiveness analysis*

Even if a strategic approach to policy-making provides considerable advantages compared with the conventional approach to policy-making, the structural investigation of policy links and the prospective analysis of party responses provide no guarantee that the desired outcome will be achieved. Therefore, it is necessary to undertake an effectiveness analysis in order to control for whether the target group's desired change of behaviour is actually observed and whether assumptions made about the likely responses of the groups involved have turned out to be correct. Such an effectiveness analysis can be conducted in three major steps through the analysis of aspects of policy related to *confining*, *structuring* and *checking* (see EEA, 1998; UNEP, 1998; OECD, 1999; Industry Canada, 2000; Burritt, 2001a, b).

*Confining* is part of a policy agreement designed to ensure that specific actions are undertaken to achieve desired outcomes. This includes support from top management and rank and file employees, as well as the buy-in of other stakeholders who will be influenced by the proposed policy. It also requires the specific identification of measurable goals, both primary and secondary, to be achieved. Finally, identification of specific sets of negative and positive drivers of desirable behaviour will be introduced in order to reward goal achievement, or penalize non-achievement.

*Structuring* relates to the general and specific constraints imposed on stakeholders associated with a proposed policy. For example, should a formal letter of intent or an agreement be signed by all the parties? Should only a certain size of company be targeted (e.g. large companies, or small and medium-sized companies)? What sectors should be constrained – a limited number of heavy polluters, or environmental innovators, all companies in a certain sector, domestic behaviour or international behaviour by companies with an overseas presence?

*Checking and monitoring* is required to provide feedback about performance achieved by stakeholders, especially environmental and economic performance of corporations in the light of established policy. It also provides an indication of whether actions have been taken in accordance with any agreements made. Included are considerations about the extent, periodicity and contents of reporting of results, giving consideration to the possible importance of public reporting. The question of independent verification of results and performance is also considered here. Questions arise of whether verification should be compulsory, whether the verifier should be internal or external, and how the verifier should be selected to demonstrate independence.

Effectiveness analysis helps to ensure that key issues are addressed before policies are introduced (e.g. as part of an ex ante integrated assessment), in the review of progress of a policy tools (as part of continual monitoring of the tools – *ex itinerere* assessment), and as post agreement analysis to check on lessons learnt that could be useful to support and improve future policy (ex post assessment) to promote corporate EMA. Consequently, careful application of effectiveness analysis to government policy choices can lead to better strategic control of policy outcomes than the conventional approach to policy-making (for an application see Burritt, 2001b).

## 14.5. Limitations and outlook

It has been argued in this paper that a strategic approach to policy-making would provide a number of substantial advantages compared with the 'muddling through' of conventional policy formulation and implementation. Such an innovative strategic approach could help to overcome the main problems of conventional policy-making and thus promises to make government policies more effective. Governments that want to promote the successfully use of EMA in companies should therefore adopt a strategic policy approach. By defining the desired behaviour of the target groups clearly (more frequent use of EMA by companies in decision-making and accountability processes) and subsequently identifying the most promising links and partners, choosing appropriate policy devices based on a multi-group perspective and finally reviewing the factual results of the policy chosen to be implemented, government initiatives to promote EMA may demonstrate considerably improved results.

However, the strategic policy approach proposed in this article also has limitations (see, e.g. Buchanan and Tullock, 1975; Hahn, 1989; ibid, 1990; Hahn and McGartland, 1989). The motivation of each actor to support the promotion of corporate EMA by the government depends very much on the institutional framework faced by the public actors. The potential guidance provided by the approach of strategic policy-making is limited by institutional incentives (see, e.g. Eisenhardt, 1989; Krueger, 1974).

# 15. Looking for Knowledge Management in Environmental Accounting[1]

*Dick Osborn*
*Green Measures (a consultancy in environmental accounting policies, programs and practices in 1999); E-mail: rosborn@greenmeasures.com.au*

## 15.1. Introduction

The Eurobodalla Shire Council represents a community of some 30,000 persons living on Australia's southeast coast. Its corporate goals include regional leadership, to be a good employer, customer satisfaction, asset maintenance, environment protection, competitive services, business investment, and sound financial management. Much (80%) of this authority's administrative area is covered by state forest or national parks, and is therefore unrateable. Tourism, timber harvesting, fishing including oyster farming, and offering quality of life as a locational advantage to residents, drive much of the community's economic activity. The local authority operates a budget of some $A50M, employing some 400 persons.

An effective environmental information system is a key element to achieving this Council's corporate goals. It has not however invested in a custom-built system to store and transform environmental data into information for its decision makers. Superficial observations of the Council's information system suggest the hardware and software configurations used are not exceptional within Australia's local government industry.

But closer examination reveals other system elements that do seem exceptional within and outside local government. The Eurobodalla Shire Council uses at least two uncommon ways to achieve its corporate goals. First, this local government sees compliance as an opportunity rather than an impost. Second, this Council gives priority to respecting, valuing and building local knowledge above other sources in meeting the diverse information needs of its stakeholders.

Operators of this local government's information system generate compliance reports for many national and state government agencies within Australia's federal system. The system must be capable of storing and analysing data from which to report information, including but not limited to:

- demonstrating that some services (and therefore specific tasks or activities) are open to market competition;
- estimating expenditures on environment protection;
- enabling comparative analysis of the authority's revenue-raising capacities and of disabilities encountered in providing services;
- maintaining accounts on an accruals basis;
- reporting the condition of public works;
- reporting environmental condition through an annual State-of-Environment-Report using Pressure-State-Response guidelines; and

---

[1] Acknowledgements: the support of colleagues in the Eurobodalla Shire Council and in the Australian Bureau of Statistics through permission to use unpublished material is gratefully acknowledged.



*M. Bennett et al. (eds.), Environmental Management Accounting: Informational and Institutional Developments,* 199–213.
© 2002 *Kluwer Academic Publishers. Printed in the Netherlands.*

- reporting how the Council's management plan seeks to improve environmental and social conditions existing within its jurisdiction.

The local authority structures the accounting frameworks and other components of its information systems under capital stock headings (natural, built, human) to store sets of physical and financial measures for assessing the impacts of:

- environment on asset;
- environment on human;
- asset on environment;
- asset on human;
- human on asset; and
- human on environment.

Examples of stakeholders sharing local knowledge include but are not limited to:

- a retired municipal engineer scoring and recording the condition of the authority's built assets;
- university students gaining practical experience while at the same time providing oyster farmers and other stakeholders with water quality measures that assist in minimising commercial and environmental risk;
- schoolchildren surveying bio-diversity;
- scientists retired from the research institutions of the nation's capital evaluating the capacity of sensitive forest environments to absorb development; and
- environmental interest groups in the local community increasing from less than 10 to more than 60 over the recent past.

One outcome then of the Council's efforts in stakeholder consultation and in building local knowledge has been to increase significantly the in-kind resources available for environmental restoration and repair. Active information-sharing has also meant that the Council has been able to increase its environmental revenues in two ways. First, the local community supported the Council's application to the state government to increase the environmental levy that it is authorised to impose on rateable property. Second, the Council has competed with considerable success for central government grants through demonstrating clear understanding of local issues, needs, and capacities in environmental and asset management.

The local government, its community and some external stakeholders participate therefore in a process of *knowledge management*. They are committed to the active sharing and transferring of explicit environmental knowledge stored mechanically within information systems, and of the tacit environmental knowledge that resides in their minds. They are achieving bottom-line and other benefits from doing so.[2]

Identifying, collecting, organizing and sharing are critical elements in knowledge management (O'Dell and Jackson Grayson Jr., 1998). Businesses are encouraged to gain competitive advantage in their markets through environmental knowledge management (Ford, 2000). Global prospects for sustainable development are also seen as benefiting

---

[2] A case study identifying the development of intelligence products in Environmental Management Accounting by this Council, and their application to decision-making, is located at http://www.greenmeasures.com.au as *How Environmental Management Accounting Supports the 'Good Government, Better Living' vision of the Eurobodalla Shire Council, New South Wales.*

from sharing experiences and ideas through environmental knowledge management (Serageldin et al., 1998).

Practices in environmental accounting *can support national income accounting, financial accounting, or internal business managerial accounting* (US EPA Environmental Accounting Project, 1995). Scholar-practitioners familiar with developments across this range and with the many disciplines involved frequently use the *macro* and *micro* prefixes which are familiar to economists, to differentiate environmental accounting practices at the scale of the economy from those at the scale of the firm.

Sharing ideas and developments between macro and micro scales of environmental accounting is consistent with arguments espousing the benefits of knowledge management. The functions of environmental accounting practice centre on informing policy and evaluating performance at both scales. But discourse on environmental accounting is usually presented in packages that separate rather than share ideas between macro and micro scales of environmental accounting (Uno and Bartelmus, eds., 1998; Bennett and James, eds., 1998; Simon and Proops, eds., 2001).

The purpose of this paper is to provide examples of how businesses can benefit from using elements of macro-scale environmental accounting. It does so through reflecting on experiences and opportunities in Australia over the recent past that should be of interest to a wider audience. For example, the European Commission is now advocating that companies uses elements of macro-scale environmental accounting practice to compile information for their annual reports and accounts (European Commission, 2001). The Commission's advocacy adds support for a macro-micro link in environmental accounting and, through the sharing of information, for knowledge management.

## 15.2. Management units contributing to Macro Environmental Accounting

### *Management Units*

Imagine that a national government uses compliance to collect data from many organizations on a financial variable important to analysing its own environmental policies. Environment Protection Expenditure (EPE)[3] is a good example. Assume that over the period of collecting EPE data the government identified organizations that, like Eurobodalla Shire Council, see compliance as an opportunity rather than an impost.

What if that government then accepted the evidence that its regulatory instrument for aggregating EPE to national scale could also become an information, education and training instrument? And what if the government considered that information, education and training associated with collection of EPE data could support programs promoting the adoption of environmental accounting practice in industry? Then what could be the size of a target population in such programs? How many management units provide their EPE estimates to statistical collectors?

A time series of national EPE estimates published by the Australian Bureau of Statistics began with fiscal 1990–91.[4] Estimates of EPE by household management units are derived

---

[3] Expenditure on environment protection activities deals with those activities *where the primary purpose is the protection of the environment, that is the avoidance of negative effects on the environment caused by economic activity.* Voorburg draft of SEEA 2000 at http://www4.statcan.ca/citygrp/london/publicrev/chv-vb.pdf.

[4] Australian Bureau of Statistics, 4603.0 *Environment Protection Expenditure, Australia (Previously Cost of Environment Protection, Australia: Selected Industries)*, Canberra, Australia.

from modelling techniques applied to data collected through a periodic Household Expenditure Survey. Farms with a gross value of annual agricultural production below a certain threshold are excluded from EPE data collections,[5] as are non-employing businesses. This leaves some 665,000 management units in Australia targeted by the statistical collector for their EPE estimates.

The probability that a target unit will be requested to compile EPE estimates varies over time and by industry class. They will vary over time between annual to periodic collections. The EPE collection in Australia will also vary between a census and a sample of the population of management units, according to industry sector.

EPE estimates for management units in general government agencies of national and state/territory governments are derived from direct interpretation of their budget documents, annual reports and other published sources. A purpose-built instrument is used to collect EPE estimates directly from a 50% sample of local authorities. For all other industries, their EPE estimates are by-products of other collections.

A distribution by industry class of some 20,000 management units in Australia compiling EPE estimates follows.

### *Macro Environmental Accounting*

Some national governments use the OECD's transaction groupings under major headings of Pollution Abatement Control (PAC) and Non-Pollution Abatement Control (non-PAC) as a framework for compiling EPE estimates. Most international government agencies have adopted the UN's 1993 System of National Accounts (SNA) as a common framework for constructing and presenting macro economic accounts. A following of that precedent is anticipated with the UN's System for Integrated Environmental and Economic Accounting (SEEA) 2000[6] becoming the international standard for macro environmental accounting. The position of the EPE account within the framework of SEEA 2000 appears in Figure 15.1.

## 15.3. Starting small in Micro Environmental Accounting: Physical or financial metrics?

Practicing knowledge management in environmental accounting involves more than breaking down barriers between macro and micro scale approaches. It is difficult to articulate any significant and practical differences between environmental accounting and environmental performance measurement.[7] Scholar-practitioners with backgrounds in Geographic Information Systems (GIS) are producing their own environmental accounting

---

[5] That threshold was around $A25, 000 at the time of preparing this paper.

[6] All sources of information on SEEA 2000 are taken from the London Group's Website at http://www4.statcan.ca/citygrp/london/london.htm, unless cited otherwise.

[7] Sikklud and Wennberg are among those noting that distinctions between practicing environmental management accounting and environmental performance measurement are difficult to identify; at http://www.lu.se/IIIEE/research/management/asa-ccc-report.html.

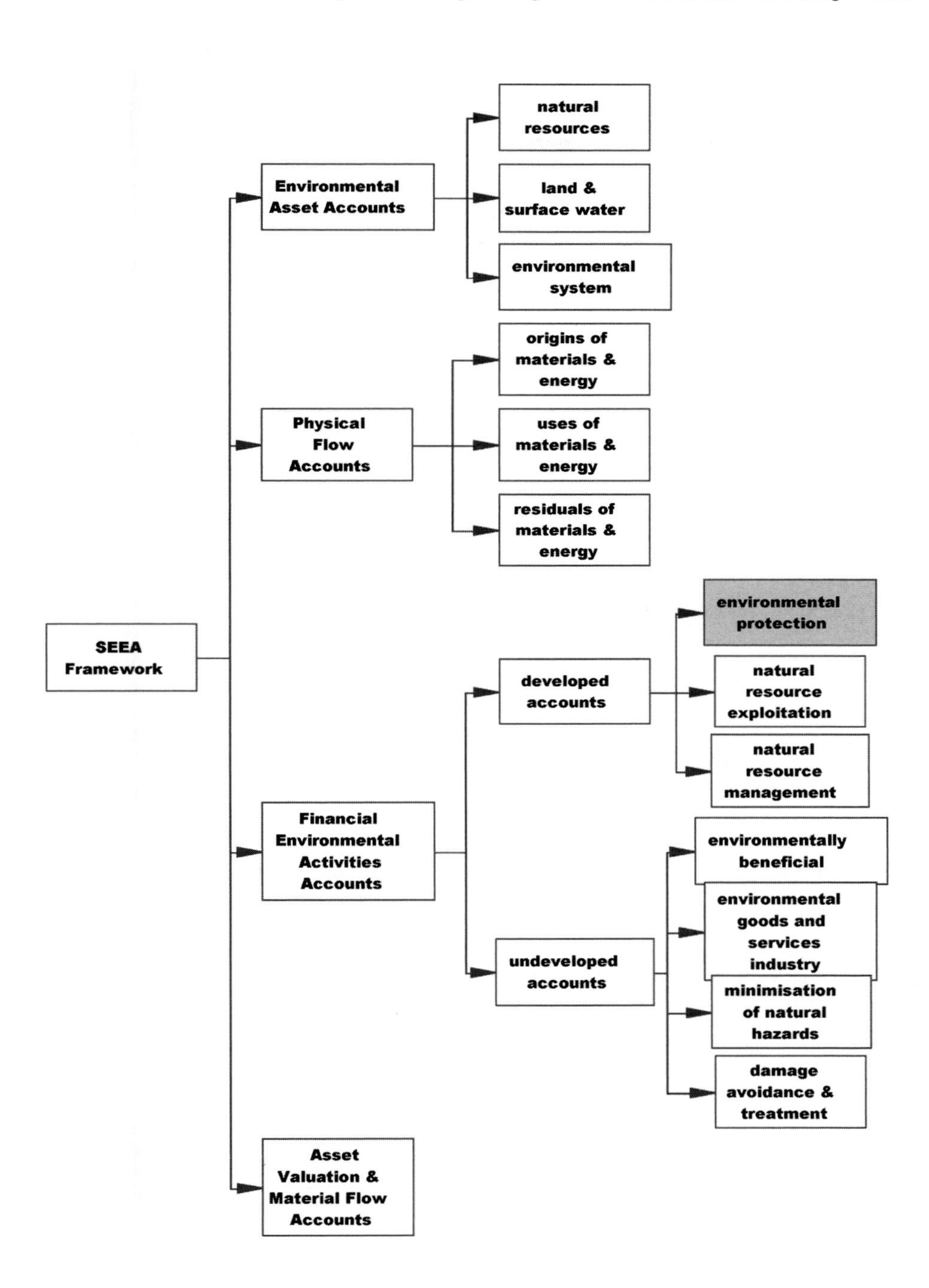


Figure 15.1. EPE within the SEEA 2000 framework.

tools,[8] as are those from, say, ecological economics. Growth in the number of environmental accounting and performance measurement tools being developed is making consumer choice more complex. In turn complexity slows down rates of adoption by firms, and inhibits progress towards more effective knowledge management.

Some individuals or firms advising entities on how to begin their environmental accounting or environmental performance measurement practice are sensitive to this growing complexity. Their advice is to start small. But with what? Should physical or financial metrics be used to embark on managing the organization's environmental performance by measuring? And what standards are available to benchmark results over time or space?

### *Emerging standards for physical metrics*

A number of national governments have initiated work on developing standard physical metrics for environmental accounting practice (environmental performance measurement).[9] Standard physical metrics suggested for measuring and benchmarking the environmental performance of key sectors within the manufacturing industries of APEC countries are identified in Figure 15.2.

### *Emerging standards for financial metrics*

Figure 15.1 identifies 'environmental protection', 'natural resource management', and 'natural resource exploitation' as sub-accounts developed to date within the 'financial environmental' activity accounts of SEEA 2000. 'Developed' in this context signifies progress on articulating 2 or 3 digit classifications within SEEA 2000's financial environmental accounts. Figure 15.3 provides a deeper perspective by identifying main elements within the most developed of these sub-accounts – the Environment Protection Account. The UN's Division of Sustainable Development has recently published a workbook on environmental management accounting principles and practice.[10] That workbook advocates using the statistical standards laid down within the financial environmental accounts of SEEA 2000 as the vehicle for an entity to assign its costs to environmental media.

## 15.4. Immediate opportunities for knowledge transfer in Australian environmental accounting

The work program of the Environment and Energy Statistics Section of the ABS necessarily centres on providing environmental information for decision-making at national and international scales. Its collection processes do however provide knowledge management opportunities for many thousands of organizations through active sharing of corporate data that has been placed in the public record.

---

[8] Knowledge management may also be described as action research processes based on collaborations between academia, business, government and community representatives. In addition to the obvious such as 'environmental performance measurement' and the work on indicators, the search terms on the Web that can lead to opportunities for knowledge transfer between environmental accounting practitioners and others with similar interests include 'collaborative spatial decision making', 'environmental informatics', 'ecoregional planning' and 'integrated environmental assessment'.

[9] A review of recent EU-developments towards standard metrics to communicate with stakeholders is Technical Report 54 (2001). A companion for developments in the Asia-Pacific region is provided by the Government of Japan's Ministry of the Environment (MOE).

[10] See chapter 3 of this book.

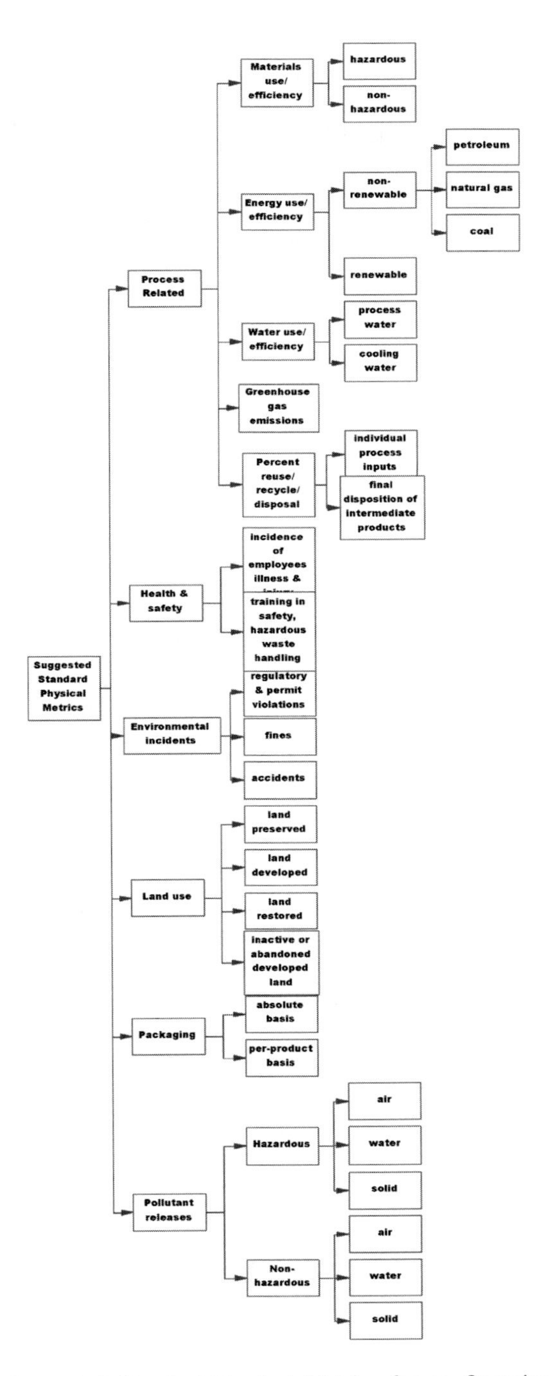


Figure 15.2. Some Suggested Standard Physical Metrics for an Organization's Environmental Accounting.

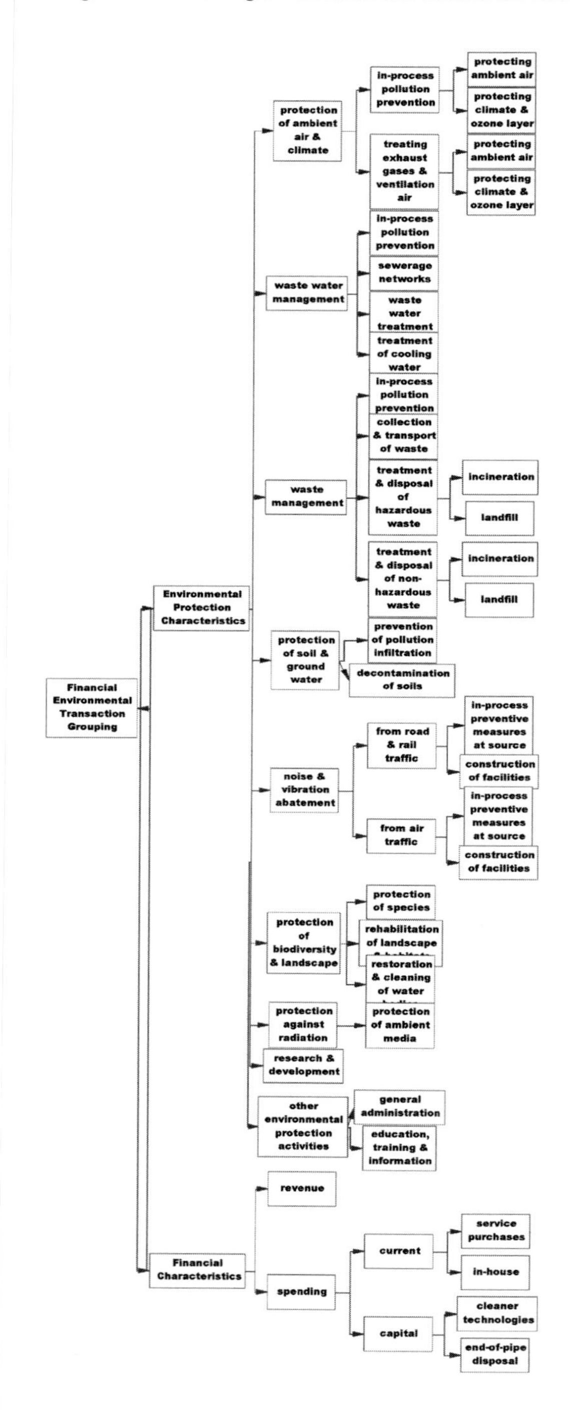


Figure 15.3. Transaction Groupings in SEEA 2000 Environment Protection Account.

### *Industry average benchmarking for manufacturing*

During the mid-1990s some 5,000 management units within Australia's manufacturing sector compiled EPE estimates, as indicated in Table 15.1. They did so by extracting data conforming to SEEA guidelines from their management accounts.

During the mid-1990s the ABS also undertook an Economic Activity Survey (EAS). Within Australia's manufacturing industry some 20,000 management units compiled estimates of their business operating expenses for the EAS, identifying some 200 variables.[11] Operating expenses incurred by management units in their purchases of materials, packaging, energy and water were among the variables identified for the EAS. Such purchases are seen as potentially harmful to the environment, consistent with concepts of sustainable production and dematerialization. Conversely, the EPE estimates would be seen as potentially beneficial to the environment.

Both harmful and beneficial environmental spending incurred by the management units surveyed can be expressed as a share of turnover to facilitate comparison over time and space. The sampling techniques employed by the ABS to collect financial environmental data from the management accounts and budget documents of management units within the manufacturing sector also enabled differentiating between processes. Information built from corporate data aggregated across a sample of Metal Products businesses appears in Table 15.2. Expenses incurred on what can be tagged as 'environmentally harmful' purchases are compared with 'environmentally beneficial' purchases.

Similar ratios for other goods-producing industries, as well as goods-distribution and service industries, can be calculated from data archived within the offices of Australia's national statistical collector. Their publication and dissemination as industry averages could be used as benchmarks for individual management units, and to compare environmental performance over both time and space. Consideration of this information by government, industry or research institutions over the five years or so of availability in the public record

Table 15.1. Management Units Providing EPE Data – Australia – *circa 1995.*

| *Industry Group & Class* | *Total Management Units ('000)* | *Management Units Providing EPE Data ('000)* |
|---|---|---|
| *Goods-Producing:* | *236* | *13* |
| Agriculture, Forestry & Fishing | 115 | 3 |
| Mining | 3 | 3 |
| Manufacturing | 42 | 5 |
| Utilities (inc. Sewerage & Drainage) | 1 | 1 |
| Construction | 75 | 1 |
| *Service Industries* | *426* | *7* |
| *General Government* | *3* | *1* |
| TOTALS | 665 | 20 |

Source: Compiled by the author from published and unpublished data provided by the ABS.

[11] Processes of collection, analysis and dissemination of corporate data from manufacturing businesses by the ABS and the Australian Tax Office are described in Australian Bureau of Statistics (1998), *8205.0 Information Paper: Availability of Statistics Related to Manufacturing.* The Paper is downloadable as a pdf document through http://www.abs.gov.au.

Table 15.2. Comparing Environmentally Harmful & Beneficial Expenses: Metal Products Industry: Australia mid 1990s.

| *Transaction Grouping* | *Expense as % Turnover* |
|---|---|
| *Environmentally Harmful:* | *46.90* |
| Material Purchases | 40.91 |
| Packaging Purchases | 0.37 |
| Energy & Water Purchases | 5.62 |
| *Environmentally Beneficial:* | *0.47* |
| Environment Protection Services Purchases | 0.15 |
| Environment Protection In-House Current Costs | 0.03 |
| Beginning-of-Pipe Capital Spending | 0.10 |
| End-of-Pipe Capital Spending | 0.19 |
| *Harmful: Beneficial Ratio* | *100:1* |

Source: Compiled by the author from published and unpublished data provided by the ABS.Ratios between potentially harmful and beneficial environmental spending in Australia for a range of manufacturing processes are illustrated in Figure 15.4.

would demonstrate knowledge management in environmental accounting. No examples of environmental policy analysis for Australian manufacturing using the statistics available have been identified.

### *Building Environmental Performance in Local Government*[12]

The ABS commenced pilot studies on the feasibility of extracting EPE estimates from budget documents and other management accounts of Australia's local governments in 1996. The studies became a major component of an Applying Environmental Accounting Frameworks in Local Government project conducted in 1996-99 by the University of Canberra's Division of Management and Technology with funding from the Commonwealth Government's National Office of Local Government.

There are some 670 general-purpose local authorities in Australia providing environment protection services. Twelve of these volunteered for pilot studies on compiling EPE estimates in the project's first year, followed by forty-five in the second year. The ABS was able to publish experimental EPE estimates for the nation's local government sector during the last year of the project as 185 local authorities volunteered data extracted from their management accounts. The ABS moved in 2000 to mandatory reporting of their EPE estimates by a sample of local authorities. This step followed extensive consultations with councils participating in pilot studies and the 1997–98 voluntary survey.

The significance of local government within the national EPE collection is illustrated in Figure 15.5. The nation's spending on Environment Protection is distributed across some seven million households and around one million businesses operating across the private and public sectors. The local government industry's 670 entities represent around 0.008% of contributors but undertake one fifth of the nation's spending on Environment Protection.

---

[12] This section of the paper draws heavily on Dick Osborn (2001), *Finding a Win-Win in Municipal Environmental Accounting*. The document is downloadable from http://www.greenmeasures.com.au.

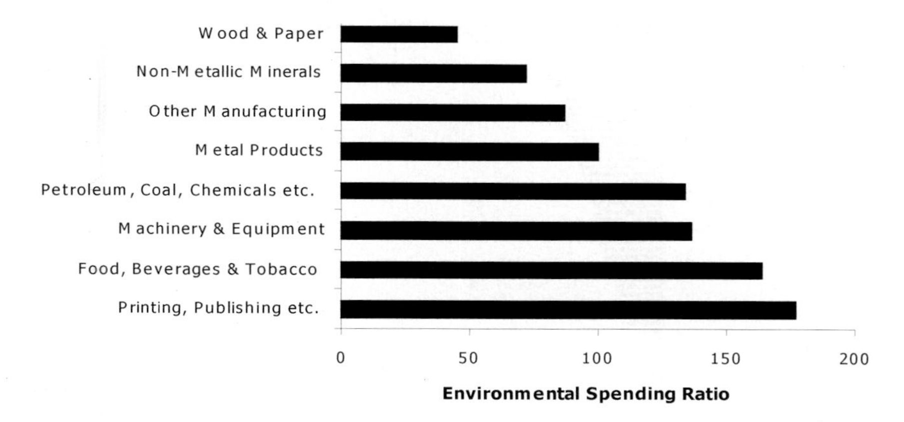


Source: Compiled by the author from published and unpublished data from the ABS.

Figure 15.4. Ratios of Environmental Spending: Manufacturing Processes, Australia, mid-1990s.

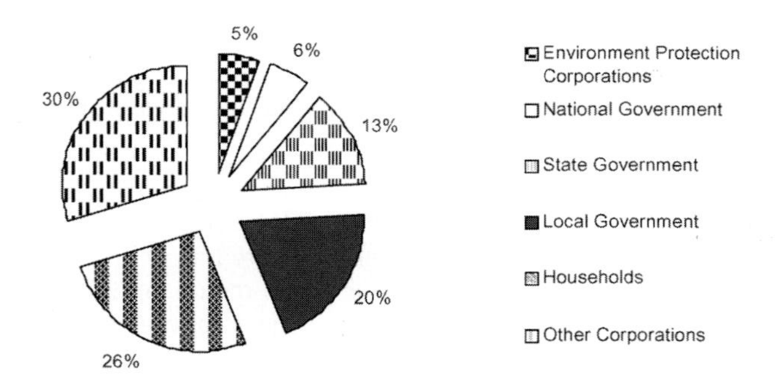


Source: Compiled by the author from data in ABS publications 4603.0 and 4611.0.

Figure 15.5. Sharing Environment Protection Spending: Australia circa 1998.

Building the capacities of local governments, and the communities they represent, to manage environmental knowledge is clearly in the national interest.

Fifty-one of the local governments collaborating with the ABS during the 1996–99 interval recently provided the author with insights into their knowledge management practices. They did so through a commission from the Commonwealth's National Office of Local Government to identify the frequency and nature of using their EPE estimates for internal decision-making.

The ABS estimates that the Australian local government sector spent some $A2.1 billion on Environment Protection during the late 1990s – equivalent to around 22% of industry outlay. A significant majority of the fifty-one councils prepared to share their experiences in compiling and using EPE were below that industry average benchmark (see Figure 15.6).

The 1997–98 collection of EPE data from Australian local governments was voluntary, but shifted to a mandatory collection the following year. This initiative led to major changes in knowledge management behaviour. In reporting for 1997–98, less than one fifth of the fifty-one councils prepared to share their experiences went beyond the collector in

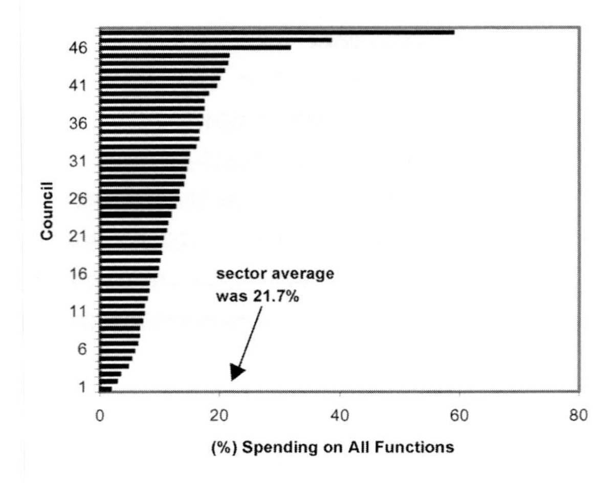


Figure 15.6. Environment Protection Expenditure in 51 Australian Councils.

reporting EPE results to stakeholders. In reporting results for 1998-99, the proportion of councils reporting beyond the collector to other stakeholders increased threefold. This significant change in behaviour signalled that the three C's of environmental performance measurement were being put in place:

- *Credibility* – the data is extracted from budget documents according to international statistical standards;
- *Comparability* – national and state governments initiated a performance measurement and process benchmarking system for all local governments in 1995. The introduction of this form of mandatory environmental reporting is seen by many councils as an opportunity to extend into benchmarking their environmental performance; and
- *Continuity* – using statutory powers for the 1998–99 survey signalled that the collectors are committed to regular publication of credible results.

Reporting of 1998-99 results by councils wishing to share their EPE Information with stakeholders beyond the collector is shown in Table 15.3.

Why did those compiling EPE estimates from budget documents and other corporate data share their information with other stakeholders? All feedback councils were able to

Table 15.3. Transmitting EPE Information Beyond the Collector (by 51 Australian Councils).

| *Stakeholder Group* | *Frequency of Transmittal* |
|---|---|
| To other staff within Council | 53% |
| To Councillors (Elected Members) | 51% |
| To ratepayers (taxpayers) | 49% |
| To local environmental interest groups | 47% |
| To state government agencies | 23% |
| To other Commonwealth Government agencies | 20% |
| To banks or other lenders | 2% |

consider and rank the worth of their EPE information against five facets of performance measurement:[13]

1. Satisfaction of the wants and needs of stakeholders;
2. Strategies put in place to satisfy the wants and needs of stakeholders;
3. Critical processes performed by the organization to execute strategies;
4. Capabilities of the organization to operate and enhance critical processes; and
5. Contributions from stakeholders for maintaining and developing the organization's capabilities.

In addition to being able to evaluate the worth of their EPE information to performance measurement, the fifty-one feedback councils identified its relevance to their fiscal concerns and to differing aspects of environmental protection and natural resource management. Choosing and ranking out of 32 possibilities among fiscal concerns and management was necessary to record their response. Figure 15.7 presents a summary image of those responses. The darkest cells in the 32-cell matrix reflect the highest weighted-average scores.

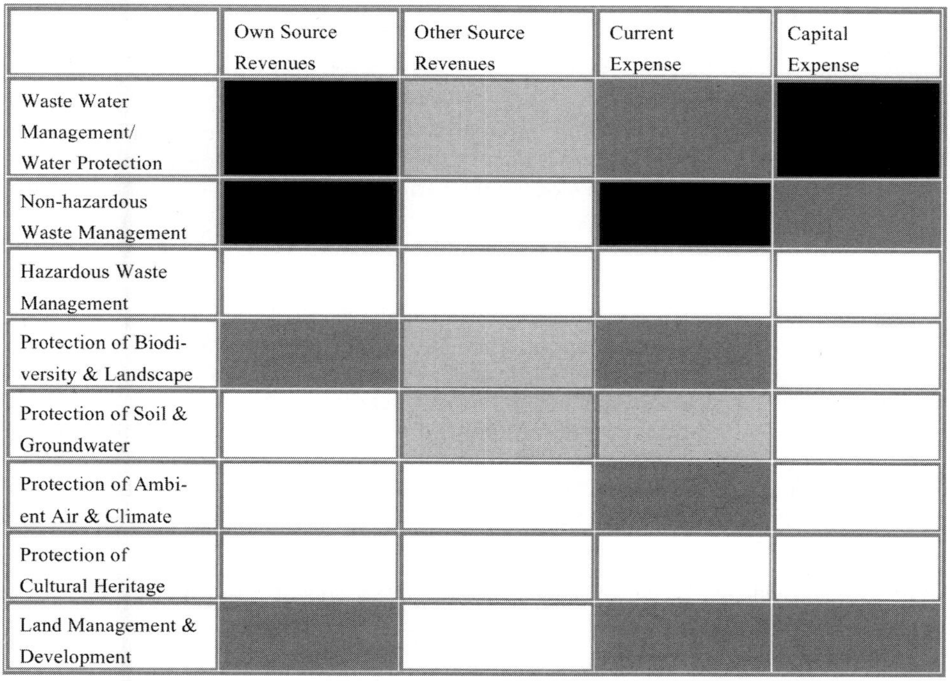

| | Own Source Revenues | Other Source Revenues | Current Expense | Capital Expense |
|---|---|---|---|---|
| Waste Water Management/ Water Protection | | | | |
| Non-hazardous Waste Management | | | | |
| Hazardous Waste Management | | | | |
| Protection of Biodiversity & Landscape | | | | |
| Protection of Soil & Groundwater | | | | |
| Protection of Ambient Air & Climate | | | | |
| Protection of Cultural Heritage | | | | |
| Land Management & Development | | | | |

Figure 15.7. Relevance of EPE Information to Fiscal Concerns & Management Functions (summary of responses from 51 Australian councils).

---

[13] Andy Neely & Chris Adams, Perspectives on Performance: The Performance Prism, Centre for Business Performance Measurement, Cranfield University; and downloadable from http://www.cranfield.ac.uk/som/cbp/prism.htm.

Active sharing of EPE information compiled by Australian local governments for contributions to environmental accounting at macro scale is evident from the responses provided by a sample of 51 councils. The evidence is based on only two years' experience in contributing to this form of environmental accounting, and from a sample whose environmental protection cost burden is lower than the sector average benchmark. Environmental knowledge management can exist at the interface between macro and micro scale environmental accounting, contrary to conventions that do little to encourage information exchange between these two types of practices.

## 15.5. Futures

Discourse on environmental knowledge management typically runs along the parallel but separate development paths found with macro and micro scale environmental accounting. The focus may be on environmental knowledge as a form of business intelligence, identifying ways and means of gaining an edge over competitors. Or it may be presented as an example of, say, transferring knowledge on cleaner technologies from developed to transitional and developing economies.

Environmental knowledge management can also be thought of as a process for promoting the use of measuring tools by management units to track their progress towards sustainability. Again two major paths can be identified. The promotion campaign may be developed around a particular tool (EMAS) or toolkit (ISO 14000). Or the promotion campaign may be developed around targeting a particular sector (households, or manufacturing, or other industry sectors). Examples that promote a particular tool or tools to a particular industry (materials flow accounting to manufacturing) can also be identified, but are less common.

Creating knowledge management in environmental accounting at a regional (subnational) scale is seen as the best way forward for a number of reasons, including:

- environments are managed as regions (watersheds, air sheds, ecoregions, soil conservation districts, river basins, representative reserves, coastal zones);
- actors in regional governments are more likely to be remain in office for the time required to implement and manage an effective campaign;[14]
- decentralisation and devolution from the centre will accelerate growth of an already high environmental protection cost burden at regional and local scale;[15]
- understanding and acting on the complexities of environmental information that can be shared through knowledge management is more practical at regional and local scale than at the centre;
- industries operate within a spatial context, locating their operations according to access to resources, to markets or for other factors that provide them with a competitive advan-

[14] For example, this paper has referred to a possible target of 665,000 management units in Australia's employing businesses. ISO 14001 certificates had been issued to some 700 of those management units at the time of preparing this paper. Probably, 100 management units voluntarily disclose the environmental impacts of their operations. Running with a central government programme that has an evaluation measure of say 0.5% of business target entities adopting an environmental accounting practice within a 3–4 year term is not likely to be politically attractive.

[15] Published ABS estimates indicate that in Australia the national government finances some 5–6% of national expenditure on environment protection. Comparable measures may be much higher in other nations, especially those with unitary systems.

tage; and some environmental accounting tools will therefore fit better in some regions than in others; and

- most universities and other research institutions have to focus at least some part of their R&D on the immediate catchments from which they draw some of their students; and are well-placed to contribute strategically to environmental knowledge management within regions.

All of the statistical processes identified in this paper are based on international standards, and can be replicated in those countries that follow those standards. National EPE, expressed as a percentage of Gross Domestic Product (GDP), is a core indicator of transition to sustainable development within the set being developed by the UN and other agencies of the international community. Any country accepting that core set has the environmental knowledge management potential identified here.

Some thirty OECD countries regularly compile national EPE estimates as an element of reviewing environmental performance. The OECD and the UN Economic Commission for Europe have worked with a number of transitional economies to also apply the OECD's environmental performance review model. They too can look for knowledge management in environmental accounting. It is time for the potential to add value to local knowledge while contributing to macro scale processes to be realised.

# 16. The Greening of Accounting: Putting the Environment onto the Agenda of the Accountancy Profession in the Philippines

*Maria Fatima Reyes*
*Chair of PICPA's Environmental Accounting Committee; E-mail: freyes@mindgate.net*

## 16.1 Inspiration

Accounting is the language of business, and business decision-makers rely heavily on the information provided by accounting to communicate and to make sound decisions. Can accounting, which provides the language of earnings and capital, also give the environment that much-needed voice to be heard in decision-making? Now more than ever that voice is urgently needed considering the state of our environment – as Maurice Strong, Secretary General of the 1992 United Nations Conference on Environment and Development, aptly put it:

> *'If we accounted realistically for current growth practices, it would be clear that much of the wealth they produce really represents a running down of earth's natural capital. Like any business that fails to provide for depreciation, amortization and maintenance, Earth, Inc. is under liquidation headed towards bankruptcy.'*

## 16.2. Introduction

The degradation of the global environment and the problems that go with it have been documented in detail. Many of these problems such as pollution, deforestation, land degradation, and resource depletion are particularly acute in countries like the Philippines, which is still struggling to solve poverty and other social inequities. The country's race for development has also put tremendous pressure for increased exploitation of its natural resources.

As the Philippines experiences disastrous floods, diseases and other consequences of environmental abuses and neglect, environmental protection has steadily become a leading cause for the Filipinos. During the past decade, environmental issues have caught the attention of policy makers, community leaders, academia, media practitioners, business and environmentalists, which is essential since only collective action by individuals, institutions, and organizations (in short, by society as a whole), can bring back the vitality of the environment.

The professions are a sector which can play an important role in these concerted efforts to save the environment. The accountant's role here is both unique and crucial, since its main purpose is to provide financial information for use in decision-making. Accounting information is constantly relied upon by a sector which is responsible for one of the biggest impacts on the environment – business. By providing accurate and reliable cost-related environmental information to managers, accountants can help companies to recognize the financial value of good environmental performance.

Through the leadership of the Philippine Institute of Certified Public Accountants (PICPA), the accounting profession is a frontrunner among the professions in the country in addressing the issue of the professional's contribution to a sustainable environment.



*M. Bennett et al. (eds.), Environmental Management Accounting: Informational and Institutional Developments,* 215–220.
© 2002 *Kluwer Academic Publishers. Printed in the Netherlands.*

## 16.3. PICPA's commitment and leadership

PICPA began to espouse the cause of environmental protection in the Philippines in 1995, when it used the theme 'The CPA: For Business and Environment' during its annual accountancy week celebrations. The event's opening ceremony saw Chair Elvira Atanacio sending out a message of combined urgency and hope when she invited all CPAs in the country to join PICPA in a campaign to promote environment-friendly business. This beginning further bore fruit in 1996 when the Accountancy Journal was published, containing articles devoted to the sole topic of business and environment. As then President Delgado Uy stated in his message, 'through the articles in this issue, we want it known that PICPA does not only talk about this matter [environment] – it also puts its words into action'.

In 1997, former President Eduardo de Guia signed the association's commitment to participate in a foreign-funded project that will teach environmental accounting skills to Filipino accountants and finance practitioners. Sponsored by the United States-Asia Environmental Partnership Program (USAEPP) and the US Council of State Governments (CSG), the project developed and delivered a multi-disciplinary course entitled 'Environmental Cost Assessment: Profiting from Cleaner Production'.

PICPA's 1998 annual national convention resolutions included a commitment by the profession to promote heightened awareness about environmental accounting among its members through education and the dissemination of materials on environmental accounting. The year also witnessed then President Mel Libre and Asean Federation of Accountants (AFA) President Antonio Acyatan leading the Philippines delegation to UNCTAD's Environmental Accounting Conference in Bangkok, Thailand. The Philippine delegation committed to working towards the integration of environmental accounting in accountancy education.

The PICPA's commitment to the environment was institutionalized in 1999 when then President Danilo Principe created the Environmental Accounting (EA) Committee headed by Fatima Reyes. Billed as a national committee, the EA Committee will lead PICPA's efforts to promote environment accounting in the country. The year also marked the completion of the PICPA/USAEPP/CSG project on environmental accounting. Highly successful environmental accounting courses continued to be conducted for different organizations in the year 2000, when President Bebe-I Gonzales was at the helm of the Institute.

## 16.4. What is environmental accounting?

The efforts of PICPA to promote environmental considerations in business is focused on spreading the practice of environmental accounting. But what is environmental accounting? How can it help provide business executives with the added impetus to improve their company's environmental performance and at the same time to safeguard the health and safety of its workers and the community?

According to the definition of the Management Institute of Environment and Business (MEB), Environmental Accounting is *a subset of accounting that deals with activities, methods and systems to record, analyze, and report environmentally-induced financial impacts of a defined economic system (e.g. a firm or a nation).*

Environmental accounting can be subdivided into various branches:

- **Environmental accounting in the context of National Accounting.** This uses physical and monetary units as measures to refer to the consumption of the nation's natural resources, both renewable and non-renewable. Sometimes referred to as 'natural resource valuation', this area is geared more towards the discipline of economics. It also includes the integration of environmental degradation cost into the computation of economic development indicators such as GNP and GDP.

- **Environmental accounting in the context of Financial Accounting.** This branch seeks to include information about the cost impacts (including liabilities) of a company's environmental performance in its financial statements. Current financial accounting standards, particularly those on contingencies, the impairment of assets, and intangibles, can apply to environmental issues; but the possibility of separate standards on environmental financial reporting is also being explored by various local and international groups including accounting communities and standards-setters.
- **Environmental accounting in the context of Management Accounting.** This deals with the use of environmental cost and savings information to improve business decisions, and is where environmental accounting in business holds much promise today. Companies who have integrated environmental accounting into their information systems use environmental cost information largely for internal purposes.
- **Environmental accounting in the context of auditing.** This covers the assessment of the company's GAAP-related issues in relation to environmental matters that may affect the financial statements. It also covers the review of a project's exposure to environmentally-related risks. As environmental matters become significant enough to affect financial performance and condition, financial auditors need to be cognizant of its impact on the financial statements.

### 16.4.1. *Hidden environmental costs*

Uncovering and recognizing environmental costs associated with a product, process, system, or facility is the core of environmental accounting. Among the growing list of internal environmental costs that need to be considered by business are costs incurred in connection with pollution reduction, waste management, monitoring, regulatory reporting, legal fees and insurance. In the midst of increasing environmental regulations affecting industry, the achievement of core business goals such as controlling costs and increasing revenues can hinge on the ability of companies to pay attention to environmental costs: current and future, actual and potential. The inclusion therefore of environmental cost data is important for good management decisions.

Conventional cost accounting systems, however, conceal environmental costs because they attribute many of these costs to general overhead accounts. This practice effectively hides environmental costs from product and production managers who receive no incentive to reduce these costs. Company executives are also often unaware of the extent of environmental costs and their impacts on operations and profits.

#### *Promoting Environmental Accounting in the Philippines*

Amidst the increasing challenge for accountants to provide business with environmental cost information, PICPA undertook various activities in understanding, publicizing, and educating its members in the basic concepts of environmental accounting, and other modern environment-related approaches such as pollution prevention (P2) and Cleaner Production (CP). The dissemination of environmental accounting concepts and tools is being achieved mainly through the following programs and initiatives:

- Continuing Professional Education (CPE);
- Integration of environmental accounting into the undergraduate accountancy curriculum;

- Dissemination of Environmental Accounting Information and Tools Through Written Materials, Conferences, Networks, etc.

**1. Continuing Professional Education (CPE)**

Through funding from the USAEPP, PICPA developed a new training course entitled 'Environmental Cost Assessment (ECA): Profiting from Cleaner Production' as part of its CPE program. The course was developed with participation from Illinois Environmental Protection Agency, Tellus Institute, and the Asian Institute of Management.

The course was designed for a mixed audience of accountants, engineers, and environmental professionals in recognition of the fact that ECA/CP requires teamwork at the facility level. The two-day curriculum was designed around a set of case studies based on Philippine businesses that have improved their financial and environmental performance through CP strategies. Topics include the following:

- Introduction to environmental accounting;
- How to estimate the true 'cost of waste' at an industrial facility;
- Basic concepts of CP for reducing the cost of waste;
- Environmental cost data collection and estimation issues and tools;
- How to perform a comprehensive profitability assessment for environmental improvement projects, particularly investments for CP;
- Case studies of CP profitability in Philippine companies;
- How to use the environmental accounting software 'E2F Philippines';
- Where to find more information and assistance.

Experienced PICPA trainers representing various regions also underwent a 3-day train-the-trainers course in order to ensure the continued dissemination of the course in other parts of the country.

Other courses dealing with environmental issues and the accounting profession will be developed and offered in the future. One of these courses is on environmental auditing which was already included in the list of courses endorsed by the Professional Development Committee to the various PICPA chapters for the purpose of continuing professional education.

**2. Integration of Environmental Accounting into the Undergraduate Accountancy Curriculum**

To ensure that future practitioners will make environmental concern an indispensable part of the practice of their professions, it is important for schools and professional organizations to review and provide their curricula with an environmental perspective. Environmental education at the college level is aimed to deepen the knowledge and develop the necessary skills for the management and improvement of environmental quality conducive to the well-being of society.

PICPA has responded to this challenge by working towards the integration of environmental accounting into the accountancy curriculum. The PICPA Model Curriculum integrates environmental accounting in the following subjects:

- *Management Accounting.* Environmental accounting topics include environmental cost analysis and capital budgeting for environment-related projects, particularly for CP investments.
- *Financial Accounting and Auditing.* Discussions of applicable financial accounting standards relating to environmental issues in companies (e.g. contingencies, liabilities and disclosures).

- *Proposed subject on Professional Ethics.* Topics for discussion include consciousness and care for the environment which is part of the social responsibility of an accountant. This also includes the integration of environmental and societal consideration into business decision-making.

**3. Dissemination of Environmental Accounting Information and Tools Through Written Materials, Conferences, Networks, etc.**

Other activities to promote environmental accounting have been undertaken by PICPA. Environmental accounting concepts and tools are being promoted via articles published in various PICPA newsletters and journals. To further spread the practice of environmental accounting to its membership, the Institute has also featured environment-related topics in conference and conventions. One prominent venue where environmental accounting was highlighted was the recently concluded convention of the Confederation of Asia Pacific Accountants (CAPA) which was held in Manila in November 2000.

Organizations like the United States Environmental Protection Agency (USEPA) and the Canadian Society of Management Accountants have sent materials to be used in environmental accounting training programs undertaken by the Institute. The Philippine Business for the Environment (PBE), and the Private Sector Participation in Managing the Environment (PRIME), a UNDP funded- project under the Board of Investments, are some of the local groups which have likewise emerged as important partners for PICPA's efforts to promote environmental accounting.

Various PICPA chapters all over the country have also initiated their own community-based environmental projects in the areas of solid waste management, recycling, and reforestation.

## 16.5. Future work

Through committed actions and the unwavering interest of individuals and groups within the organization, PICPA has accomplished much in putting the environment onto the agenda of the accountancy profession as well as of the business sector in the Philippines. But much work still has to be done in sustaining the various programs that are already in place, particularly in the area of educating professionals and students in the practice of environmental accounting.

The Institute can also look forward to elevating its commitment to the next level by looking at the following partnership avenues that can further the cause of environmental consideration in business.

### *Partnership with the Government to Promote Environmental Management Accounting*

The United Nation's Division for Sustainable Development (UNDSD) has launched a project to explore policy pathways for national and local governments to support and promote the practice of environmental management accounting. PICPA can pursue initiatives along this line to help the government encourage business to choose more environmentally-friendly options in their operations, such as the switch to cleaner production and the installation of environmental management systems.

### *Promotion of Environmental Accounting in Other Parts of the World*

Through its active membership in various regional and international organization, PICPA is in a good position to assist other accounting organizations to begin their own environmental accounting initiatives in their respective countries.

## 16.6. Conclusion

The greening of accounting is a challenge ably taken on by the accounting profession in the Philippines through the efforts of the Philippines Institute of Certified Public Accountants (PICPA). The organization promotes environmental accounting which is a part of a larger universe of accounting tools that are necessary for good decision-making. Environmental accounting is particularly helpful because it not only improves business decisions but also links environmental care to core business interests.

And yet environmental accounting, just like conventional accounting, is only a tool, and one which is only as good as those who use it. It should be remembered that the greening of the accounting professional does not end with environment accounting. What is more important is to ensure the formation of holistic values and a green ethic that would enable current and future accountants, and the decision-makers who rely on their work, to count the environment as a legitimate stakeholder in the choices made in society.

# PART IV

## DIFFERENT EMA PERSPECTIVES

# 17. Environmental Performance Measurement[1]

*Edeltraud Günther and Anke Sturm*
*Professor of Business Management resp. researcher, Dresden University of Technology, Dresden, Germany*

## 17.1. Introduction

The Chair of Business Management (in particular Environmental Management) at Dresden University of Technology (TUD) in Germany has executed a research project on how corporate environmental performance can adequately be measured, evaluated and assessed (Environmental Performance Measurement (EPM)). An ideal-typical model has been developed for this purpose, which includes the five steps that are explained as follows.

1. The model sets out to determine the objectives (or even very concrete targets) that have to be reached by EPM, taking into account the interests of the company's major stakeholders.
2. It is necessary to focus the measurements on the environmental effects as the basis of environmental performance (ecological success). To prevent the confusion of effects and impacts, environmental effects are here referred to as 'environmental-influence factors' (Etterlin et al., 1992, p. 19).[2] Measurement of these factors will be based on a particular classification of ecological results (cf. Figure 17.1 below). The principle behind this 'ecological breakdown of ecological results' is essential to the model.
3. The company's environmental-influence factors have to be evaluated, as these at the operational level define environmental performance. As a result of this, the effects[3] that the company has on the environment can be made explicit.
4. To draw conclusions as to how well the company is performing from an environmental perspective, it is necessary to compare the measured values of company influence on the environment with the target values. Distance-to-target indicates to what extent the objectives of environmental management have been achieved.
5. These objective achievement degrees form the basis to derive the relevant action recommendations in the company. Additionally, decision making should be accompanied by checking the established EPM objectives.

---

[1] With thanks to Ms Katrin Pönisch-Pörschke from the Language Centre of Dresden University of Technology and Stephan Schöps, who committed themselves greatly to the English translation of this paper.

[2] Environmental influences are emissions, e.g. of carbon dioxide or sulphur dioxide. In the following scientific discussion, we shall use the term *environmental influences* instead of the term *environmental effects* contained in the EEC Environmental Audit Directive (and in the Questionnaire, cf. Annex), since, however, environmental effects are substantially referred to as environmental influences in accordance with the EEC Environmental Audit Directive, but they could be easily confused with the term environmental impacts (cf. footnote 3).

[3] Environmental impacts are caused by environmental influences. The first designate impacts (emissions) on flora, fauna, human beings and materials, cf. also Section 3 (1) and (2) of the Federal Emission Control Act (BimSchG).



*M. Bennett et al. (eds.), Environmental Management Accounting: Informational and Institutional Developments*, 223–229.
© 2002 *Kluwer Academic Publishers. Printed in the Netherlands.*

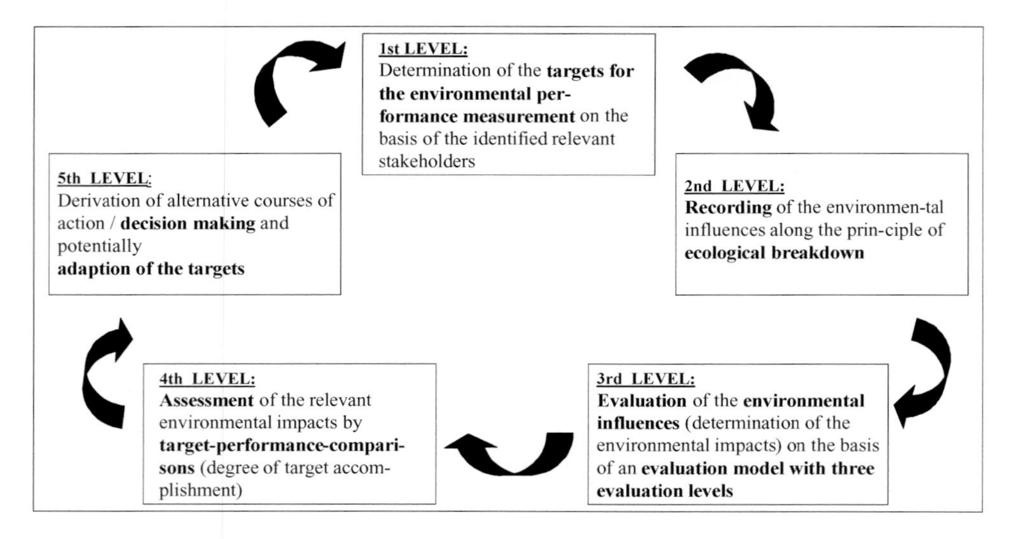


Figure 17.1. Ideal-typical model for the Environmental Performance Measurement (EPM) (scheme designed at the TUD Environmental Management chair).

The particular features of this model are (Günther and Sturm, 2000):

- **Strategic nature of the environmental objectives and targets.** There should be a close link between environmental performance measurement at an operational level and the environmental objectives at a strategic level, i.e., the environmental performance parameters should reflect the company's stakeholder-oriented strategy.
- **Ecological breakdown of ecological results.** The ecological breakdown of ecological results (cf. Figure 17.2 below) is applied to differentiate between environmental effects (influence) that relate to continuous processes, and those that have an incidental character ('ordinary' versus 'extraordinary'). Moreover, a distinction is made between intended environmental results versus unintended environmental side-effects of certain measures ('goal-oriented' versus 'non-goal-oriented'). In this way, it is possible to link measurements with strategic goals and to identify opportunities for improving the company's environmental management.
- **Process and control orientation.** Application of the above-mentioned ecological classification make apparent the potential internal control mechanisms that make it possible to reach environmental targets, and leads to a different way of looking at the company's processes.
- **Target-oriented evaluation model with three evaluation levels.** Basically, this approach distinguishes between three types (and levels) of targets related to operational environmental objectives that allow the company to evaluate its environmental effects (influence):–
  1. the targets derived from compulsory regulations (e.g. limit values);
  2. the targets that go beyond the EEC Environmental Audit Directive and DIN ISO 14001, which may involve new scientific findings and innovative measures;

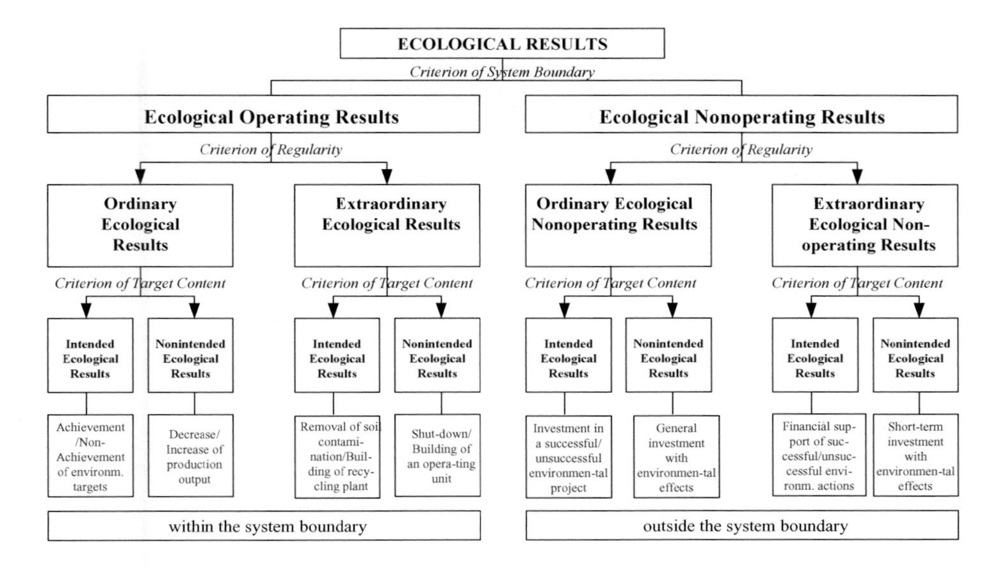


Figure 17.2. The concept of ecological breakdown of ecological results (scheme designed by the TUD Environmental Management Chair).

3. targets of sustainability with the three dimensions of economy, ecology and social issues.

These three ways of evaluation do not exclude each other; instead, they build on each other.

## 17.2. The research design

The five-step model provided the conceptual basis for carrying out empirical work on Environmental Performance Measurement. When defining the research population, two criteria were used:

- **Reference to industrial sector.** The Environmental Performance Measurement model is designed to be sector-specific, assuming that the environmental problems differ between sectors due to differences in production processes, activities and materials used. The research was carried out in the German mechanical engineering industry (Batschari et al., 1995), although it could also have focused on one or more other sectors.
- **EMAS and DIN ISO 14001.** Only those companies were selected that officially complied with *EMAS (EEC Environmental Audit Directive)* or that had been certified under *DIN ISO 14001*. It was assumed that these companies had collected comparatively reliable environmental data. However, a first analysis of the environmental reports of the EMAS companies showed that the available data was not sufficient to quantify the conceptual model. Therefore, a special questionnaire was developed for this purpose.

Eventually, the population was made up of 111 company sites. 18 percent of these sites had been certified according to DIN ISO 14001 only. In February 1999, the researchers interviewed these companies. The participants were asked to fill in the questionnaire (primary research) and the EMAS sites were also asked to send their environmental reports (secondary research).

The collection of data was closed early in September 1999. Out of the 111 companies that were involved, 52 responded (a response rate of 46.8%); 45 of these returned a properly filled in questionnaire (a return rate of 40.5%). Out of the 91 EMAS sites (82.0% of the population) 83 environmental reports (including simplified statements) of 65 companies could be processed (return rate: 71.4%). The analysis of the questionnaires was done by site, not by company. In particular, it focused on finding out whether indeed the companies were capable of providing the data needed to quantify the conceptual model.

## 17.3. Empirical investigations – results

### 17.3.1. *Ecological breakdown of ecological results*

Comparable to the breakdown of the management performance,[4] the ecological breakdown of ecological results is intended to help to identify success factors and thus new possibilities of internal control. Therefore, as a first step the environmental effects (influences) are measured on the basis of the criterion *Within the Company's System Boundary*. These are influences that occur within the company and therefore can be controlled by the company itself (ecological operating results). A difference has to be made between the latter and the results of company investments outside its system boundary.

Only 8.9% of the companies appeared to have made investments that contributed to the company's **ecological financial performance**. All companies that reported an ecological financial success referred to other projects than those that were key in the questionnaire (it must be added that frequently these financial successes were based not on commercial activities, but instead resulted from intra-company support or support from an affiliated group). Just one company indicated to have sponsored a regional project. Thus, the investment categories named *Participation in ecological or environmental funds* and *Sponsoring of supra-regional or regional ecologically oriented projects* seem not to be feasible in practice.

With regard to **ordinary and extraordinary ecological performance (in terms of burden on the environment)**, the influence factor *production increase* appeared to be dominant. For example, 53.3% of the companies indicated that over the last five years their burden on the environment had increased together with a growing production volume. From an accounting point of view, it can be concluded that so far as **ordinary ecological performance** is concerned, the companies have collected relevant and verifiable data.

Further influence factors are (in order of importance – see also Figure 17.3):

- *other factors* (such as increased inspection or overhaul of machinery: 13.3% of the number of items mentioned);

---

[4] The management breakdown differentiates between the operational and the financial performance in accordance with the criterion *Employment*, and the ordinary and extraordinary performance in accordance with the criterion *Ordinariness*. The operational and financial performance of the enterprise are regarded as ordinary performance, while the extraordinary performance comprises 'all extraordinarily occurring, i.e., extraordinary and extra-period performance/success components', cf. Coenenberg (1997, p. 337).

- *extraordinary economic measures* (such as construction of a new operating unit: 11.1% of the number of items mentioned);
- *accidents* (4.4% of the number of items mentioned);
- *extraordinary ecological measures* (such as sealing of the soil by building an own sewage treatment plant: 2.2% of the number of items mentioned); and
- *failing to meet the environmental goals set by the company* (2.2% of the number of items mentioned).

The findings suggest that it is economic factors that are seen to be responsible for a company's environmental performance rather than the ecological influence factors as such.

Concerning **ordinary and extraordinary ecological performance,** 82.2% of the companies could provide the solicited information according to the ecological breakdown of ecological results. The other respondents could not classify their environmental effects (influence) on that basis.

Of the companies that did not classify the environmental influences, 50.0% indicated that classification was not possible; 37.5% indicated other reasons (e.g., in-house restructuring).

When looking at the major factors responsible for the reduction of the environmental burden, we get a different picture. The economic influence factor *extraordinary economic measures* (e.g., close-downs of operating units) was mentioned in 14.4% of the cases and the ecological factor *achievement of relevant environmental goals* was mentioned in 71.2% of the cases).

The *achievement of environmental goals*, therefore, can be regarded as the main factor influencing the reduction of the company's environmental burden. This means that environmental performance is within the reach of the company by means of feasible

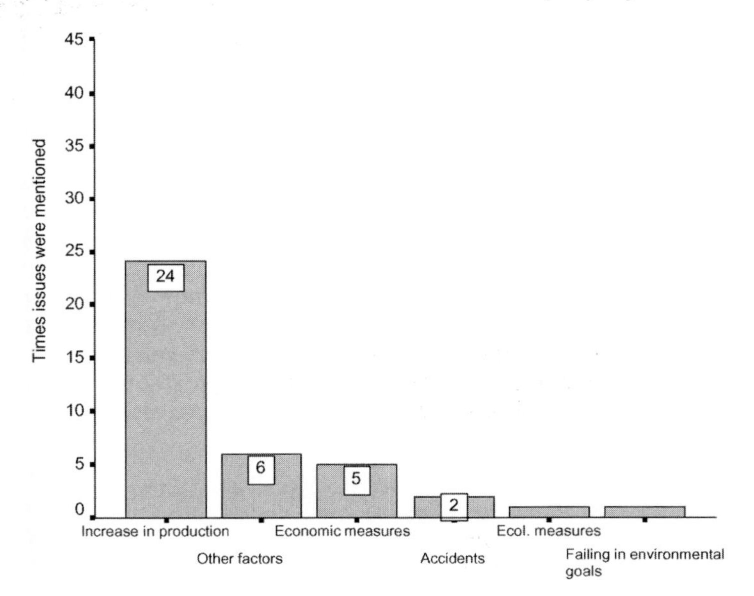


Figure 17.3. Are the environmental influences covered separately (in the sense of environmental burden) following the principle of the ecological breakdown of ecological results? (random sample: 43 company sites; multiple answers possible).

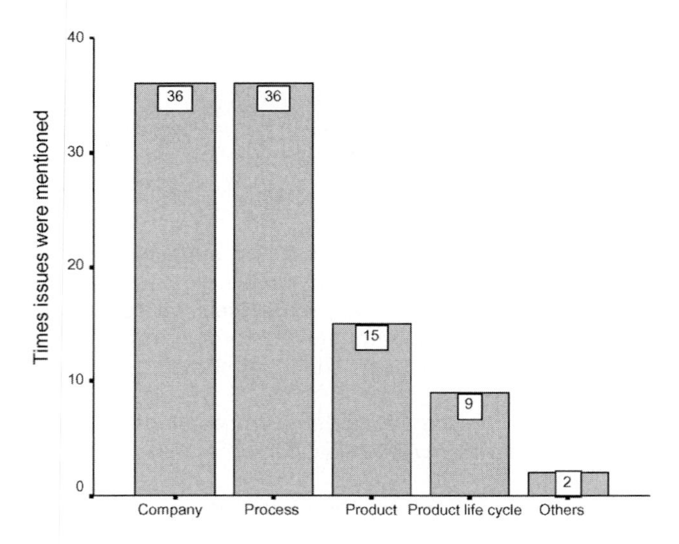


Figure 17.4. Can the environmental influences be classified with the production processes and activities (process and control orientation)? (random sample: 45 company sites; multiple answers possible).

environmental management practices. It can be assumed that the companies involved have a database that is sufficient to be able to measure their actual performance against the targets (**scheduled ecological performance**).

## 17.3.2. *Process and control orientation*

The questionnaire (see Figure 17.5) aimed at relating the strongest environmental influences to the company's production processes and activities. A majority of the companies (73.3%) could classify their major environmental effects (influences) to the various materials/substances, production activities and plants. 6.6% of the companies could do this for only a limited number of environmental effects. 20.0% of the companies were not able to provide any details that the questionnaire required.

The high percentage of companies that could provide the figures according to the classification used can be explained by the process orientation of most of the companies (80%). However, a product-centred consideration of environmental effects still plays a minor role (product consideration: 33.3%, product service life: 20.0%). This is a weakness that needs to be considered when evaluating the information needs of the companies.

## 17.3.3. *Evaluation basis*

As to the evaluation basis that the companies use, it appeared that 97.8% of the companies apply company-specific environmental goals. This is the most common procedure of comparing to the other scores: legal regulations (e.g., limit values, 62.2%), sustainability goals (31.1%), and other evaluation procedures (e.g., company-specific procedures, orientation towards environmental compatibility tests or external measurements, 24.4%). The high relevance of the company-specific environmental goals probably depends very much on the population chosen. The figures indicate that those main ecological fields are

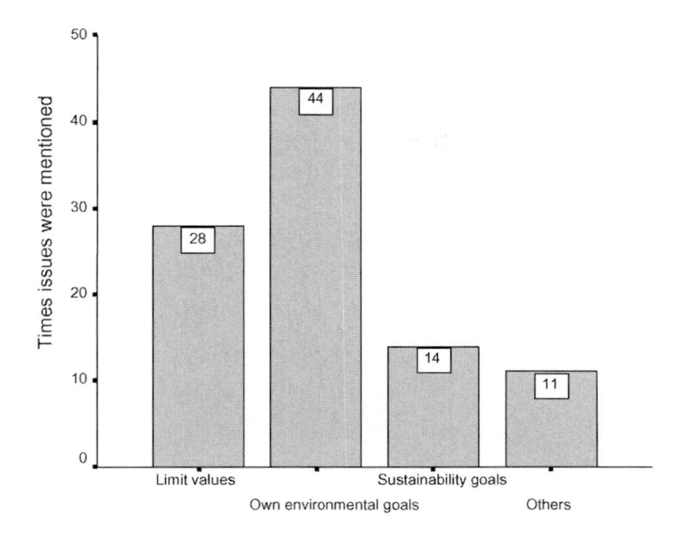


Figure 17.5. What is the evaluation basis on which the environmental influences are identified? (random sample: 45 company sites; multiple answers possible).

defined in the companies on the basis of limit values. The companies will confront different economic-ecological problems in these fields that they need to address. The presence of EMAS and/or ISO 14001 is likely to have inspired the companies to define company-specific objectives that go beyond compulsory regulations.

## 17.4. Conclusion and outlook

The descriptive evaluation and representation of the results of the study *Measurement of the Corporate Ecological Results* is a first major step towards the empirical substantiation of the conceptual model (cf. Figure 17.1). The results show that the companies are conversant with the concepts behind the model and avail themselves of the data needed to quantify the model. Essential elements are:

- Process orientation, to be supported by suitable tools, such as process balances;
- Abilities to carry through the ecological breakdown of ecological results to support the process and control orientation;
- Company-specific environmental goals that are oriented towards ecological problem areas within the company. Distance-to-target measurements will indicate whether the goals are achieved.

Companies that cannot comply with these criteria need to determine their relative environmental performance by external benchmarking. The research reported in this paper has shown that, for the production processes of the German mechanical engineering industry, external benchmarking is possible. This is illustrated by the fact that for most companies in this industry set similar priorities for the main environmental issues in the context of their overall environmental objectives.

# 18. Towards Sustainability Indicators for Product Chains

## *With Special Reference to an International Coffee Chain*

*Teun Wolters*
*Director, ISCOM Institute for Sustainable Commodities, Utrecht, The Netherlands; E-mail: twolters@iscom.nl*

*Myrtille Danse*
*Consultant, CEGESTI, in San Pedro, Costa Rica; E-mail: myrdanse@racsa.co.cr*

### 18.1. Introduction

This paper is about sustainable development and how it can be promoted by strategies for product chains. It focuses on how to develop indicators that adequately measure corporate and chain-oriented performance in reaching sustainable business.

Sustainable development means making the world a better place to live, for both present and future generations. It entails the preservation and protection of the earth's wealth-creating resources by bringing about the social and economic conditions for a transformation in that direction.

The title of this paper refers to three concepts that have become part of the terminology that belongs to the world-wide quest for policies and practices which conform to sustainable development: sustainability, indicators and product chains. Rather than attempting to provide a comprehensive overview of the literature, this paper focuses on a selection which are particularly relevant to the development of indicators that can support sustainable development at corporate levels. The focus on indicators means that this paper is in the domain of Environmental Management Accounting (EMA), which itself may evolve over time into accounting for sustainability.

### 18.2. Sustainable development

#### *Sustainable development and its indicators*

Although sustainable development is relevant to all walks of life, it is the business community that has a crucial role to play in making it a reality. Its common definition implies a long-term view – that is, considering the interests of future generations – and makes it apparent that 'business as usual' is not sustainable. There is a need for corporate sustainability strategies that lead to the necessary transformations. 'Sustainability' here means being consistent with a path towards sustainable development. Sustainability indicators are indicators that reflect progress (or a lack of it) in reaching sustainability.

Of course, since sustainable development concerns the whole of society, the corporate level alone cannot determine how to head for it. Society at large, both in civil and political terms, needs to be involved. Interdependencies between different institutional levels also require policies at the meso and macro level. Therefore, sustainable development is relevant to all the systems that together bring forth a society's wealth, justice and care.

Although co-ordination and priority setting are crucial elements, it is an undeniable fact that paving a path towards sustainable development is associated with all kinds of uncer-



*M. Bennett et al. (eds.), Environmental Management Accounting: Informational and Institutional Developments,* 231–247.
© 2002 *Kluwer Academic Publishers. Printed in the Netherlands.*

tainties. This explains why sustainability strategies cannot be taken for granted; they require vision and entrepreneurial courage.

The concept of sustainable development is 'a star affording a general direction' (as in the nineteenth century Edgeworth characterised his theory of bilateral monopoly) but it is impossible to define it in terms of final attainment levels. Therefore, as it is now, sustainable development is primarily a matter of movement by incremental steps. What these steps actually entail and how fast they are taken will depend on a company's tacit and explicit strategies whereby stakeholder pressures and market opportunities may be decisive factors. Given the uncertainties, it is natural that these strategies differ substantially between companies. Companies will at the least wish to consider their 'licence to operate' (both literally and metaphorically), but may also take creative steps in the field of sustainable business that give them a special profile.

### *Product chains*

This paper looks at a particular area of sustainability, namely the consideration of entire product chains. Rather than focusing exclusively on the environmental and social aspects of those economic activities that take place within the boundaries of one's own company, it is generally recognised that companies should accept responsibility for the fact that they belong to entire product chains. These go from the production of raw materials through the different stages of processing and distributing up to and inclusive of final consumption and the handling of waste. The product chains frequently cross borders, which gives their management global dimensions. It is understandable that when companies become aware of their product chains and wish to take them into account when defining their corporate strategies, they will first consider the chain-related aspects of their core activities and final products.

### *Accountability*

Considering that corporate sustainability strategies need to be developed within the companies themselves, one may ask what role public policies may have in them. Of course, there still are the unfinished tasks of policy-making and legislation to show society's interest in realising sustainable development. As such these may give positive incentives to taking sustainability into account. Likewise consumer concerns about health and safety as well as public awareness of the adverse social and ecological consequences of certain economic activities may induce a change of course. However important these external forces may be, as soon as the 'low hanging fruits' have been reaped, sustainability will become a matter of strategic decision-making that to a great extent remains hidden from the public eye. The link to the broader society lies in the concept of accountability. This concept implies that a company has the right to make its own decisions in the field of sustainability but has to give account of what decisions are (to be) made to those stakeholders who do not directly participate in the decision-making processes (external stakeholders).

## 18.3. Developing the desired indicators

### *External and internal stakeholders*

In the conventional domain of accounting, it is financial accounting that has the explicit function of informing external stakeholders of a company's financial bottom line under the assumption of going concern. Sustainability, however, requires a much longer-term view than 'going concern' normally implies. Moreover, it would involve a 'triple bottom line' by adding environmental and social accountability.

Distinct from financial accounting, there is management accounting, which is one of the most important information tools used by managers. It directly supports internal strategic and decision-making processes, gives feedback on results to internal stakeholders, and acts as an instrument of management control. In principle, management accounting can be a crucial instrument in designing and operating corporate sustainability strategies. This can happen only if top management shows a genuine interest in explicitly incorporating sustainability criteria into its strategic decision-making. Moreover, management accountants and controllers need to be flexible enough to broaden their perspectives and amend their accounting practices accordingly. These aspects touch upon the behavioural side of accounting relative to diverse cultural, institutional and other contingent factors that this paper does not further discuss.

Above, along the lines of financial and management accounting, a distinction was made between external and internal stakeholders. External stakeholders do not participate in the corporate decision-making processes (although they may influence these processes in different ways), whereas internal stakeholders do participate in them (in particular line managers and senior staff who organise the various planning processes).

Although the information needs of the two groups of stakeholders differ, both share an interest in basically the same bottom-line results. Moreover, the dividing line between the two groups may change case by case. For instance, banks or governmental bodies may in a particular case be closely involved in a company's investment planning whereas usually they remain on a distance. This kind of involvement is likely to influence the disclosure of relevant information to these two stakeholders.

Within the framework of sustainable product chains, a distinction between internal and external stakeholders may turn into a sliding scale of distinctions. When a particular situation conforms to chain management based on the collaboration of partners in the chain, these partners share a common interest in the results of their collaboration. However, the power distribution within a chain also influences how the chain partners relate to one another and how decisions are made. The two extremes here are: (a) a chain totally controlled by one powerful company, and (b) a chain consisting of a number of small companies that are fully interdependent.

### *Towards a core set of indicators*

Whatever the situation is, an analysis of the information needs associated with the development of a sustainability strategy for a particular product chain will at least be helpful in defining a core set of sustainability indicators that is required for both internal and external purposes. Such a set of indicators is to be based on a common understanding of the basic issues to be addressed. In addition, certain indicators may be essential to one particular stakeholder whose participation is crucial to the whole endeavour. This is an additional reason for adding certain indicators to the core set, serving the purposes of both decision-making and accountability.

The formulation of the core set of sustainability indicators is a much needed step in implementing a sustainability strategy for a particular chain. As indicated above, it is preceded by a broad analysis of the information needs. Given the multi-faceted nature of chain-related sustainability and the many stakeholders involved, such an analysis requires a suitable methodology. A few shortcuts in defining the major issues in the chain may be useful as a way to generate quickly some preliminary results, but since conclusions on where the hotspots are located in the chain may have a substantial impact on the strategic priorities and responsibilities in that chain, it is necessary to have a well thought-out methodology which will lead to results that cannot easily be challenged.

As a first step such a methodology could prescribe the collection of available information on the chain in question. By putting together the different pieces of literature and other documents, it must be possible to get a general picture of the main environmental and social issues over the respective links of a product chain.

A second step is to study different frameworks that inform the discussion on what should be seen as relevant aspects of sustainability, in both a general sense and when focusing on a particular case. Such frameworks may comprise standards like ISO 14001 and SA 8000 as well as different codes of conduct and good practices. Instruments such as LCA and environmental-cost accounting models can also be helpful here. This step will create awareness of the things that matter and help to compose a list of potentially relevant aspects of sustainability.

A third step entails a careful process of communication and interaction with the various stakeholders in order to find out what items need to be considered as part of the indicators that have to be implemented and regularly updated. To understand the issues that stakeholders tend to emphasise, it may be helpful to find out who are the stakeholders' own stakeholders, in particular who are the most important ones. Partners in the chain have to respond to their own specific political environments with a view to their own 'licences to operate', or perhaps to certain strategic goals that they have adopted in the past.

As a fourth step, based on the three previous steps, the total set of sustainable indicators can be composed. Strategic goals and action plans can be expressed in terms of target figures that match with the indicators that measure performance. A core set of indicators can be selected and given priority in the process of implementation.

Finally, there will be issues of how to integrate and present the various economic, social and environmental indicators to the different stakeholder groups. Next, procedures are needed that prescribe how one should update the indicators. The theory and practice of accounting can be used to define the quality criteria that the indicators have to comply with.

## 18.4. Indicators, accounting and environmental management

Sustainability indicators for product chains as outlined above are still imaginary entities. However, as chain-oriented approaches increase in importance, the need for such indicators will grow. Here lies a prominent task for those engaged in EMA to lead the way. Their development can benefit a lot from the knowledge within accounting about how to generate effective indicators of high quality. To be able to do so, it is important to realise that management accounting is to play an integrating role. This is particularly important in the case of the larger companies that have several management systems in place. As the ECOMAC-project has shown (Bartolomeo et al., 1999), for these companies it is not just their environmental management system that plays a central role in environmental management, but the other systems as well. Table 18.1 reports on the percentage of 84 European companies in 1998 which perceived a particular management system to be of considerable importance or crucial to environmental management.

Table 18.1 shows that the way in which companies organise their environmental man-

Table 18.1. Management systems: percentage of firms that perceive them as of considerable importance or crucial to environmental management in the future (percentage of number of firms in size class, based on number of employees).

| *Size of firm: (number of employees)* | *5–50* | *50–250* | *250–500* | *> 500* |
|---|---|---|---|---|
| Quality Management Systems | 30 | 54 | 73 | 67 |
| Health & Safety Systems | 40 | 58 | 73 | 72 |
| Environmental Management Systems | 20 | 71 | 82 | 74 |
| Materials Requirements Planning Systems | 30 | 29 | 55 | 62 |
| Process/Job Management Systems | 30 | 38 | 45 | 49 |
| Management Accounting Systems | 50 | 71 | 91 | 77 |
| Financial Accounting Systems | 0 | 42 | 82 | 46 |

Source: Bartholomeo et al., 1999, p. 105.

agement may take different forms but frequently involve diverse operational management systems. It is striking that management accounting systems have the highest scores and exceed even environmental management systems as systems which are significant and crucial to environmental management.

Table 18.2 subdivides management accounting into a number of accounting activities so that the future importance of management accounting can be differentiated.

Table 18.2 indicates that companies with 250–500 employees make the most intensive use of management accounting as a significant tool of environmental management. For the very largest companies there is some decline in importance for management accounting as a tool for environmental management, other than for capital budgeting. This tendency could be explained by the fact that the largest companies continue to implement management systems for different purposes so that the weight of a single management system eventually diminishes.

It seems that expectations as expressed in Table 18.1 and Table 18.2 are that management accounting functions will develop further in order to play a prominent future role in environmental management. Part of this development has to be the integrational role of environmental accounting. All management systems reflected in Table 18.1 have data collection and accounting systems, whether simple or complex. All these data sets in principle belong to the data sources of management accounting.

Table 18.2. Management accounting activities: percentage of firms that perceive them as of considerable importance or crucial to environmental management in the future (percentage of number of firms in size class, based on number of employees).

| *Size of firm: (number of employees)* | *5–50* | *50–250* | *250–500* | *> 500* |
|---|---|---|---|---|
| Bookkeeping | 33 | 75 | 70 | 54 |
| Budget setting | 57 | 72 | 80 | 78 |
| Budget control | 57 | 78 | 90 | 81 |
| Capital budgeting | 50 | 85 | 82 | 89 |
| Product costing | 63 | 47 | 71 | 71 |
| Performance measurement (financial) | 33 | 56 | 89 | 82 |
| Performance measurement (non-financial) | 20 | 54 | 90 | 77 |

Source: Bartholomeo et al., 1999, p. 105.

This also applies to the product chain aspects of environmental management. All aforesaid management systems have linkages with the product chain as they all, in one way or another, relate to the flows of materials and energy going through the company. However, none will bring automatically the desired perspective. Environmental management has usually focused first on the emissions and waste streams directly related to a company's internal production processes. Next steps involve the suppliers and then further links in the chain can be considered (Klinkers et al., 1999). Although in general, environmentally purchasing as a systematic way of acting is rare so far (Wolters and Hoeben, 2000), it is expected to increase in importance in the near future. Interesting enough, it is governments as purchasers at different levels that have initiated exemplary greener purchasing programs. In the private sector pioneering efforts in this area have been reported (Halwood and Case, 2000) that have to a great extent been strategically motivated.

An integrating role of EMA in chain-related environmental management could be argued to require a comprehensive framework such as integrated eco-control (Schaltegger and Burritt, 2000). However, given the mostly incremental nature of chain-oriented environmental management, such a framework will be hard to accomplish. However, EMA could also acquire an integrating role by crossing the borders of the different management functions, so as to be able to combine data from different sources and generate relevant chain-related information.

The chain-related environmental management that a company may embark on can develop along two lines: either as company-based product development or as integrated chain management. Product development uses chain-related environmental and social information only to shape its policies without making any arrangement with other actors in the chain. Integrated chain management, by contrast, focuses on arranging common policies with partners in the different links in the chain whereby all steps in the chain-wide production process are taken into account. When it comes to accountability and credibility vis à vis a broader audience, both types of chain-related environmental management require largely the same indicators. In the case of integrated chain management, however, other elements such as the distribution of the costs and benefits of the common policy over the different partners along the chain need specific accounting procedures in order to produce figures that are generally acceptable to all partners. Composite indicators may meet controversy because of different weightings that chain partners can be expected to advocate, depending on their own social and business environments.

These two directions in which chain-related environmental management can develop are relevant to current business practices. Progress in the field has been and will continue to lead to different blends of the two. It is important that EMA follows this up and is in the forefront when it comes to producing the right indicators.

## 18.5. Illustration of sustainability indicators under construction

### 18.5.1. *Introduction*

To illustrate what developing sustainability indicators entails in terms of policy making, this paper looks at two cases that are quite different in nature. The first refers to the initial efforts of a group of Dutch companies in the food industry to reach a common understanding of sustainable food production and to develop corresponding sustainability indicators. The second case is of a consortium of Costa Rican co-operatives of coffee farmers which is making efforts to produce and bring to the European market sustainably produced coffee, and part of whose management structure is a set of relevant indicators.

The first case involves a group of firms that have in common their Dutch backgrounds, and that all their businesses involve agro-business food chains. They feel the pressure of Dutch society to produce sustainable products, and as large companies with well-established names realise that they have a high profile in the public eye which requires a certain degree of pro-activity. Although each company has its own chain-related policies, working together underlines that they share certain interests. For a part, this is a matter of public relations, though it is also valuable in helping through joint co-operation to fill gaps in know-how in this field.

The second case tells the story of a group of co-operatives in Costa Rica that not only share commercial interests but actually work together in developing and selling sustainable coffee. The implementation of sustainability indicators forces the co-operatives to be much more transparent towards each other than they might otherwise have been inclined to in terms of their current policies, future goals and governance.

### 18.5.2. *Sustainability indicators for the food industry*

Since 1995 a group of 15 Dutch companies in the food industry have been working together in the Foundation of Sustainable Foodstuff Chains (Stichting Duurzame Voedingsmiddelenketen DuVo) to promote sustainability in foodstuff chains by building up a relevant body of knowledge through research, generating a compendium of relevant concepts, exchanging company-based experiences, and encouraging a dialogue. In 1998 the foundation decided to launch a cyclical process to arrive at measurable criteria for the most prominent sustainability issues in the Dutch foodstuff chains. Along with this, the foundation intends to stimulate a structured but otherwise informal dialogue between companies and non-governmental organisations (Stichting DuVo, 1999 and 2000).

In December 2000 the foundation organised a conference on the development of sustainability indicators. The organisers presented a list of sustainability aspects of the primary sector that came out of interviews and documents (see Table 18.3).

Blonk and Dutilh (2000) report that in most cases the respondents focused strongly on typical Dutch agricultural themes such as eutrophication and energy consumption. On a global scale, however, issues such as water consumption and erosion are as least as important. Also, the use of ozone-layer depleting coolants were not frequently mentioned, although most coolants are used in foodstuff chains. Priorities may vary by region – for instance, genetically modified ingredients may not cause consumer problems in the USA whereas they do in the European Union.

Table 18.3 provides a comprehensive picture of relevant sustainability issues over entire foodstuff chains. When it comes to particular projects, the list of relevant issues can be adapted in terms of priorities, as in most cases projects do not aim to address all aspects of sustainability.

Besides the specific context and scope, the use of indicators is also contingent on the kind and level of ambition. Blonk and Dutilh (2000) *inter alia* mention three dimensions:

- visionary vs. relative;
- incremental vs. big strides;
- integrated vs. specifically-focused.

These categories are of interest but are not a substitute for the generation of more distinctive images of what a sustainable foodstuff chain might be. It was striking that in a

Table 18.3. Inventory of measurable sustainability aspects.

| *Environmental* | *Social* | *Economic* |
|---|---|---|
| Use of mineral and fossil materials/fuels | Safety | Financial performance of farmers (income, profitability, future perspectives etc.) |
| Use of chemicals that affect higher ozone-layer | Health | Compliance with requirements concerning food safety and hygiene |
| Manure (balance) | Animal well-being | Meeting consumer concerns: risk assessment, communication and ethics |
| Heavy metals (balance) | Working conditions | Expenditure on local community |
| Non-energy contribution to eutrophication and nutrification | Child labour | Product innovations and product development |
| Drying out of soil | Human rights | |
| Water consumption | Working hours/wages | |
| Use of pesticides and deterrents | Education/training | |
| Non-energy contribution to global warming | Gender issues | |
| Bio-diversity | Participation in decision-making/organisational capacities | |
| Use of space | Local community development/ capacity-building | |
| Erosion | | |
| Soil ecology | | |
| Risks | | |
| Environmental profile of product (LCA) | | |

Source: Blonk and Dutilh, 2000.

conference organised by the Foundation, in discussions on Table 18.3 only two images of agriculture were referred to: conventional agriculture and organic agriculture. The latter was most frequently mentioned to express the view that organic agriculture cannot be the only legitimate alternative to conventional agriculture, but no other stylised images of a sustainable chain were obviously available. Sustainability strategies require that choices are made and priorities are set, which cannot be out-sourced; however, it should still be possible to build a number of consistent images of sustainability that companies can opt for and communicate to their stakeholders, and a set of sustainability indicators can then be applied within an comprehensible framework. Such a framework is particularly needed when the purpose of indicators is to measure either distance-to-ambition levels or concrete targets.

To sum up, the case shows the relevance and importance of the discussion within the Dutch food industry. The companies involved have made different steps that support sustainability. The need to develop indicators to monitor the process towards sustainability increases awareness of the need for different images of sustainable product chains that will help and stimulate companies to make strategic choices and to shape their operations accordingly.

### 18.5.3. *The Sustainable Coffee Project*

#### *Introduction*

The sustainable coffee project (SUSCOF) is a common endeavour of the Consortium Suscof RL which consists of six Costa Rican coffee co-operatives, the Institute for Sustainable Commodities (ISCOM) and the Costa Rican Centre for Technology and Innovation (CEGESTI).[1] The project is based on a chain-oriented approach aimed towards reaching continuous improvement in the production of coffee. Continuous improvement is in this project defined as a process whereby management utilises all human resources and relevant information to produce a constant stream of improvements in all aspects of consumer value.

To be able to apply the chain-oriented approach to the Costa Rican coffee sector required a rethink of environmental issues in industrial processes. Environmental problems are traditionally regarded as technical problems. By prescribing a set of technologies the coffee sector was believed to be capable of curtailing and controlling its environmental impacts. Although technology is important, it is only part of the solution.

#### *A managerial approach*

In order to come to integrated solutions, which are preventative in nature and cost-effective, environment had to be seen as a managerial problem and responsibility in the first place. To recognise that, it was considered important to have a comprehensive insight into the environmental effects of the coffee-growing and milling processes, so as to assess the latter's environmental strengths and weaknesses. This led to elaborate Initial Environmental Reviews which provided recommendations on how the most pressing environmental (including human health) problems could be addressed.

#### *Towards ISO 14001*

Managerial experience with documentation may be slight at best in this sector, as witnessed by rudimentary or non-existent accounting and production related documentation systems. For such firms the drive for competitiveness had to begin with the most fundamental approaches to workplace improvement and housekeeping. However, ISO 14000, if implemented in a prudent and cost effective way, can be a useful guide when beginning a program of continuous improvement of product quality and production processes. For this reason, it was decided to implement Environmental Management Systems based on the ISO 14001 norms in the coffee mill of each of the seven co-operatives.

The implementation of the EMS started in February 1999. Already the first implementation activities have allowed the co-operatives to obtain a better insight into their envi-

---

[1] The project is financed by funds based on the Sustainable-Development Agreement between Costa Rica and The Netherlands (Ecooperation in the Netherlands and Fundecooperacián in Costa Rica) and by the Netherlands Ministry of Environment (VROM).

ronmental performance related to water, air and soil pollution and plan activities to improve themselves.

### *Good agricultural practices*

In order to improve environmental impacts along the whole chain, environmental management activities were also defined at farm-level. The six co-operatives together cover an area of more than 18.000 hectares of coffee plantations, being owned by 9.600 farmers. ISO 14001 includes procedures pertaining to the environmental aspects of purchasing. In the case of coffee, the coffee producers (farmers) are by far the mills' main suppliers. This justifies a monitoring system based on different criteria related to quality, environment and human health. For a large part these criteria will be derived from recently defined 'Good Agricultural Practices' (GAP) that have been formulated by a group of European food companies (although GAP has by no means the same status as does GAAP in the accounting profession, they indicate what the market is likely to require in the coming years). The short-term objective of the implementation of such an monitoring system (the so-called 'Eco-monitor') is to get an insight into the social and environmental performance of the coffee producers and into the data available for monitoring. The medium-term objective is to improve the social and environmental performance of the producers by further implementing the GAP. In addition, a small group started this year with the implementation of organic coffee production as an alternative, in order to meet the demands of certain niche markets.

### *Strategic components involving the market*

The management aspects of environmental care are not only operational in nature but also have clear significant strategic components. In particular, the application of advanced environmental technology and processes is likely to be more rewarding if this can be communicated to the market which then either increases sales or and ensures a premium price. Originally, this aspect was considered within the framework of discussions with one big retailer in the Netherlands. The advanced idea implied a broader market approach by reaching potential clients in different market segments, and the building up of a sales organisation at the level of a consortium. In the past the co-operatives in this project had not generally been involved in the commercial side of the coffee business but simply sold their coffee to middlemen as soon as they had processed it, being satisfied with the certainty of immediate cash flows. However a growing interest in the origins of the coffee which has developed amongst overseas customers had already made coffee farmers and their organisations more sensitive towards the prevailing environmental and social conditions under which they produced their coffees. A strategic answer to this requires a direct link with overseas clients in order to present one's own special product qualities and to directly respond to their preferences. This idea is being taken up in this project, in order to develop it into the commercialisation of coffee that is produced under a consistent and verifiable sustainability regime.

The sustainable chain management approach that the SUSCOF project pursues implies a substantial increase in the importance of environmental accounting for both internal and external purposes. A first major step in acquiring basic environmental information was made by conducting elaborate Initial Environmental Reviews of the coffee mills. To illustrate this, two environmental aspects of coffee processing will be briefly discussed: wastewater, and energy consumption.

### *Waste water in the processing of coffee*

In producing coffee, the beans are first 'de-pulped' (their outer skins are removed) in the coffee mills, followed by the removal of subsequent shells by fermentation and husking. After the de-pulping process, the coffee is left in basins for at least 8–12 hours to ferment, then washed twice in order to remove the acid juices which are created during the fermentation process. These juices form part of the ensuing waste-water stream and are detrimental to the environment, as are the organic substances in the waste-water. Table 18.4 shows some data on the volume of waste-water generated yearly from processing coffee.

Table 18.4 in the first place shows the conspicuous absence of waste-water figures. Thanks to the environmental management system that has been implemented since then, this situation will change so that waste water can be subject to explicit environmental improvement programs. Before the 1990s, this waste water was discharged directly into a nearby river, which caused, due to the acidity of the water and the volume of water and pulp, an important environmental problem in the surroundings of the coffee mill as well as elsewhere in the same river basin.

In general, it can be observed that each co-operative has succeeded in the construction of a waste-water treatment plan. However, most of them still have problems with managing the open-air basins which are used for the waste-water to settle out and purify in a manner that meets legal requirements throughout the year. The implementation of the EMS enables the co-operatives to improve the monitoring needed to control the treatment of waste-water as a basis for reaching the required standards. To meet future legal requirements, additional changes have to be implemented.

### *Energy consumption*

Two major sources of energy for the milling processes are electricity and fire wood. In the past, energy use was never seen as an environmental issue. This also explains why electricity consumption is substantial (see Table 18.5).

As part of the implementation of ISO 14001, a monitoring system has been designed which allows the co-operatives to measure their electricity consumption per working area.

Table 18.4. Volume of waste water generated in $m^3$ per co-operative per year.

| *Harvest period/ Co-operative* | *Unit* | *A* | *B* | *C* | *D* | *E* | *F* |
|---|---|---|---|---|---|---|---|
| 1992–93 | $m^3$ | n.d. | 33,183 | n.d. | n.d. | n.d. | n.d. |
| 1993–94 | $m^3$ | n.d. | 37,332 | n.d. | n.d. | n.d. | n.d |
| 1994–95 | $m^3$ | n.d. | 38,362 | n.d. | n.d. | n.d. | n.d. |
| 1995–96 | $m^3$ | 15,621 | 36,788 | n.d. | n.d. | n.d. | n.d. |
| 1996–97 | $m^3$ | 6,548 | 3,584 | n.d. | n.d | n.d. | n.d. |
| 1997–98 | $m^3$ | 3,531 | 7,505 | 7,5026 | 81,045 | 28,934 | 15,282 |
| 1997–98 | W/F | 0.32 | 0.53 | 1.2 | 1.14 | 0.4 | 1.1 |
| 2000–01 | $m^3$ | 1,995 | 7,849 | 37,496 | 20,365 | 58,590 | 60,410 |
| 2000–01 | W/F | 0.17 | 0.43 | 0.73 | 0.50 | 0.81 | 0.50 |

n.d. = no data available. W/F = volume wastewater ($m^3$) per unit of coffee processed (fanegas).
1 fanega = 46 kg. F = total amount of fanegas.
Source: Chacon (1999) and EMS updates.

Table 18.5. Total electricity consumption (kWh) per co-operative per year.

| *Harvest period/ Co-operative* | *Unit* | *A* | *B* | *C* | *D* | *E* | *F* |
|---|---|---|---|---|---|---|---|
| 1992–93 | | n.d. | n.d | n.d. | n.d. | n.d. | n.d. |
| 1993–94 | | n.d. | n.d. | n.d. | n.d. | n.d. | n.d. |
| 1994–95 | | n.d. | n.d. | n.d. | n.d. | n.d. | n.d. |
| 1995–96 | kWh | n.d. | n.d. | n.d. | n.d. | n.d. | 1,886,500 |
| 1996–97 | kWh | n.d. | 123,578 | 573,633 | 915,546 | 915,546 | 1,196,875 |
| 1997–98 | kWh | 91,787 | 107,544 | 656,594 | 654,040 | 993,174 | 1,500,076 |
| 1997–98 | E/F | 8.5 | 7.7 | 10.5 | 9.3 | 14.1 | 10.5 |
| 2000–01 | kWh | 88,037 | 128,925 | 459,572 | 446,950 | 896,962 | 8,700,001 |
| 2000–01 | E/F | 7.46 | 7.12 | 9.02 | 11.00 | 12.50 | 7.23 |

n.d. = no data available. E/F = electricity consumption (kWh) per unit of coffee processed (fanegas). 1 fanega = 46 kg. F = total amount of fanegas.
Source: Chacon (1999) and EMS updates.

The analysis of these data will make it possible to define next year's adjustment programs, which will decrease the electricity consumption.

*Firewood consumption*

After the de-pulping and rinsing process, the coffee beans have to be dried. The ovens that are used to dry the beans, are heated by burning fire-wood. The co-operatives buy part of this wood from their members (both abandoned coffee trees and wood from other trees on the coffee farm that are there to provide shade to the coffee trees, the so-called 'shadow trees'). However, in part it comes also from other (forest) areas in the region of the coffee mill though its exact origins are not known (yet).

The yearly amounts of wood the co-operatives require may create a serious threat to the local forests as supplies from local plantations are insufficient. Moreover, wood as fuel creates local health problems. Table 18.6 indicates the wood consumption.

One of the solutions which has been devised to decrease the volume of firewood required, is to use the husk of the coffee as an alternative energy generator. The co-operatives have implemented monitoring systems as a means to regulate the balance between husk and wood such that the correct temperatures are reached while decreasing the amount of fire-wood needed.

An alternative (but in fact traditional) solution refers to substituting the drying machines by open-air platforms where the coffee beans are dried by the sun. This alternative has been applied by co-operative A, which explains the relatively low figures of this co-operative. For the largest co-operatives, however, this may not be a feasible solution.

The environmental improvement programs that the co-operatives have adopted clearly require indicators that can show the ultimate results of reduced usage of energy, in particular of the less desirable types of energy. First, the individual co-operatives need such indicators to control and, as need be, to correct their processes. Secondly, the seven co-operatives participating in the SUSCOF consortium have an interest in each other's results as these have an impact on the entire group. Thirdly, overseas companies that will buy the SUSCOF coffee will be interested in whether the sustainable aspects of the coffee, that they may communicate to the consumers, are based on real improvements.

Table 18.6. Total wood consumption (m3) per co-operative per year.

| Year/Co-operative | Unit | A | B | C | D | E | F |
|---|---|---|---|---|---|---|---|
| 1992–93 | $m^3$ | 244 | n.d | n.d. | 2,840 | n.d. | 10,000 |
| 1993–94 | $m^3$ | 243 | n.d. | 4,390 | 4,000 | 7,161 | 7,500 |
| 1994–95 | $m^3$ | 252 | 800 | 3,803 | 2,800 | 5,902 | 6,800 |
| 1995–96 | $m^3$ | 274 | 1,000 | 4,359 | 5,000 | 7,467 | 6,500 |
| 1996–97 | $m^3$ | 227 | 1,200 | 2,900 | 2,200 | 5,460 | 7,000 |
| 1997–98 | $m^3$ | 217 | 1,000 | 3,029 | 4,670 | 7,178 | 4,065 |
| 1997–98 | WC/F | 0.08 | 0.07 | 0.04 | 0.06 | 0.10 | 0.02 |
| 2000–01 | $m^3$ | 176 | 1.04 | 4,741 | 1,222 | 2.80 | 5,755 |
| 2000–01 | WC/F | 0.07 | 0.06 | 0.09 | 0.03 | 0.04 | 0.04 |

n.d. = no data available. WC/F = wood consumption ($m^3$) per unit of coffee processed (fanegas).
1 fanega = 46 kg. F = total amount of fanegas.
Source: Chacon (1999) and EMS updates.

### *Developing indicators*

The entire SUSCOF-project leads to a consistent set of management control systems that are mutually related and supportive to one another. The management at the consortium has a co-ordinating role to play. It can receive information from the individual co-operatives, evaluate it against common goals and benchmark the outcomes so as to find out what organisation has the best performance and why. The various control systems are reflected in Figure 18.1.

There is a form of 'hierarchical' structure which is involved here. The Eco-Monitor is management by the co-operatives. That means, each co-operative monitors its own farmers. It is impossible to include all the farmers in the monitor at once. Therefore, the system begins with a limited number, but this number is to be increased step by step. It will make it possible to have indicators with respect to the use of chemicals and workers' health & safety.

Each individual co-operative will manage the EMS according to ISO 14001 for the processing plants. However, the programmatic goals and the way the monitoring will be executed can be partially co-ordinated at the level of the consortium.

At the consortium level, further co-ordination is possible by common strategic decision-making and by setting up a system of performance indicators that allows for comparison and integration. This is necessary both for the internal evaluation of the process through continuous improvement and for informing the stakeholders, in particular the overseas clients, about how the consortium is performing. By including the consortium's common goals in the environmental programs under ISO 14001, the regular verifications to be done by external auditors provide a major means to build up credibility.

To get a reliable picture of the whole, efforts have been and will be further made to carry out an LCA of the coffee chain as the basis to build up indicators that take the chain into account. Here, we touch upon different intriguing methodological questions, particularly concerning the weighting of environmental impacts that occur in quite different regions. The consortium itself cannot be expected to deal with this. By working

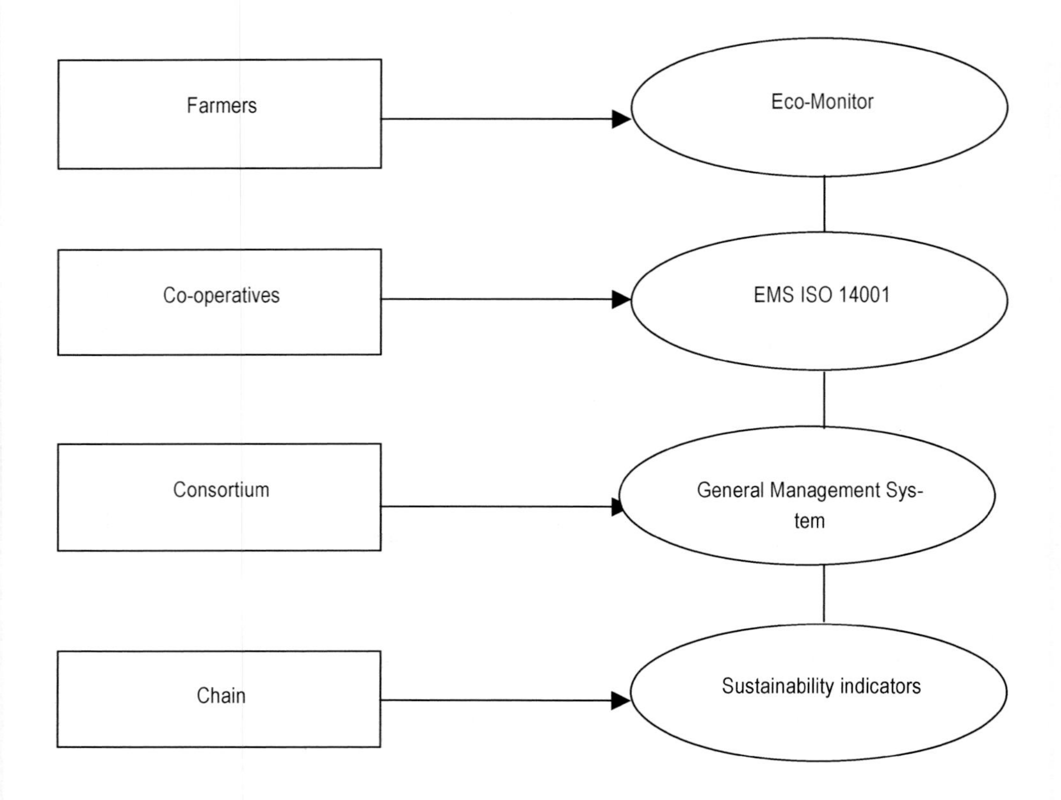


Figure 18.1. Management control for sustainability.

together with different network partners in different countries, including university-based researchers, research capacity has been available for this.[1]

Another aspect is the need to measure distance-to-target, at least a limited set of core indicators in the field of use of chemicals, waste water and energy. To realise this, it is necessary to define a longer-term improvement program for the coming five years with concrete targets. Advisers assist the consortium, but the actual adoption of such targets needs to be based on a decision that the co-operatives involved have to make together.

The most promising approach in the early stages of implementation is to develop simple, readily understood measures that reflect environmental priorities and make maximum use of available data. In addition, it is important to interview the staff of the organisation about non-environmental issues by discussing general business concerns. This basic level of environmental performance evaluation (EPE) mainly relates to, what may be called, defensive or reactive environmental policies, involving measures that are imposed by external stakeholders or that are easy to take, i.e. that do not require substantial organisational change or heavy investments. In the case of the co-operatives, this kind of EPE has supported the improvement of environmental performance already during the imple-

[1] Among them are CEGESTI and CINPE in Costa Rica, the Wageningen University in the Netherlands and the Australian National University in Australia.

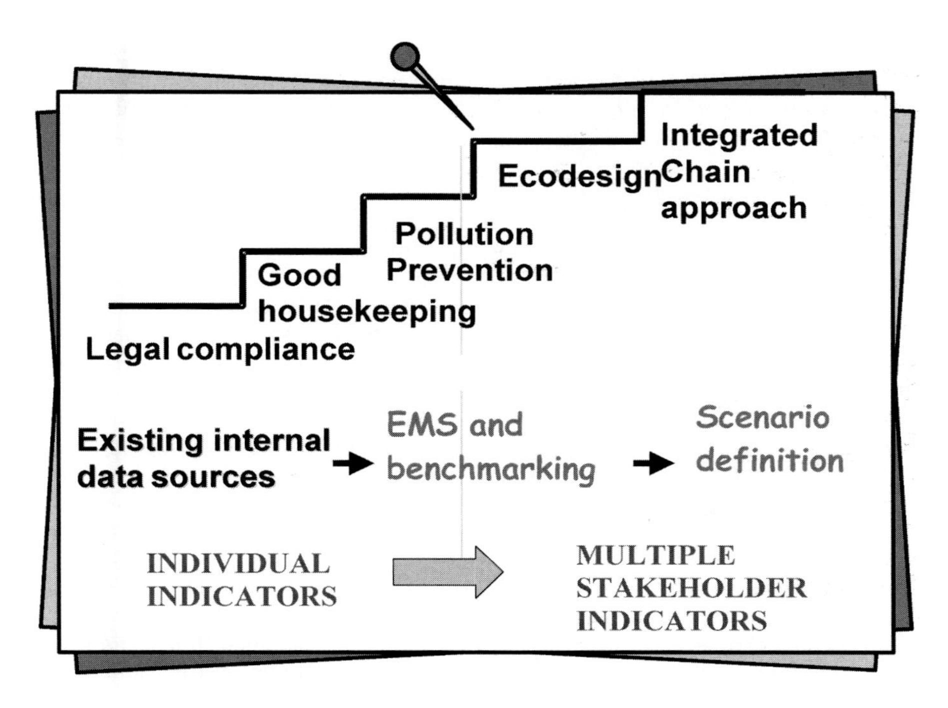


Figure 18.2. Stimulating continuous improvement by multilevel data collection and definition of environmental performance indicators.

mentation of environmental management systems. The information required to define the environmental performance indicators (EPIs) has for a great part been derived from the information obtained by the Initial Environmental Reviews of the coffee mills. These indicate where without substantial investments environmental improvements are possible, in particular 'good housekeeping' measures. For instance, savings in the use of energy and water have been realised by monitoring more carefully the production process. The information obtained through the monitoring process is used to define work instructions and training activities, and resulted into a better control of peak loads of electricity and water. Husk, a waste created by the processing of coffee cherries, can be used as a fuel, partly substituting the use of firewood. For the entire process – from harvesting to the dispatch of green beans – a quality-control protocol could be set up and accounting tools can be developed that help to identify cost-saving opportunities.

This basic level of EPE, which is based on general environmental condition indicators (ECIs) and environmental performance indicators (EPIs), is largely related to risk management and dealing with environment-related costs – such as high-energy consumption, high waste rates or emissions – that require potentially expensive pollution control measures. Major risks are non-compliance with legislation and financial liabilities. To a large extent, EPE occurs within the framework of considering audit results by environmental managers and their senior managers.

However, on the medium term, the SUSCOF Consortium has to develop a more advanced EPE and related indicators in order to satisfy external stakeholders, particularly in the overseas market. Main aim of EPE at this level is to stimulate and support continuous improvement, particularly through awareness and identifying appropriate measures. The organisation has to move towards a pollution prevention approach if it wishes to be in an advanced position in terms of ecological sustainability. This requires more detailed data on the materials, water and energy flows. This data is also crucial to increase resource productivity through waste minimisation and other initiatives.

For the co-operatives, this level of EPE will make it possible to define investment opportunities in the area of cleaner technology. It is expected that on the medium term (two to four years), parts of the existing equipment will be eligible for replacement. For instance, a number of electrical engines are over-dimensioned and otherwise fairly old. Replacement by newer and more efficient equipment could save energy and costs. In certain cases, more advanced technologies could be considered, at least as partial substitutes for existing machinery. For instance, photovoltaic energy equipment could be used in the drying and further processing of the coffee beans, particularly if the investment is subject to a short payback period. Also the use of worms in the composting of pulp can be a major contribution to replacing chemicals at the farm sites. This implies, however, considerable operational adaptations in the coffee mills. The investments concerned need to be paid back within a fairly short period of time by means of costs savings or opportunities to sell the obtained results such as the compost to the farmers who can use it as fertiliser.

Finally, for the long term (more than 5 years), it is necessary to extend the horizons of the co-operatives in order to be able to design integrated production systems that consider the entire chain. There is no quick jump into such systems, since they require careful scenario definition and planning processes based on the interactive participation of the different stakeholders in and around the chain. Interactive learning can enhance the information level about the long-term key factors in terms of technology, logistics and commercialisation. Scenario development is a necessary means not only to build up consistent pathways towards sustainability, but also to reach consensus among the major actors in the coffee chain as to how the future should look like.

One aspect of EPE is that of 'unintentional environmental improvements'- the improvements in environmental performance that result from non-environmental measures – such as investments in new equipment or general cost reduction programs – can be a major cause of environmental improvement (Bennett and James, 1998). Therefore, it is vital to extend EPE into areas such as product development and capital budgeting as these are likely to influence the eco-efficiency of future processes and products.

Altogether, the fine-tuning process of EPE is accompanied by a shift in emphasis from individual indicators to the development of scenarios based on multiple-actor indicators. Within these, indicators that relate data from different units to each other become more important for comparative analyses, for example emissions per thousand dollars of turnover, or waste per unit of production. Related to the SUSCOF project, the consortium – that consolidates the strategic alliance between the six co-operatives involved – is the appropriate level for the creation of the overall EPE framework. This implies that there should be a core set of uniform indicators that are used in each of the six co-operatives. Moreover, uniformity allows different kinds of comparison and benchmarking. Of course, along with the core set, individual co-operatives could use other indicators, which are attuned to their own specific situation, as well.

## 18.6. Concluding remarks

The development of sustainability indicators at corporate and chain levels is still in its infancy. There is still much work to be done in this area involving different disciplines. It is important that environmental management accountants take this development very seriously as sound accounting is of major importance to the cause of sustainable development. EMA specialists can help to keep a fair balance between theoretical soundness and the need for pragmatic solutions.

Although indicators as such are neutral to policy choice, developing them induce those involved to make explicit choices that put their commitment to the subject to the test. For instance, if key indicators are expected to measure distance-to-target, then to implement such indicators it is indeed necessary to define concrete targets. Discussions about them may bring to light a lack of clear images of to what the road towards sustainability looks like, both technologically and technically, and how fast the pace should be. Working on sustainability indicators, therefore, promises to be challenging, intriguing and rewarding.

# 19. Towards Transparent Information on the Environmental Quality of Products – LCA-based Data Production for the Finnish Foodstuff Industry

*Torsti Loikkanen and Juha-Matti Katajajuuri*
*Group Manager resp. Research Scientist, The Research Group of Industrial Environmental Economics (IEE), Environmental Technology, Espoo, Finland, VTT Chemical Technology; E-mail: torsti.loikkanen@vtt.fi, juha-matti.katajajuuri@vtt.fi*

## 19.1. Enhancing competitiveness through green innovation

As well as economic and technological competitiveness issues, the environmental issue has become an important element in current industrial competition. According to several observers, the response of industry to environmental problems may even be a leading indicator of its overall competitiveness. Moreover, a truly competitive industry is more likely to take up the new issue as a challenge and to respond to it innovatively (Porter and van der Linde, 1996; Loikkanen, 1999). As well as environmental issues, ethical and social aspects will also be of increasing importance to company success, when considering current and foreseeable future trends in financial and stock markets (Schmidheiny and Zorraquin, 1998).

From the perspective of a foodstuff company, the ultimate reason for producing data on the origin and different properties of foodstuffs, including environmental aspects, relates to the intense international competition in current and future food markets. The players in the food and foodstuff business value chain largely recognise that a purposeful quality policy, integrated with the principles of sustainable development, will be a necessary condition for their future competitiveness in globalizing markets.

A customer and stakeholder-oriented approach requires the production of transparent and traceable data on product origin and quality which also encompasses environmental aspects. In the case of agricultural and food products, data on health, safety, environmental and hygienic properties is of especial importance for consumers. On the other hand, both economic and related technological competitiveness is crucial for the success of companies today. Life cycle assessment (LCA), in combination with other environmental management methods and tools, supports the identification of improvement options of individual firms and of the entire supplier and business value chain of food industry. Subsequently, this analysis may bring forward ideas for innovative product and process development and related R&D to support sustainable development (Loikkanen and Hongisto, 2000). It is important that companies from the whole business value chain participate in such a common effort, because the quality of the final product will be determined by the weakest links in business chains.

## 19.2. Life cycle data of Finnish foodstuff production

So far, only limited LCA-based data exist on the environmental burdens and impacts of the entire business value chain of foodstuff production. LCA methodology is a systematic



*M. Bennett et al. (eds.), Environmental Management Accounting: Informational and Institutional Developments*, 249–252.
© 2002 *Kluwer Academic Publishers. Printed in the Netherlands.*

and comprehensive method of producing data on the environmental burdens of products and related processes. According to a recent Finnish study, the use of LCA is gradually expanding among other environmental management tools in companies (Loikkanen et al., 1999). By relying on this method, the Finnish foodstuff industry launched the production of LCA-based data about agricultural and foodstuff products, which was considered to be a necessary element in the long-term development of its quality management and accounting systems. As the LCA method had not previously been widely applied to the assessment of agricultural products and complex biology-related processes, several specific issues and development needs had to be dealt with.

The production of LCA-based environmental data was launched through a pilot study of barley production, extending from ploughing in farmers' fields up to the food troughs in cattle houses. The study also included the production of fertilisers and other inputs into barley production, as well as their packaging and transportation. The intention of this pilot study was to deepen the understanding of special problems of applying LCA to biological and environmental phenomena in the special case of agricultural processes, in order to be utilized in LCA studies on food products. The initiative for the pilot study was made by the Association of Rural Advisory Centers, and the research consortium consisted of the companies Suomen Rehu (feed industry), Valio (dairy industry), Kemira Agro (fertilizer industry), Broilertalo (meat industry) and UPM-Kymmene Forest (forest industry). The research partners were the Industrial Environmental Economics Group (IEE) from the Technical Research Centre of Finland (VTT) and the Agricultural Research Centre of Finland (MTT). As well as companies, the pilot study was supported by the National Environmental Cluster Research Program and was completed in 1999 (Katajajuuri et al., 2000).

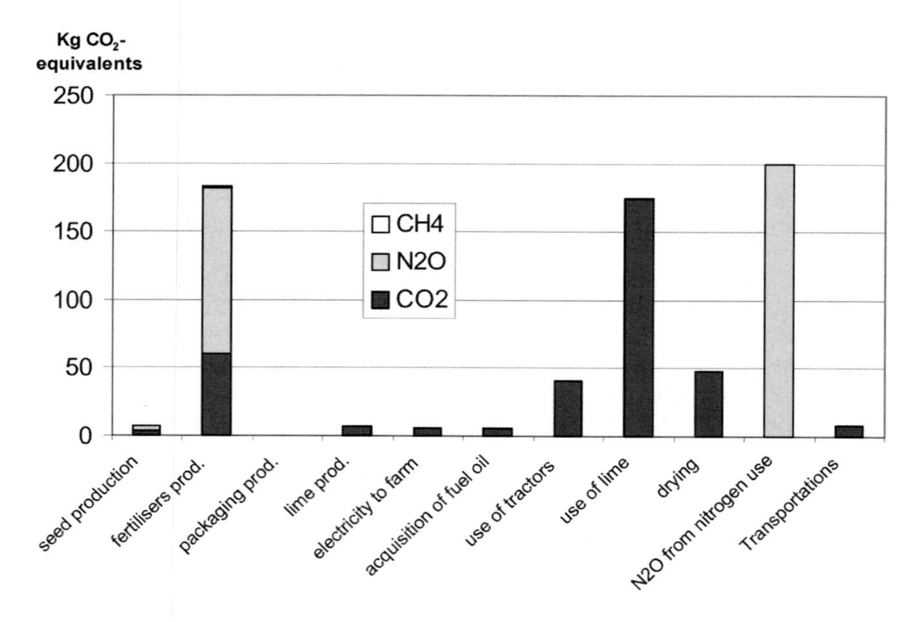


Figure 19.1. The potential contribution to global warming by life-cycle phases and emissions according to pilot study (Katajajuuri et al., 2000) (kg of $CO_2$-equivalents per 1000 kg barley, time horizon of 100 years).

The pilot study turned out to be quite cumbersome, and data production to be exceptionally labour-intensive, in particular because of complicated biological processes such as nutrient cycles and crop rotation, and because of the lack of suitable input data for LCA modelling. The results demonstrate well the steps of barley production which are the major sources of environmental burdens and impacts.

According to the results of the life-cycle inventory and impact assessment, the environmental load from arable land was concluded to be the primary cause of environmental impacts in feed barley production. As expected, washouts dominated eutrophication impacts. The main sources of global warming potential appeared to be nitrous oxide, emissions from agricultural soil, and carbon dioxide emissions from the use of lime. There are however uncertainties in those emission estimates due in particular to incomplete data and also to simplifying assumptions which were made in the impact assessment methodology. The production of fertilisers was also found to be an essential contributor, since its relative share of primary energy consumption and global warming potential appeared to be relatively high. Other important pollution sources turned out to be the drying of grain and the use of agricultural machinery. The energy consumption in agricultural processes was almost 50% of total energy consumption. Correspondingly, the share of the production of fertilisers appeared to be over 40% of the total energy consumption. In addition to these energy-intensive phases, ammonia volatilisation from fertilisers contributed essentially to acidification.

On the basis of the results of the LCA exercise, the main options for improvement seem to relate to efforts to optimise the use of inputs into the agricultural process in question. The opportunities for decreasing the use of energy relate especially to the drying of barley grains. Moreover, the cultivation practices and related behaviour of farmers may affect many other environmental issues. Estimation of the effects of using pesticides, as well as the changes on landscape and on bio-diversity, turned out to be difficult and complex and will be studied in more detail in forthcoming research efforts.

The LCA-based data on the barley process in respect of the environmental burdens and impacts of this pilot study can already be used by participating companies to support the continuous improvement of agricultural processes and products, both within firms and in their business value chains. The results also provide support for further R&D and innovation, again both in individual companies and in their business value chains. Moreover the LCA-based environmental data which was produced can also be utilised for external purposes, as information to support sustainable choices by customers concerning the product in question. In forthcoming research efforts the aim is to extend consideration also to issues of environmental costs. All these methodological approaches are aimed at supporting more sustainable investment decisions, cost-effective strategic management, and environmentally benign production and product choices.

## 19.3. Towards interactive communication in an information society

Due to the rapid current development of information and communication technology (ICT), environmental information on products and processes may be stored, maintained and disseminated by ICT tools such as the Internet. Access to ICT will increasingly be essential to full participation in society and citizenship, and it has even been proposed that Internet access should be considered to be a fundamental right (Information Society Forum 2000). In this way, in the context of a current and future information society, consumers and other stakeholders can become informed and aware about environmental, hygiene and other quality characteristics of products and processes in production chains (see Figure 19.2 for a view of different actors and stakeholders). Consequently ICT will play an important role in the development of sustainable industrial production. Interactive information systems, based on the dissemination of transparent documented and traceable environmental data, will support more environmentally friendly choices by consumers, and give comprehensive information to companies, authorities and other stakeholder groups for their different needs (Bouma, 1999).

Life cycle assessment is presently amongst the most systematic and accurate methodologies in the assessment of the environmental performance and properties of products and processes. This methodology can be used both in individual companies, and also among companies in business value and supplier chains, upstream and downstream. Combined with other environmental management tools and indicator-based monitoring and accounting systems, LCA-based data will play an important role in the search for eco-efficient solutions for products and production processes. Recent and foreseeable trends related to the information society offer increasing opportunities for the dissemination of the origin and quality characteristics of products, including environmental and hygiene aspects. In this way the interactive communication between different actors and stakeholders – one of the key factors in the coming information society – can create conditions for the more sustainable future of foodstuff products as well as of other consumer goods.

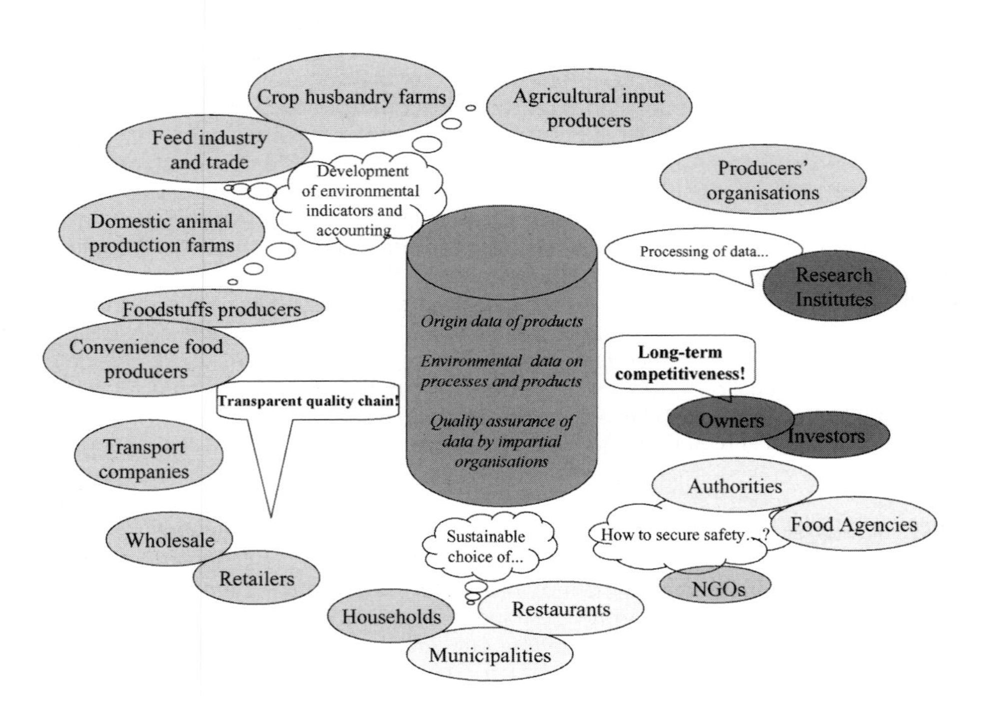


Figure 19.2. A simplified presentation of different actors in the food production chain producing environmental data and of the different stakeholder groups using this data.

# 20. Prospective Analysis for Implementing an Environmental Management System in Pig Farms: Likely Role of an Environmental Management Accounting System[1]

*Bertrand Montel*
*Assistant professor, Institut National Agronomique Paris-Grignon, Paris, France; E-mail: bmontel@inapg.inra.fr*

## 20.1. Introduction

In 1970, French authorities set up a national plan to rationalise pig production because current levels of output were insufficient in comparison with levels of consumption. One of the main reasons for this shortfall was the type of pig production systems which were normal at that time, which were mainly unspecialised systems each with only a few pigs – in 1968, there were 800,000 pig-holdings which each farmed an average of only 12 pigs (Teffène et al., 1998).

This plan for the rationalisation of pig production was structured around two points: (i) the modernisation of pig farms, and (ii) the organisation of the French pig market. The first point was pursued by providing support for technical improvements such as in the training of farmers and the genetic improvement of animals, which led to the specialisation of pig farming systems, notably farrowing-fattening systems. As a consequence, pig production is now concentrated in fewer farms, mainly in Brittany. In 1995, there were 90,000 pig farmers which held an average of 157 pigs per herd (1,223 pigs per herd in specialised farms, compared with 37 in 1968) (Tefféne et al., 1998). The second point relied on producer groups, on a payment scale and on an auction system (Marché du Porc Breton). At the same time, firms which were either upstream of farms (feedstuffs industries) or down-stream (slaughterhouses, food-processing industries) developed themselves, once again mainly in Brittany. Socially, this model of development found its legitimacy in the Fordist compromise (sharing of productivity gains) which prevailed in French society up to the 1980's, which emphasised productivity as an achievement. In Brittany, this has led, among other consequences, to a huge increase in pig density and thus in the amount of nitrogen (N) spread. One effect of this has been water pollution by nitrates and phosphates.

In a social context where environmental concern has arisen as a key value, the question of a new development pattern has come up more and more acutely. Authorities have intended to regulate environmental impacts of pig production using statutory means such as the national Classified Installations for Environment Protection Act (CIEP) or the European Directive on Nitrates (DN). The implementation of these regulations has not yet achieved its goals, so it is necessary also to consider other approaches to support them.

Environmental management systems (EMS), as defined by ISO 14000 standards, justify consideration. In France, no pig farm has implemented an EMS yet, so this paper will use a systems approach to develop a prospective analysis. First, it is necessary to understand the environmental issues of pig farms. The paper will then briefly present what an EMS

[1] Acknowledgement: This work is part of a PhD thesis project supported by the Bretagne Eau Pure 2 program (convention n° 97/09-071).



*M. Bennett et al. (eds.), Environmental Management Accounting: Informational and Institutional Developments*, 253–263.
© 2002 *Kluwer Academic Publishers. Printed in the Netherlands.*

is, and then go on to analyse what are the consequences of the implementation of an EMS in a pig farm, and to show that information management is a key issue and that environmental accounting can be a worthwhile support tool.

## 20.2. What are the environmental problems of pig farms?

### 20.2.1. *Pig farm as a controlled system*

Since a systemic approach has been chosen for consideration as a way to cope with environmental issues of pig farms in Brittany, the pig farming system needs to be defined. Pig farms are controlled and open systems; using Le Moigne's system approach [1984], such a system can be considered to be made up of a *decision-making* sub-system, an *informational* sub-system, and an operating sub-system. The latter can be divided into two other systems. The first (the *resources system*) includes the labour force, the technological and financial resources; the second (the *biotechnical system* – Landais and Deffontaines, 1991) is the set of biological processes controlled by the farmer. Analysis of the decision-making process will use Simon's IMC (Intelligence-Modelling-Choice) model (Le Moigne, 1974). Figure 20.1 illustrates the pig farm seen as a system, for a typical Breton pig farm.

Generally, an environmental problem of a system can be defined as any problem posed by changes in its relations with its environment. These may be due to changes in either

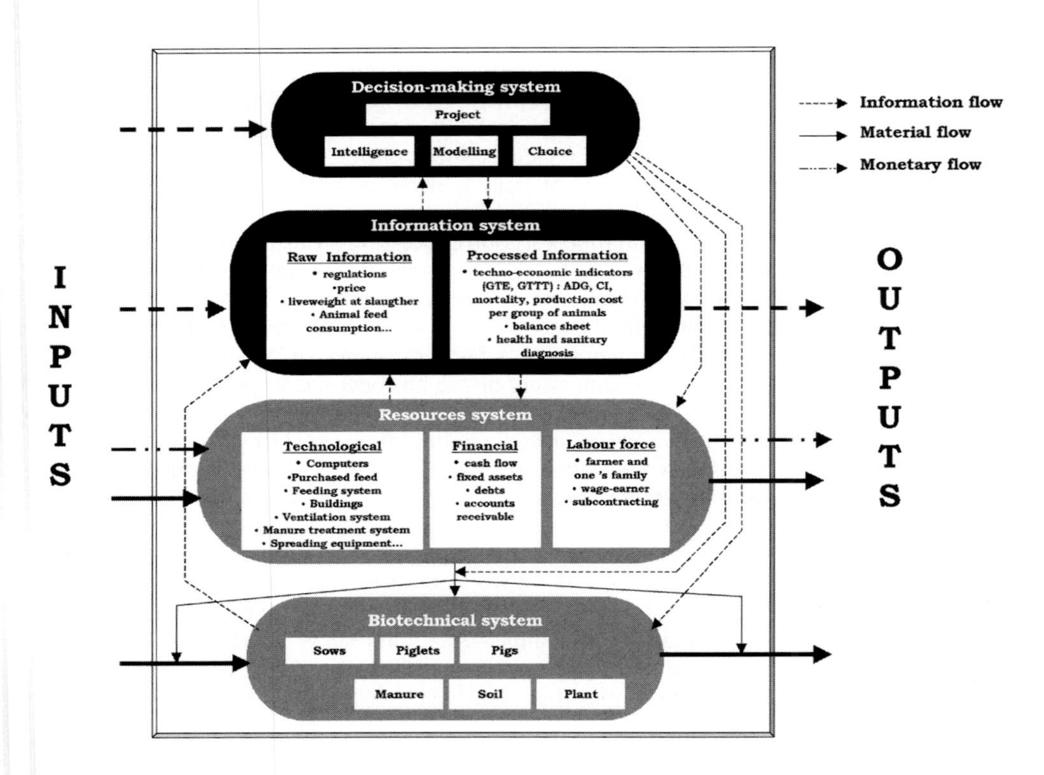


Figure 20.1. Pig farm seen as a system.

the environment and/or in the system itself, so we will analyse how the environment of pig farms has changed.

### 20.2.2. *Changes in the environment of pig farms: from pollution to illegitimacy*

Since pig farming is a human activity, we will consider the anthropogenic environment (society) and the non-anthropogenic one (the eco-system, but its anthropogenic elements) for which this paper will use the term 'natural' (although it is recognised that nature and society cannot be opposed).

Regarding the latter (the eco-system), the major change has been pollution. A pig farm is linked to its natural environment by various flows; among others, organic matter, nitrates, phosphates, and ammonia. So long as some kind of balance is maintained, these flows do not alter the integrity of the natural environment, so one may then consider that there is no change in the natural environment. However as soon as these flows begin to modify the integrity of the environment, there is pollution. In the case of pig farming in Brittany, the increase of animal density has implied an increase of the amount of N to be spread per unit of area. Because the natural cleansing capacity of the soil has been exceeded, leaching of nitrates and phosphates has occurred and has created modifications of the continental water eco-system such as eutrophication. This pollution also has another consequence, since continental water is used as a source of drinking water. In Brittany, some sources are closed because the nitrate drinkability standard (50 mg/l of $NO_3^-$) has been exceeded. At this point, the ecological problem becomes an health problem and thus comes within the anthropogenic environment of pig farms.

One of the main relations between a system and its social environment is the relation of *acceptance*. This relation is determined by the conformity of the system to a set of values prevailing within the society. Such conformity gives legitimacy, without which no activity can sustain its functioning since not only are some in-flows dependent on other people's goodwill (financial receipts, for instance), but also its very existence if it breaks the law. Recent legal cases show that pig farming in Brittany, by making water undrinkable and because of nuisances such as odours and noise, has reached the limits of its legitimacy. As society changes, so do its values – environmental concern has increased for nearly twenty years as a key value, particularly in western societies. Consequently, we conclude that the environmental problem of pig farms in Brittany is due to pollution which alters the ecological integrity of its natural environment and its social legitimacy.

When we look at the process leading to this pollution (Capillon, 1992), it appears to result from an interaction between the natural eco-system of the field (air-water-soil-plant), and the amount of nitrates and phosphates spread, resulting from the implementation of spreading practices. Generally, a practice is a set of co-ordinated actions modifying the state of a system in order to meet the farmer's aim. At its very first step, a practice is built at the decision-making level. Le Menestrel and Panes (1996) have proposed a distinction between three levels of legitimacy in the decision-making process: *consequential* legitimacy, *procedural* legitimacy and *cognitive* legitimacy. We will use this analysis pattern to assess how pig farmers have integrated the environmental factor in their decision-making process.

### 20.2.3. *Environment and decision-making in pig farms: an assessment*

Clear understanding and knowledge (intelligence) of the environmental situation of the farm was almost non-existent until the early 1990's. Indeed, the only widely used tool which enabled farmers to assess their environmental impacts was the fertilisation balance, which is basically designed not to minimize leaching but to supply sufficient amounts of nutrients to plants. At the end of 1993, French authorities and farmers association concluded an agreement to support a national plan of agricultural pollution control (PMPOA). Among the various measures, there was an environmental diagnosis of the farm, the DeXeL® (Dockès et al, 1995). However this diagnosis method is only partial since it deals only with nitrates and phosphates pollution attributable to farm wastes (animal manures, dirty water), so the intelligence step of the decision-making process remains mainly focused on technical and economic issues and cognitive legitimacy is lacking. Besides, use of DeXeL is linked to PMPOA which is to last only until 2001, and there is concern that once PMPOA is achieved, DeXeL will no longer be widely used.

The modelling step is generally barely implemented by the farmer himself (Bourgeat, 1999). Moreover, this step is often not made formally in regulation and steering decisions since farmers tend to place greater reliance on routine procedures (Cerf and Sebillotte, 1997). One must note that these procedures have not been 'greened' yet because few references exist; they are under construction. For planning decisions, the modelling of the various alternatives is seldom environmentally sound. Since these decisions bind the farm for a long time (say, at least five years), pig farms may continue to be controversial because they cannot achieve procedural legitimacy.

Regarding the final step, choice, we note that the environment is not taken into account and the choice is made on either technical and economic criteria or on previous experiences. We underline here that, in France, there is no environmental accounting system implemented in pig farms nor in any other kind of conventional farms.

At this point, we should conclude that pig farming systems need at least to make deep changes to their decision-making systems in order to solve their environmental problems. It should be emphasised that nitrates and phosphates are not in themselves polluting – pollution occurs only when there is an excess. So far as this excess is due to decisions made by farmers, pollution is likely to be controlled if their decision-making systems can be changed. According to Llerena (1996), this is an organisational failure of the system.

In 1996, following the Earth Summit held in Rio de Janeiro in 1992, ISO set up a new family of standards (ISO 14000) designed for any human activity from chemical plants to forest management. These standards define what an environmental management system (EMS) is and provide guidelines to implement a 'greened' decision-making process.

## 20.3. An environment management system in a pig farm: a prospective analysis

### 20.3.1. *What is an environmental management system?*

According to ISO 14000 (AFNOR, 1996a [p. 15] and 1996b [p. 31]), an EMS is a set of procedures whose aim is to enable a firm to achieve its environmental goals. Priority is given here to the more satisfactory method to reach an environmentally sound decision over the best *ex ante* alternative. This is justified by the complexity and singularity of environmental issues (Llerena, 1996). As shown in Figure 20.2, an EMS relies on an environmental audit and five points, and is driven by continuous improvement.

### 20.3.2. *Implementing an EMS in a pig farm. What could it mean?*

Firstly, the system in which the EMS will be implemented has to be determined, in particular the decision-making system. At a first glance, this appears to be quite simple as the individual farm seems to fit, though variations in business practice may cause some

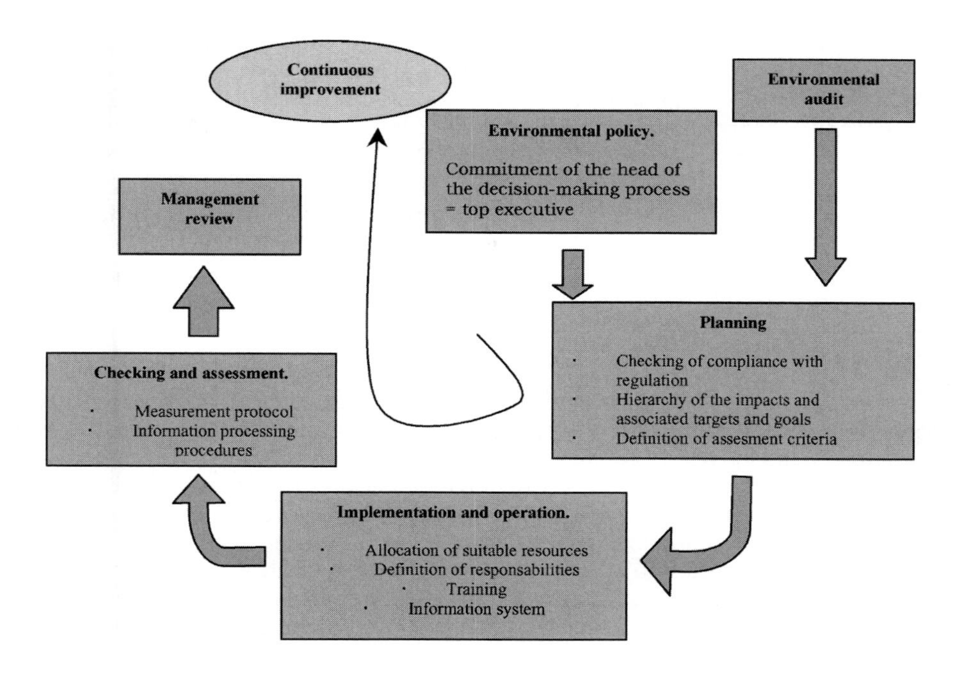


Figure 20.2. Principles of an environmental management system.

complications of definition. For example, if a farms were to subcontract fattening to an outside supplier, decision power would then lie in the hands of both parties so that it may not be easy to determine where the main centre of the decision-making system lies – this would depend on the particular contract.

Once the centre of the decision-making system has been identified, an EMS can be implemented only if there is an actual environmental commitment, which is the first condition (Bezou, 1997; AFNOR, 1996a, b; IISD, 1996). For such a commitment, a farmer must change his/her system of values by including environmental concern. This (r)evolution may help the farmer to rebuild legitimacy. To achieve this, information and education are necessary.

Planning is the next stage, following the commitment stated in the environment policy to implement an EMS. The farmer must check that the farm is in compliance with all environmental regulations (DN, CEIP, medical wastes, etc.), which implies some basic juridical training. Then, using an exhaustive environmental audit, the farmer must define a hierarchy of the farm's impacts on the environment, and the corresponding targets and goals according to its environmental policy (Table 20.1). In order to follow up the functioning of the EMS, specific indicators have to be created, which supposes joint work with research and extension services. From the previous elements, the farmer can establish a plan of priorities and target dates for achievement.

The implementation itself requires that the various resources that are required (labour force, money, technologies) are adequately allocated. This means that a formal investment plan has to be designed, which is not a widespread practice among pig farmers (Bourgeat, 1999, communication personella). Each worker's responsibilities (from head-farmer to

Table 20.1. Some examples of environmental impacts and associated targets and goals in a pig farm.

| *Impacts* | *Targets* | *Goals* |
|---|---|---|
| Non-point pollution by nitrates | Nitrate amount spread per area unit | Balance supply-demand in the soil-plant system |
| Ammonia emissions | Manure storage and spreading | Minimise emissions from storage buildings |
| Soil pollution by heavy metals (Cu, Zn) | Cu and Zn content of feedstuff | Decrease total Cu and Zn content of feedstuff |

part-time wage-earner) must be defined, which demands some formalisation of labour organisation in the farm. Such a change is likely to involve an increasing functional specialisation, though not a Taylorian labour organisation. The question of decision-making autonomy arises at this point: to what extent is each person autonomous, and then responsible, in the decision-making process? Depending on the answer to this question, a training program is needed.

Since an EMS is a means to reach environmental aims, it needs a measurement protocol and an information processing system. For example, in order to balance fertilisation the farmer needs to know how much nitrates remain in the soil, so soil analyses are required. Once these have been done, the information has to be processed and registered, following which the farmer (or whoever is responsible for this task) must assess the environmental performance of the farm in relation to the defined goals. Keeler and Schiefer (1997) have underlined that this can be formalised within an environmental information operational system, based on both internal and external information and using computer technologies. Once again, the consequent need for training and in research (expert system) must be noted.

The first section of this paper argued that pig farms need social legitimacy. Certification of products or processes is one of the ways leading to this, so the implementation of an EMS is likely to solve part of their environmental problem if it is certified by a third party. This will involve non-agricultural players such as certification companies, or firms conducting their own supplier or compliance audits.

Before considering the economic side of the implementation of an EMS in a pig farm, we summarise the main points of our analysis:

1. an important need for farmers' training;
2. enhancement of the functional division of labour between managerial and technical functions;
3. improvement of the information management system;
4. broadening of farmers' professional network to more non-agricultural players.

## 20.4. Economic assessment: interest in environmental management accounting

This paper has previously addressed the organisational aspects of implementing an EMS in a typical pig farm. We ought now to look at the economic issues for a pig farmer, who (like any manager) is also interested in costs and revenues. First, we will consider the annual operating cost of an EMS, excluding both specific costs of implementation (measurement devices and computers, for instance), and costs induced by specific investments required by the EMS (waste treatment facilities, for example). This limit on the scope of the EMS cost is necessary since there is little information available as no pig farm has yet implemented an ISO 14001 EMS. Since the usual accounting system in a pig farm is an annual analytic accounting system, and in order to provide useful findings to inform discussions with pig farmers, an annual time scale has been taken.

### 20.4.1. *Estimate of the operating cost of an EMS in a pig farm*

This operating cost is determined mainly by the required labour-force and by the certification cost (in cases when an EMS is in fact certified). The appropriate labour force is at least three full-time workers, since there are two technical functions (farrowing and fattening), and a managerial function which requires a work-load equivalent to at least one-third of a full-time person. This may require that an extra worker be hired, on either a full-time or a part-time basis. In any case, we can allocate up to a third of an annual qualified labour unit (in Brittany, about 200% of the legal minimum wage) (Table 20.2).

Training is one of the keys of success in EMS implementation, so each farm worker should attend a training course about every five years. We have estimate the training cost according to the average cost of training courses designed for farmers.

The certification cost includes the cost of the certification itself and the cost of two audits to follow it through. Since certification expires every three years, we have divided the whole cost by three.

Related to the information requirements of an EMS, we have to assess measurement costs. The main ones are analyses to be made outside the farm – for instance, the analysis of a single soil sample to monitor nitrogen and phosphate soil content costs about 70 euros, and several analyses may be needed for a single land plot. In order to follow up the EMS, some audits may be necessary in addition to the initial environmental assessment. The other costs refer mainly to the maintenance cost of measurement devices.

If we consider that these costs are almost structural, we can assess how this affects pig production costs, assuming that pig production is the only source of revenues for the farm. Here we must expect some economies of scale as shown by Figure 20.3. This leads to the conclusion that an ISO 14001 EMS may be successfully implemented in pig farms which produce more than 6000 pigs a year, on the assumption that 0.06 euros per kg of carcass is a tolerable additional cost. The Danish Kvamilla® system experience underlines the damning effect if it is not possible to trade enhanced value in order to compensate

Table 20.2. Estimate of the annual functioning cost of an EMS in an typical pig farm.

| | *Estimated cost per year (euros)* |
|---|---|
| Labour force | 0 to 10,000 |
| Training | 1,000 to 2,000 |
| Measurement | 3,000 to 7,000 |
| Initial environmental assessment | 1,500 to 4,000 |
| Audit (not included in the certification process) | 1,000 to 3,000 |
| Certification | 2,500 to 4,000 |
| Others | at least 1,000 |
| Total cost | 10,000 to 32,000 |

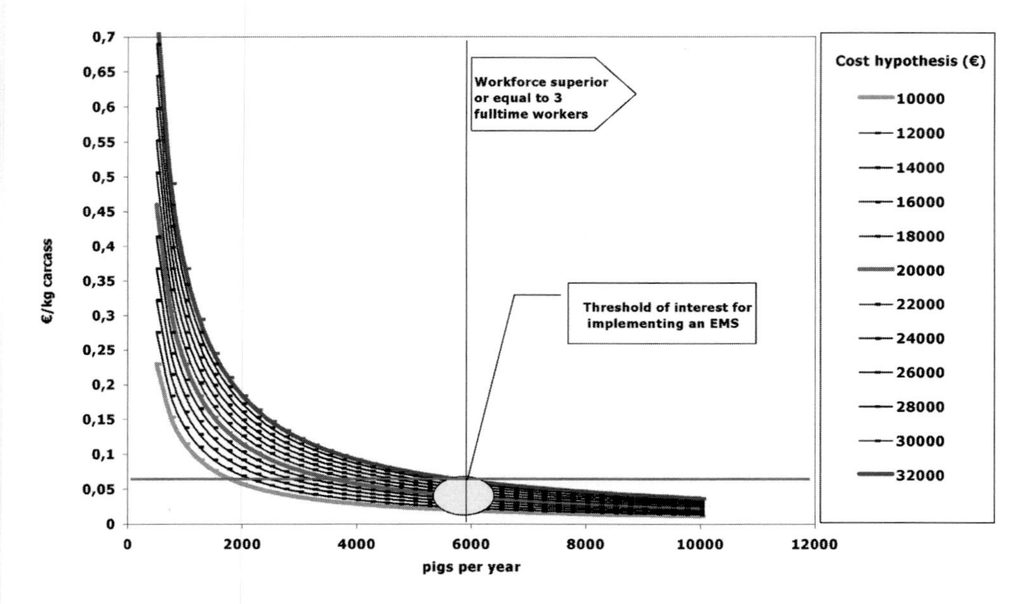


Figure 20.3. Effect of EMS implementation on annual pig production cost.

this additional cost (Petersen, 1999), and also points out the importance of the labour force, since an EMS requires both time and competencies.

Nevertheless, the lack of information means that it is not possible to draw definite conclusions to be made, which itself highlights the need to improve information management systems within pig farms. Moreover, we can provide only some items of knowledge on EMS operating costs, but none about environmental costs themselves, so we are not able to identify possible opportunities for savings which may be a key argument for implementing an EMS from a pig farmer's point of view. This situation raises the question of the possible role of environmental management accounting as a support tool for information management in an EMS framework.

### 20.4.2. *Interest in environmental management accounting*

This section is based on an introductory document on environmental accounting (EPA, 1995). An environmental management accounting system (EMAcS) aims at identifying, collecting and analysing information about environmental costs for internal purposes. These environmental costs are of five main types: conventional costs, potentially hidden costs, contingent costs, communication (relation/image) costs and societal costs (externalities).

If we consider an typical pig production site, we can notice that management accounting is almost limited to 'GTE' (Gestion Technico-Economique, or technico-economic management). GTE is a procedure in which technical and economic parameters are collected and processed in order to provide technical and economic indicators to the farmer about the fattening process. Some of the main outcomes are related to the technical efficiency of animals (e.g. feed conversion ratio, average daily gain). The main economic outcome is what may be called feed-cost margin (sale price per kg of carcass, feed cost per kg of carcass). GTE can be made on a three-month, six-month or annual basis.

We can state that current management accounting within typical pig farms is characterised by prominent overhead costs which cannot be allocated to specific points in the production process. This reduces management efficiency since it does not facilitate the targeting of corrective actions. Therefore, as environmental issues are deeply challenging for pig farms, setting up EMAcS may improve the whole management of pig farms by fostering the implementation of management accounting procedures.

As an illustration, a possible EMAcS for nitrogen management will be presented, by addressing the various kinds of costs mentioned above. As already mentioned, nitrogen management is one of the key issues of environmental management in a pig farm. Figure 20.4 shows the main flows and pools.

Because nitrogen is necessary for the growth of both animals and plants, it is not so easy to define costs as 'environmental costs'. We could assert that nitrogen-related environmental costs are induced by misuse and structural surpluses (i.e. regarding the natural assimilation capacity of the farm land). Some of these costs are presented in Table 20.3.

We can see here that the environmental accounting approach is likely to bring a substantial quantity of information to pig farmers about their production and management systems, but this approach also requires substantial changes in not only their management knowledge and skills but also in their behaviour. A current survey (Devienne and Montel, 2000) of the effects on Dutch dairy farming of the enforcement of the MINAS system (Mineral Nutrient Accounting Systems) appears to confirm this.

## 20.5. Conclusion

The potential for the use of EMAcS is clear, for several reasons. First, it will provide information about cost and benefits related to environmental performance, thus making it easier to convince pig farmers to commit themselves to the implementation of ISO 14001 EMS by demonstrating the economic advantages of greening their farms. Secondly, an EMAcS suits EMS as a support tool to set up a relevant information system. EMAcS will help to enhance the accountability of pig farms, and thus their legitimacy.

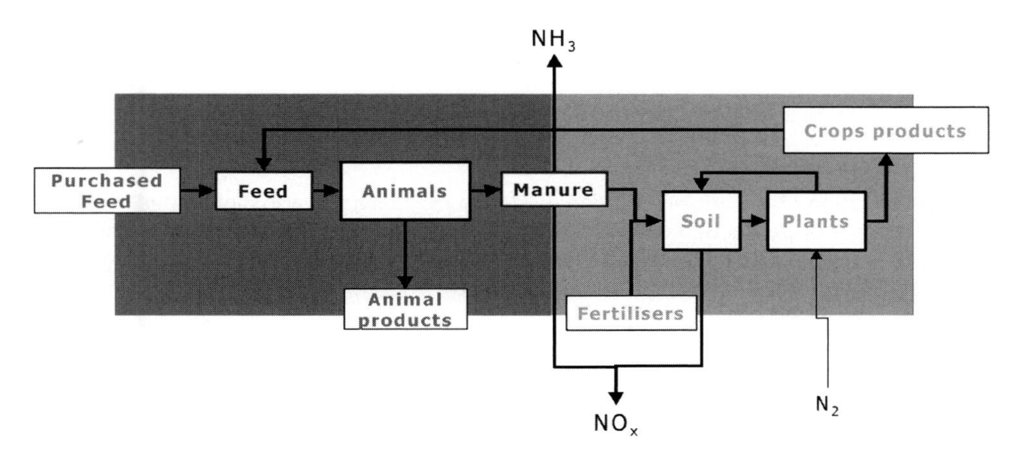


Figure 20.4. Nitrogen pools and flows in a livestock farm.

Table 20.3. Some of the environmental costs related to nitrogen management.

| *Private costs* | |
|---|---|
| **Direct environmental costs** | |
| – Explicit environmental taxes (General Tax on Polluting Activity, Water Pollution Tax)<br>– Manure treatment facilities (initial investment and maintenance) | |
| **Potentially hidden costs** | |
| • *Regulatory* | – record keeping for fertilisation plan (CIEP act)<br>– spreading periods and areas (CIEP act) |
| • *Up-front* | – fertilisation advice<br>– feeding advice |
| • *Conventional costs* | – feed: multi-phase diet.<br>– feeding system: multi-phase diet distribution system.<br>– fertiliser<br>– manure and fertiliser spreading equipment: injection device. |
| • *Back-End* | – maintenance of manure storage facilities<br>– replacement of spreading equipment |
| • *Voluntary* | – adherence to a fertilisation improvement program.<br>– training course on nitrogen management.<br>– implementation of an environmental management system |
| **Contingent costs** | |
| – Loss of crop yield because of nitrogen surplus in the soil (wheat for example)<br>– Health hazards for piglets in case of nitrates overload in drinking water<br>– Legal expenses in case of water pollution or illegal spreading | |
| **Communication costs** | |
| – Participation in an open house program (to improve the image of pig farming, mainly damaged by bad nitrogen management) | |
| **Societal costs** | |
| – Water treatment to reach drinkability standards<br>– Closure of water catchment points<br>– Health hazards for new-born babies (controversial from a scientific point of view) | |

Consideration of the implementation of EMS in pig farms has also highlighted that both new standards of production and new management principles are emerging, as shown in this paper. It would be interesting to look more closely at how farmers will change their know-how in an information-rich context, and how their local professional networks will evolve (Darré, 1996).

Consumption behaviour has changed over the past twenty years, and also the social values underlying the legitimacy of human activities. All these elements seem to give evidence of a deeper change in pig production. The whole development pattern is changing. As underlined by Boyer and Durand (1998), after any internal difficulties, changes in productive systems meet barriers within the society itself: to set up new professional relations, new education systems and even new patterns for state intervention.

There is a dilemma here. The industrial nature of these emerging pig production systems is quite clear, especially in respect of the changes in management principles which are required in order to deal with environmental issues. At the same time there is a societal disapproval of industrial agriculture, which is becoming widely considered to be intrinsically polluting. One of the consequences of this dilemma may be to strengthen the position of the biggest (and ironically therefore, often the most unpopular) pig farms, if these also accept their industrial nature and commit themselves to environmental (and also quality) management. In this respect it could also be interesting to address how current environmental regulations are leading, in some cases, to agriculture becoming even more industrialised (for instance, dairy farming in the Netherlands: Kempf and Chotteau, 1996).

# 21. Environmental Management Accounting and the Opportunity Cost of Neglecting Environmental Protection

*Stefan Schaltegger and Roger Burritt*
*University of Lueneburg, Germany and The Australian National University, Canberra*

*Reflections on how much voluntary expenditure a company should make on environmental-protection measures are dominated by the discussion of relevant direct internal costs. Compulsory spending on environmental protection (e.g. expenditure forced by regulations) is not taken into account here, because compliance with environmental laws and regulations is a minimum requirement in order to continue in operation. When making decisions on future investments, a comparison of the direct and indirect costs of corporate environmental protection with other potential commercial investments is, without doubt, economically highly relevant. However, from an economic point of view, a comparison based on opportunity costs is even more important (see, e.g., Hirshleifer, 1980, p. 265; Kreps, 1990). Economists consider that the economic cost of undertaking any activity should be interpreted as being the cost of the best alternative opportunity foregone. This paper addresses the question of why environmental management accounting is a necessary foundation in order to obtain opportunity cost information as a foundation for making an informed choice on whether a business should pursue or neglect environmental protection.*

## 21.1. Introduction

Environmental management accounting (EMA), and in particular environmental cost accounting, have recently gained importance in business (see Bartolomeo et al., 1999, 2000; Bouma and Wolters, 1999; Bennett and James, 1996, 1998b, 1999; and Schaltegger and Burritt, 2000). A better understanding of the scope and boundaries of EMA is now becoming apparent (Bennett and James, 1998a; IFAC, 1998; and Schaltegger et al., 2001). In particular, EMA tools may focus on the provision of either regular or ad hoc information for decision-making, and may emphasize past or future data, or the short-term or long-term, as the basis for analysis. Decision-making involves looking at future possible alternatives and choosing the best course of action for the organization. In this area, economics and accounting are very close indeed, because a set of tools has been developed by economists to compare alternative courses of action so that the most efficient use of resources results from the actual choice made. Accounting information can provide the basis for the prediction of the costs and benefits of alternative courses of action. This paper examines such situations, and their link with EMA as the provider of environmental management information to help guide the selection of *efficient* decisions about expenditures that may reduce the environmental impacts of an organization. Managers are concerned about issues of effectiveness, efficiency and equity, and so it must be recognized that the following analysis has a narrow perspective, ignoring as it does issues of effectiveness and equity in order to focus specifically on efficient outcomes and the type of EMA information that can drive efficient decision-making.



*M. Bennett et al. (eds.), Environmental Management Accounting: Informational and Institutional Developments*, 265–277.
© 2002 *Kluwer Academic Publishers. Printed in the Netherlands.*

Opportunity cost is a notion that economists use to help to guide the choice between alternative courses of action. Here, the focus is on the opportunity cost of either pursuing, or neglecting, environmental protection. These two courses of action can be appraised on the assumption that other things remain equal (e.g. that production levels and production costs, etc., remain constant.). Opportunity cost has long been claimed to be relevant to businesses in making a choice between available alternatives. A strong British contribution to this aspect of cost analysis emerged in the 1930s (Edwards, 1938; Coase, 1938; Baxter, 1938) and has been further developed (Baxter and Oxenfeldt, 1961; Gould, 1974; Arnold and Scapens, 1981). It has to be recognized that the treatment of costs in conventional accounting is not entirely in line with how costs are treated in economics. Conventional accounting does not record opportunity costs, but accounting costs remain of interest when a business is planning future courses of action, especially as a way of predicting opportunity costs for ad hoc decisions that involve significant amounts of resources (see Arnold, 1980, p. 103).

This paper discusses the notions of the opportunity cost of environmental impacts[1] and of environmental protection (section 2); the internal environmental costs of a company (section 3); and the opportunity cost of pursuing or neglecting environmental protection (section 4). It looks at the juxtaposition between the concept and use of opportunity cost in environmental decision-making, and why environmental management accounting is a necessary foundation in order to obtain opportunity cost information about the choice of whether a business should pursue or neglect environmental protection.

## 21.2. Opportunity costs of environmental impacts and environmental protection

According to neoclassical economic theory, the cost of producing a particular product is represented by the value of the other products that the resources used in its production could have been used to produce instead (Mansfield, 1997). In other words, because of limited resources, a product can be produced only at the expense of another product that could otherwise have been produced using the same resources. If the other product is not produced, then the opportunity of obtaining value from the other product is given up, or foregone. The best of these alternative values given up is known as the opportunity cost of the course of action that the business actually follows. The on decision whether to establish and use an EMA system in order to help guide the use of scarce business resources also represents an opportunity cost, since the resources used for the accounting system might alternatively have been used in other more profitable ways.

Using this concept of the opportunity cost of environmental information, an environmental manager who has been granted a particular budget and who has to determine on which projects the money should be spent, will invest in an EMA system until the marginal costs are equal to the marginal benefits expected from the investment. The following analysis assumes two simple choices for a business – to invest in environmental protection, or to not invest in environmental protection. All other factors are assumed to be constant and therefore do not affect the decision to be made. The costs of these two alternatives can be examined either in terms of total costs (upper part of Figure 21.1) or

[1] Environmental impacts are, following ISO 14001, taken to be any change to the environment, whether adverse or beneficial, wholly or partially resulting from the activities, products or services of the organization.

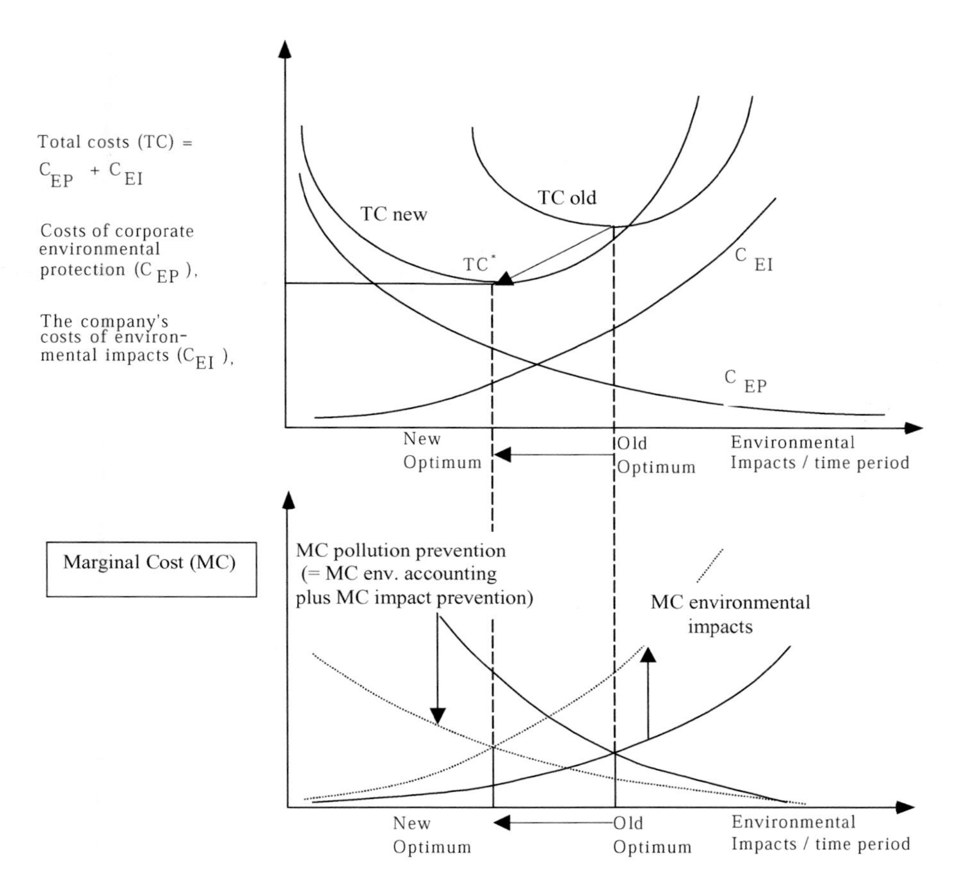


Figure 21.1. Total (upper part of Figure) and marginal (lower part of Figure) cost curves for environmental reduction (including the costs of environmental accounting) and environmentally driven financial impacts of environmental impacts.

of marginal costs (lower part of Figure 21.1), which are referred to as 'incremental costs' in conventional management accounting. Information about marginal costs is useful because it focuses on the differences between available alternatives. Its role is to cut down on the amount of computation necessary to compare alternative plans, by eliminating elements that are not affected by the decision on hand (Gould, 1974, p. 101). In contrast, total cost information shows the overall projected costs of each alternative when a certain number of pollution prevention investment alternatives are available. The relationship is examined below, and total costs are then examined for the remainder of the paper so that all cost information is made transparent.

First, consider the total costs shown in the upper part of Figure 21.1. Most companies face some costs of environmental impacts ($C_{EI}$) as well as costs of environmental protection ($C_{EP}$). The costs of environmental protection ($C_{EP}$) decrease with increasing environmental impacts (because prevention and reduction costs proportionately less with higher pollution levels), whereas the costs of environmental impacts ($C_{EI}$) increase (because of higher fees, fines, etc.) with increasing environmental impacts of business activities, products or services per period of time. The costs of environmental protection ($C_{EP}$) include the environmental accounting costs of collecting and analyzing the information to make good pollution prevention decisions, as well as the costs of technical and organizational measures to reduce and prevent environmental impacts.

The *total costs* (TC) include the sum of the costs of corporate environmental impacts ($C_{EI}$) and the costs of environmental protection ($C_{EP}$). Since TC is equal to the sum of $C_{EI}$ and $C_{EP}$, the point at which total costs ($TC_{old}$ or $TC_{new}$) are lowest are mostly not where the costs of environmental impacts equal the costs of environmental protection (intersection of $C_{EI}$ and $C_{EP}$). If the company seeks maximum profit it will aim for corporate environmental impacts associated with the lowest point of the total cost curve ($TC_{old}$ or $TC_{new}$).

As decisions are taken on an incremental basis, however, setting $MC_{EI}$ equal to $MC_{EP}$ will lead to lowest total costs. The *marginal costs* are shown in the lower part of Figure 21.1. As with the total costs, marginal costs can be separated into the marginal costs of pollution prevention ($MC_{pollution\ prevention}$, left curve sliding to the right) and the marginal costs of environmental impacts ($MC_{impacts}$, right curve increasing to the right). The marginal costs of pollution prevention ($MC_{pollution\ prevention}$ in Figure 21.1) are the organizational, technical and accounting costs of preventing environmental impacts. The marginal costs of environmental impacts ($MC_{environmental\ impacts}$ in Figure 21.1) include any failure costs (e.g. fees, fines, additional administrative and lawyer costs, etc.) that are expected to arise if the organization fails to prevent unacceptable environmental impacts.

Now consider the dashed curves in the lower part of Figure 21.1. They illustrate the shift in marginal cost curves that has occurred during the last decade. The marginal costs of pollution prevention have decreased (i.e. the cost curve $MC_{pollution\ prevention}$ has shifted downwards to the left) due to two developments. Firstly, costs have fallen because of the development of more advanced information systems and environmental accounting systems, and the associated skills. The same has happened to the marginal costs of organizational and technical measures of pollution prevention because of the development of advanced pollution prevention and abatement technologies. Thus, the marginal costs of pollution prevention (including environmental information management) have been decreasing over time (dashed line $MC_{pollution\ prevention}$).

On the other hand, the marginal costs of environmental damage caused by the company ($MC_{environmental\ impacts}$ in the lower part of Figure 21.1) increase with growing environmental impacts. Because of stricter regulations (e.g. fines and fees), this cost curve has shifted upwards during the last decade, i.e. the marginal costs of environmental impacts have been increasing (dashed curve $MC_{EI}$ and arrow line).

As a result of this development (i.e. the shift of the MC curves in the lower part of Figure 21.1) the optimal economic point has been sliding to the left on the 'environmental impacts' axis (i.e. from the 'old optimum' to the 'new optimum'). The optimal point from an economic perspective is shown in the lower part of Figure 21.1 by the intersection of $MC_{pollution\ prevention}$ and $MC_{environmental\ impacts}$, that is, by the balance between the marginal costs of pollution prevention including environmental accounting information (appraisal) on one hand, and the costs associated with the failure to avoid environmental impacts on the other ' '. With the change in marginal costs, the total costs also change. This is shown by the shift in the total cost curve in the upper part of Figure 21.1 from $TC_{old}$ to $TC_{new}$.

The implication is that from an economic point of view it might be expected that more companies would introduce environmental management accounting systems and invest in the prevention of environmental impacts occurring, in order to reduce their total environmentally relevant costs (i.e. to realize $TC_{new}$ instead of remaining with $TC_{old}$).

It appears, however, that while these predicted costs have shifted the expected optimum to the left, many companies in environmentally sensitive situations have not yet become aware of these changed relationships between the expected costs of environmental protection (including provision of environmental accounting information) and the expected costs of environmental impacts (see, e.g., Fischer et al., 1997; Schaltegger and Müller, 1997; and Bartolomeo et al., 1999, pp. 190–210). Furthermore, they may not even be aware that it could be worthwhile to implement suitable information systems to measure the economic potential of greater environmental protection. However, for companies that have been ignoring the potential cost to them of not preventing environmental impacts, there has been a considerable increase in the opportunity cost of such ignorance. If management is not aware of the shift in marginal and total costs (the arrow from $TC_{old}$ to $TC_{new}$ in the upper part of Figure 21.1), and if it does not invest in more environmental protection measures, then by remaining at the old optimum, the total costs (on curve $TC_{old}$) and environmental impacts each period will be higher than necessary (total costs on curve $TC_{new}$). The costs of foregone opportunities to invest in additional environmental protection measures (and environmental accounting) have been increasing. These opportunity costs will be discussed further in section 4.

The next section reviews the crucial aspect of the development of internal environmental cost accounting which must be understood by environmental managers and accountants in order to identify and measure the change in total and marginal environmental costs discussed in Figure 21.1. Conventional management accounting records some of the costs of corporate environmental impacts. These are examined in the next section. They provide actual cost data that forms the basis for predicting the costs of prevention, implementation and operation of environmental management accounting systems, and the costs actually incurred because the organization has failed to avoid certain environmental impacts. In short, they provide the basis for an assessment of the costs associated with neglecting opportunities to undertake further investment in environmental protection.

## 21.3. Internal environmental costs of a company

Conventionally, internal environmental costs of companies (also called 'private environmental costs') have been defined and managed as *costs of corporate environmental protection* (e.g. costs of sewerage infrastructure and waste water treatment plants; also, see Fichter et al., 1997). Internal environmental costs derived from a conventional management accounting system are seen as normal costs of doing business and can be divided into ordinary and extraordinary costs, direct and indirect costs, and potential future costs (see the left side of Figure 21.2).

Among the most obvious costs related to environmental impacts are *ordinary* costs such as capital and operating costs for clean-up facilities. For example, environmental costs associated with the production of cars are ordinary costs for a car manufacturer (e.g. costs to treat the waste water from production). An unexpected, exceptional accident, however, results in *extraordinary* costs (e.g. clean-up costs caused by an unexpected explosion at Esso's Longford gas plant in Melbourne).[2]

[2] ASC defines 'extraordinary' and 'ordinary' in IAS 8, §6, as follows: Extraordinary *items* are income or

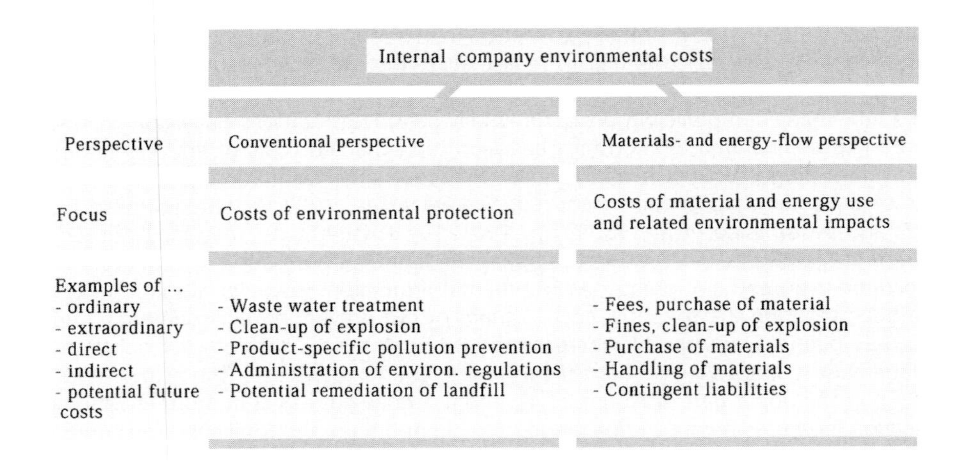


Figure 21.2. Two different perspectives on internal company environmental costs.

*Direct* environmentally driven costs could be, for example, the costs of scrubbers which are linked linked directly to the production of a specific type of car. Costs of joint clean-up facilities, such as a waste-water treatment plant, are *indirect* costs since they have to be allocated specifically to cost centres and cost objects. *Potential future clean-up costs* include, for example, costs of future remediation of landfills and the potential costs of oil spills on site (Burritt and Gibson, 1993, p. 18). Table 21.1 provides some examples of external and internal costs.

However, in contrast with this conventional perspective, environmental costs can be defined as the sum of all costs which are directly and indirectly related to materials and energy use and their resulting environmental impacts (see right side in Figure 21.2, also see Fichter et al., 1997; Schalteger and Burritt, 2000). These environmentally driven costs include all costs which are incurred because materials and energy flows are not reduced such as, for example, fees (ordinary costs); fines (extraordinary); purchases of materials (direct); or administrative costs which are caused by environmental regulations (indirect costs such as, e.g., reporting costs), and contingent environmental liabilities (potential future costs). Both the conventional and the materials and energy flow-based definitions of internal company environmental costs can be reflected in conventional management accounting systems; however, the opportunity costs tend to be excluded.

Internal company costs can, moreover, be distinguished according to their *measurability* and their *visibility in the accounts*. Conventional ordinary and extraordinary direct costs are usually quantified and included in management accounting. Indirect ('hidden') costs (US EPA, 1995) are, however, often not explicitly recognized in management accounting but are instead considered to be part of general overhead costs. Less tangible costs include negative effects on the goodwill of a company.

---

expenses that arise from events or transactions that are clearly distinct from the ordinary activities of the enterprise and, therefore, are not expected to recur frequently or regularly. *Ordinary activities* are any activities which are undertaken by an enterprise as part of its business and such related activities in which the enterprise engages in furtherance of, incidental to, or arising from these activities.

Table 21.1. Examples of external and internal environmental costs (Source: Adapted from: Whistler Centre for Business and the Arts. Environmental Accounting. Prepared by Berry and Failing, 1996; IFAC, 1998, § 29).

| *External and Internal Environmental Costs* | |
|---|---|
| *External Environmental Costs* | |
| Examples:<br>– Depletion of natural resources<br>– Noise and aesthetic impacts<br>– Residual air and water emissions | – Long-term waste disposal<br>– Uncompensated health effects<br>– Change in local quality of life |
| *Internal Environmental Costs* | |
| Direct or Indirect<br>Environmental Costs<br><br>Examples:<br>– Waste management<br>– Remediation costs or obligations<br>– Compliance costs<br>– Permit fees<br>– Environmental training<br>– Environmentally driven R&D<br>– Environmentally related maintenance<br>– Legal costs and fines<br>– Environmental assurance bonds<br>– Environmental certification/ labeling<br>– Natural resource inputs<br>– Record keeping and reporting | Contingent or Intangible<br>Environmental Costs<br><br>Examples:<br>– Uncertain future remediation or compensation costs<br>– Risk posed by future regulatory changes<br>– Product quality<br>– Employee health and safety<br>– Environmental knowledge assets<br>– Sustainability of raw materials inputs<br>– Risk of impaired assets<br>– Public/customer perception |

Potential future (contingent) costs have to be estimated. They are sometimes included in accounts as provisions or charges on income, but in general these costs, which are needed for predicting opportunity costs associated with not protecting the environment, are not provided in the past records of an EMA system. Likewise with intangible costs, large measurement problems can occur (e.g. a loss of reputation) and these external costs are not usually included in the accounts. Nevertheless, these costs are important when predicting opportunity costs.

For example, in the 1960s, the asbestos industry sold products that caused tremendous health damage in the eighties and nineties. Today, asbestos as a product has mostly been phased out, and it is insurance companies that often have to foot the financial bill. Financial liabilities for pollution, illnesses such as asbestosis, clean-up liabilities and related claims have all to be borne by the insurance industry and today's payers of insurance premiums. Insurance claims related to asbestos have been estimated at US$2 trillion in the USA alone, but only US$11 billion of these are covered by reserves and provisions (Knight, 1994, p. 48).

Typically, these costs have not been measured in the accounting systems of those parties responsible for them, although years later the negative financial consequences have had to be internalized. In general, the focus of environmental managers has not been on opportunity costs associated with such problems, and management accounting systems have not been able to provide the full set of data upon which decisions about alternative courses of action can be based. The following section considers these opportunity costs.

## 21.4. Opportunity costs of pursuing or neglecting corporate environmental protection

The reason for considering opportunity costs is that they illustrate that no decision is without cost, even if no direct internal or external costs arise or are recorded in the conventional accounts. The *opportunity cost of an investment in environmental protection* (see Box 1 below) is equal to the benefit of the most attractive alternative investment foregone (e.g. the return that could have been earned in the financial marketplace for the same level of risk). In turn, the *environmentally relevant opportunity costs of non-environmental investments* are the unrealized benefits of the most beneficial investment in pollution prevention. From this perspective, environmental costs include the costs of the purchase and handling of materials that become waste at a later stage in the production process (see Bouma and Wolters, 1999b; Fichter et al., 1997; Schaltegger and Mueller, 1997).

*Box 1. Opportunity costs of pursuing and neglecting environmental protection*

**Opportunity costs** are the costs which arise from unrealized opportunities whenever an alternative is chosen from a set of available alternatives.

The **opportunity cost of an investment in environmental protection** equals the benefits of the most attractive alternative investment given up (e.g. in a production device or in the financial market).

**Opportunity costs of neglected environmental protection** are the unrealized benefits of the most beneficial investment in pollution prevention. From this economic perspective, environmental costs are defined as the costs related to the best unrealized pollution prevention alternative. Like external environmental costs, opportunity costs are not recorded in management accounts in a regular, systematic way. Instead, they are provided to managers when a set of alternatives is being considered in a particular decision setting.

The implication is that the decision on whether a company should voluntarily spend more money on pollution prevention should be based on the choice that has the lowest opportunity cost. Other things being equal, investment will take place until the Net Present Value of all implemented projects, including environmental protection projects, is equal to zero. EMA systems are designed to help provide relevant information for these ad hoc investment decisions. To arrive at the correct investment decisions, according to this criterion, requires knowledge of the benefits of voluntary corporate environmental protection. Yet, as pointed out above, many benefits from a company's environmental protection are of course not quantified, because they are intangible or external. Possible ways of including these benefits in investment appraisal are discussed in Schaltegger and Burritt (2000), although the details are not important here.

Although environmental protection cost is both an internal cost and measurable within a conventional management accounting system, the possibility of reducing the opportunity cost of unrealized environmental protection is usually not highlighted in management accounting, or considered by business. For example, pollution prevention that is *not* carried out costs business money, if it could in fact have been introduced using techniques that would have helped overall income to rise. It is suggested that the *opportunity cost of neglected environmental protection (OCNEP)* represents the potential foregone income arising from environmental protection. From an economic perspective, the opportunity costs of unrealized corporate environmental protection by a company are shown in Figure 21.3.[1] Figure 21.4 is based on the upper part of Figure 21.1 and focuses on total costs.

The horizontal axis in Figure 21.3 shows the environmental impacts of a cost centre's activities (e.g. a production process or site), a cost object (e.g. a product or product group), or a company. Total conventional costs for the activity, cost object, or company are depicted on the vertical axis. To understand the concept of opportunity costs of neglected environmental protection, the factors which influence the curves in Figure 21.3 are briefly repeated.

As mentioned in the discussion of the upper part of Figure 21.1, many internal costs of environmental impacts (relating to the environmental damage caused by the company) ($C_{EI}$) increase more than proportionally in relation to increases in the number of environmental interventions (e.g. fees, fines, liabilities, and administrative costs to comply with regulations). Some fees or regulations become relevant only when threshold amounts of hazardous waste or materials are used. In these cases, special administrative activities, with their associated costs, become mandatory. Education costs can also arise or increase more

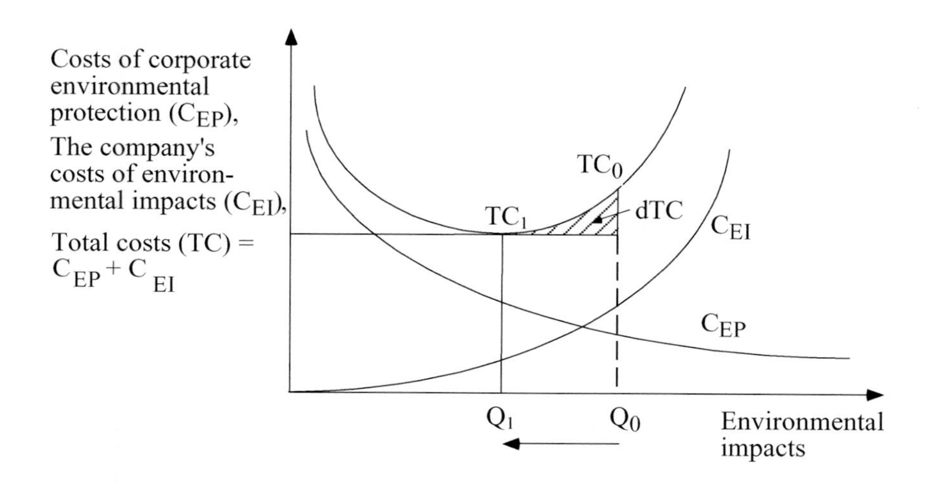


Figure 21.3. Opportunity cost of neglecting environmental protection (OCNEP).

[3] Note that Bartolomeo et al. (1999) use a different classification of relevant cost categories that is linked with total quality environmental management. This appealing approach is based initially on a classification of prevention costs, appraisal costs, internal and external failure costs that is compacted into three categories – prevention, correction and failure costs.

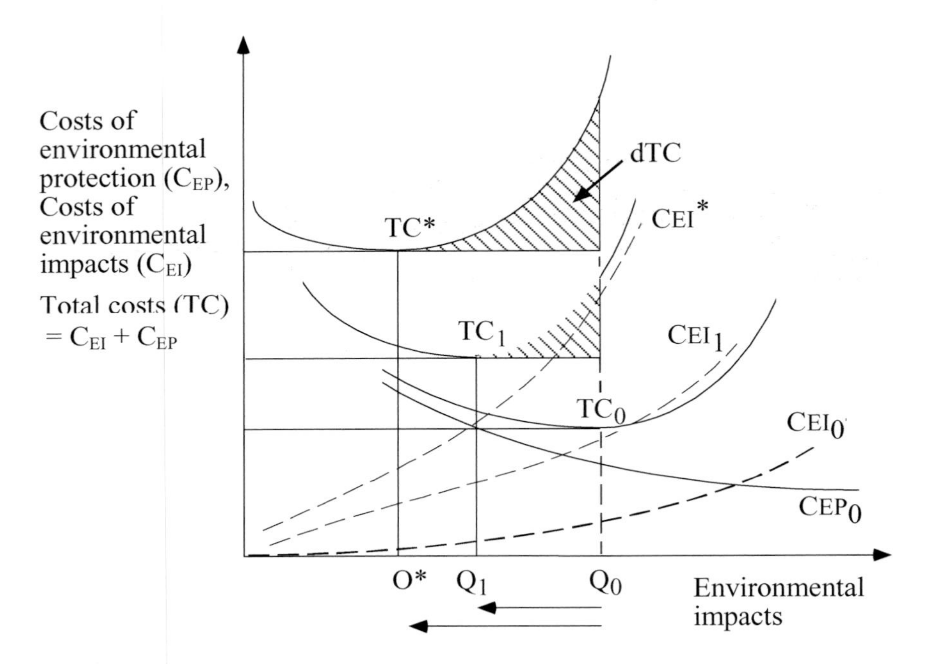


Figure 21.4. Emergence of the opportunity cost of unrealized environmental protection.

than proportionally when staff need special education because a certain minimal amount of waste is expected to be exceeded.

On the other hand, the total cost of environmental protection ($C_{EP}$) decreases with a higher incidence of environmental impacts because the fixed costs of environmental protection are spread out over many impacts that share the fixed element of cost. Hence, the optimal amount of environmental impacts from a company is where the total costs (TC, i.e. the sum of these two costs) are minimized ($Q_1$). This local optimum is where the sum of $C_{EP}$ and $C_{EI}$ is minimized and is not related to the intersection of $C_{EP}$ and $C_{EI}$ as Figure 21.3 shows *total cost figures*. It is, however, the same place as the point where the marginal costs of pollution prevention equal the *marginal* costs of environmental impacts (as shown in the lower part of Figure 21.1).

Starting from the initial point of environmental impacts of $Q_0$ (which is to the right of $Q_1$ with the lowest total costs of $TC_1$ in Figure 21.3), unrealized environmental protection causes opportunity costs for a company because the total costs of $TC_0$ could be reduced with more environmental protection (and lower environmental impacts of $Q_1$) until the point of the lowest possible total costs of $TC_1$. These opportunity costs are shown by the difference between minimal total costs ($TC_1$) and actual costs ($TC_0$) of pollution. Area dTC, therefore, shows the total opportunity cost of unrealized environmental protection.

The question remains why a company would wish to take the opportunity cost of unrealized environmental protection into account in its management accounting decisions, as this could only reduce income. There are two main reasons. *Firstly*, most accounting systems have hitherto not been adjusted to recognize foregone income in their investment in environmental management decisions. *Secondly*, in order to judge whether a company has incurred an economic loss, all alternative investment opportunities have to be considered and the best of these chosen. In the limited context of environmental information decisions, the opportunity cost of unrealized environmental protection has to be compared with the Net Present Value of the planned alternative investments. It is equal to the Net Present Value of the most economic pollution prevention measure (the foregone alternative).[4] Looking at the first reason, the emerging importance of the opportunity costs of unrealized environmental protection is illustrated in Figure 21.4.

Section 1 shows how the *marginal* costs of environmental impacts have been increasing for business in the last decade because of stricter regulations and stakeholder pressures. This has lead to an upward shift of the *total* cost curve, so that the local optimum point in Figure 21.4 has been sliding to the left of the 'environmental impacts' axis. However, because historically environmental costs have not been treated as an important class of cost in management accounting they have not been separately accounted for in conventional information systems.

Figure 21.4 shows the development of opportunity costs of unrealized environmental protection based on the inclusion, in sequence, of indirect costs, intangible costs (e.g. loss of reputation), and the internalization of what were external costs.

Cost curves $CEI_0$ and $TC_0$ illustrate the perceived cost situation if only the direct financial consequences of the environmental impacts of a company are considered. In practice, in many companies environmentally driven indirect costs, such as administrative costs incurred to comply with regulations, are often treated as overhead costs and are thus not explicitly considered in decision making, i.e. in investment appraisal of pollution prevention technology. If these indirect and internal costs are included, the total cost curve shifts upwards to the left ($TC_1$). The cost curve would shift even further to the upper left if liabilities caused by formerly external costs, for instance costs of dumping toxic waste, were internalized ($TC^*$). Both of these types of indirect cost have increased substantially in the last decade, thus shifting the optimal level of environmental impacts from $O_0$ to $O_1$ and $O^*$. While economic analysis recognizes their importance, few of these costs have been recognized in accounting systems. Exploration of the significance of such costs and the ways that they can be reflected in management accounting have been explored (see IFAC, 1998; Ditz et al., 1995; Parker, 1999; Schroeder and Winter, 1998).

The above analysis shows the result when incomplete consideration is given to environmentally driven monetary impacts on a company. Effect number one is that the total costs of many profit and cost centres (e.g. polluting production processes, equipment) and cost objects (e.g. environmentally harmful products) are underestimated ($TC_0$ instead of $TC^*$). The second effect relates to the presence of opportunity costs because corporate environmental protection is not at its optimal level (area dTC represents the total opportunity cost of under-investment in environmental protection from a social rather than a private decision maker's perspective), something that environmental managers need to champion. Thirdly, this leads to a lower level of environmental protection than would be optimal in economic terms ($Q_0$ instead of $Q^*$). Evidence supporting the view that many economically beneficial measures of environmental protection are not realized is provided in the results of a large survey conducted in the US State of Washington (DOE, 1992b, c, 1993). As a result, the less-than-optimal level of company eco-efficiency also means that

---

[4] It may be added that if a company was positioned to the *left* of $TC_0$, i.e. if more environmental protection was realized than would be economically optimal, opportunity costs of environmental protection would be larger than the opportunity costs of neglecting environmental protection measures. However, given the historical change of curves discussed in Figure 21.3, opportunity costs of neglected environmental protection can be considered to be relevant more often and for more companies.

eco-efficiency levels for the whole economy are too low, with X-inefficiency[5] being encouraged (Leibenstein, 1966).

## 21.5. Summary

This paper can be summarized as follows: opportunity costs relating to unrealized environmental protection have been neglected for too long by many companies. Public pressure and increasing government legislation is making managers consider strategies for internalizing these opportunity costs, before companies are 'forced' to do so. Once these costs are identified and recognized, as illustrated in Figure 21.4, managers will tend towards higher levels of environmental protection. Once the opportunity cost of unrealized environmental protection is recognized by managers, they will focus on ways to reduce these opportunity costs by lowering environmental impacts in a cost-efficient manner. In short, in anticipation of having opportunity costs of unrealized environmental protection forced upon them, managers will implicitly adopt the concept of corporate efficiency where they look for a reduction in environmental impacts, while maintaining or improving profitability. In Schaltegger and Burritt (2000, p. 102), an example is provided illustrating how the opportunity costs of unrealized environmental protection can be calculated as part of an investment project. Of course, it should be noted that an environmental management accounting system does not record opportunity costs on a regular basis as these are not cash-outflows or expenses that are actually incurred by an organization. Unless internalized, these will be recognized only through ad hoc provision of information for investment appraisal as part of a decision to select one course of action rather than another. The two main questions relating to opportunity costs of neglected environmental protection (OCNEP) can therefore be summarized in the following way:

- Why do environmental managers need to know the opportunity cost? At the time that they make a decision they need to know the best alternative course of action in order to make sure that it is not as good as, or better than, the alternative they are taking.
- When do they need to know opportunity cost? At the time that they make a decision, not after a decision is made, because the opportunity cost (cost of the best alternative not taken) could have changed at a later date.

Given the potential effects of neglecting the opportunity cost of unrealized corporate environmental protection, in particular the resulting corporate losses associated with being told what to do by a regulatory body rather than choosing the best course of action in the ordinary course of business, and given the growing stakeholder concerns over environmental impacts, it is hardly surprising that some stakeholders, such as the United Nations Division of Sustainable Development (UNDSD, 2000) or the US State of Washington (DOE, 1992a) have been exerting their influence to try and ensure that better consideration is given to environmental issues by companies in their management accounting systems.

If corporate management cannot (because it lacks the information) or will not (because it fears competitive repercussions) consider the opportunity costs of neglecting environmental protection, then government legislation will be the only way forward. An alter-

[5] The difference between the best possible efficiency ratio and the efficiency ratio actually achieved is described as *X-efficiency* (Leibenstein, 1966). The concept of X-efficiency is useful because it suggests that, in practice, organizations do not appear to be cost minimizers (using latest technology); instead, they are more inclined to imitate the policies of their rivals and to follow industry norms and targets.

native is that policy can be brought to bear to influence the voluntary take-up of environmental management accounting systems by companies in order that the opportunity costs of neglecting environmental protection can be made transparent to management through improved environmental management accounting classification, recording, and internal reporting such that the information is taken into account in decision-making. The conclusion is, therefore, that the concept of opportunity cost is useful for conceptualizing what it is that environmental management accounting should measure. Furthermore, it would seem useful for environmental managers to draw upon and appropriate the term when holding discussions with treasurers, conventional accountants or finance staff about the potential problems if investments in environmental protection are neglected.

# 22. Wanted: A Theory for Environmental Management Accounting

*Jan Jaap Bouma*
*Visiting professor in environment management at the University of Ghent and associate professor at the Erasmus University Rotterdam; E-mail: bouma@fsw.eur.nl*

*Mark van der Veen*
*Researcher at the University of Amsterdam, Economics Faculty; E-mail mvdveen@ua.nl*

## 22.1. Introduction

Most research in environmental management accounting is prescriptive, contributing to the further development of tools, and often based on a limited number of case studies (e.g. Bennett and James, 1998). Empirical research in environmental management accounting (e.g. Bouma and Wolters, 1998) is scarce and is focused more on describing the current state of implementation than on analyzing or critically evaluating the effectiveness of the new tools. Our understanding of environmental management accounting could therefore be enhanced by extending our insight into the spread of environmental management accounting practices across different countries and industries. A second option for the further development of environmental management accounting theory could come from applying it to explain the adoption and effectiveness of environmental management accounting practices. This paper will first discuss the current state of environmental management accounting research; secondly, will use both contingency and institutional theory to formulate hypotheses and to develop a questionnaire on the adoption and effectiveness of environmental management accounting, and finally report on the feedback that was received from a company and draw conclusions for further research.

## 22.2. The development of environmental management accounting

Environmental management was first related to management accounting in the 1980s, when the experience of several us firms in particular was that pollution prevention could often be profitable. Management accounting expertise was necessary to assist in determining the exact costs and benefits of pollution prevention. In addition, a more theoretical contribution which was sought from management accounting was to explain the obvious inefficiencies of producing waste: why had so many seemingly profitable pollution prevention opportunities remained undiscovered? How could engineers and controllers in firms that operate in competitive industries have overlooked the cost saving potential?

To evaluate environmental investments, new management accounting tools such as calculation methods and checklists for the inventory of environmental costs and benefits, were developed (Freeman, 1990). Often the development and adoption of these techniques were initiated by governmental agencies or NGO's (White et al., 1991), but have nevertheless been noticed at a business level and seem to have been adopted to some extent (Bartolomeo et al., 2000). These practices became known as 'environmental management accounting' (EMA) (Bennett and James, 1998; Van der Veen, 2000). However,



*M. Bennett et al. (eds.), Environmental Management Accounting: Informational and Institutional Developments*, 279–290.
© 2002 *Kluwer Academic Publishers. Printed in the Netherlands.*

there is very little empirical research that underpins the effectiveness of environmental management accounting in promoting change towards sustainable business (Boons et al., 2000). The question arises of the extent to which EMA facilitates the achievement of strategies that integrate environmental objectives. Clearly the congruence between environmental management within firms and the achievement of sustainable development remains a difficult hypothesis, but the effect of using management accounting systems and techniques with respect to environmental management is also less obvious than many case-based studies and policy makers have claimed.

If the special issue on environmental and social accounting of the European Accounting Review (2000) can be taken to provide a representative overview of the current state of research in environmental management accounting, then we have to conclude that most of this is not strongly related to other research in management accounting. Much of this EMA research does however provide useful insight, like the conclusion from the Ecomac project (Bartolomeo et al., 2000) that only a small percentage of interviewees considered that '*the benefits of introducing environmental management accounting were sufficiently high in themselves to compensate for the high costs of changing accounting systems*'. But when going through the texts and references of most of the contributions, it looks as if EMA is developing as a new field of research separate from the much wider body of management accounting research generally. In 'The Green Bottom Line', which provides an overview of EMA research, it even looks as if some are deliberately seeking for such an insulated and unique position for EMA. Bennett and James (1998) defined environmental management accounting as the generation, analysis and use of financial and non-financial information in order to optimise corporate environmental and economic performance and achieve sustainable business. Such an exclusive positioning may be understood from a perspective that new tools have still to be developed, and that we should not waste time by recycling old concepts.

## 22.3. Understanding the adoption of EMA

To understand why EMA is adopted and when it is effective, we believe that it is necessary to relate our research to similar developments in the history of the management accounting literature. It is not unique for EMA that advanced tools are not immediately adopted; we know that even the penetration of basic concepts such as activity-based costing, net present value and balanced scorecards is only modest. The purpose of this paper will be to explore how we can improve our understanding of why companies implement EMA tools. As we expect that the application of EMA is rather limited, the paper will focus on forms that we expect are less scarce, in order that sufficient cases may be found for analysis. It is assumed here that having some economic figures on the cost of environmental management is one of the least advanced forms of EMA. These economic figures are also relevant because the literature on environmental management accounting contains many case studies which suggest that a low level of cost allocation is one major reason why organisations ignore the potential of pollution prevention measures to cut down on production costs. Four costs categories in particular were selected that are especially important for environmental management: waste management costs, energy costs, water costs and waste water cost.

In our selection of environmental management accounting practices a similar criticism may be anticipated as was expressed about much of the research in management accounting generally – that there is too much focus on formal practices (Chapman, 1997) and financial information alone. We accept this limitation to our research, but defend our choice in the light of the very limited knowledge until now on the implementation and effectiveness of EMA.

In this paper we will explore how far two different theoretical perspectives can be used to understand the adoption of EMA. Clearly, other theoretical approaches could be applied to study aspects of the adoption of EMA, but the scope of this article has deliberately been limited. Following the selected perspectives of contingency and institutional theory, an empirically based research prospect will be developed. *Contingency theory* was selected because much of the work based on the adoption and effectiveness of management accounting refers to this theory, and because it is used to analyse the relationship between business strategy and management accounting. Contingency theory provides major insights into these relationships, especially at firm-level. *Institutional theory* has the benefit of analysing relationships at a more meso and macro level (for example, the theories addressed as *new institutionalism* explore organisational change at the meso level). Although management theories seek to understand firms' behaviour by pointing to motives such as efficiency improvements and to profit-increasing motives such as improving the competition position, the new institutionalism theories broaden the unit of analyses by looking at the organisational field.

## 22.4. The contingency approach

The relationship between business strategy and management control systems is a major issue in many contingency studies of management control. Contingency theory is based on two major claims: (1) that there is no single universally best way to organise, and (2) that each way of organising is not equally effective (Galbraith, 1973). Contingency studies aim to identify the determinants for the selection and effectiveness of organisation form. However although the basic assumptions of contingency theory may be plausible, there have been many critical reviews that question even whether it should be considered to be a theory at all. In 1979 Burrell and Morgan concluded that the present state of contingency theory at that time could best be formulated as a loosely organised set of propositions which in principle are committed to an open systems view of organisations. These are committed to some form of multivariate analysis of the relationship between key organisational variables as a basis for organisational analysis, and which endorse the view that there are no universally valid rules of organisation and management. Theoretical confusion begins when the contingencies are to be listed that affect organisational design. A large number of empirical studies aim to explain the organisational arrangements with respect to accounting systems and techniques. Relevant contingencies that are found (see Pugh and Hickson, 1976) for these arrangements are: market environment; competitive strategies; history; size; ownership and control; technology; location; available resources; and interdependence with other organisations.

In empirical research, a wide range of control mechanisms such as incentive schemes, budget controls, cost controls and goal setting, have been related to organisational effectiveness. From this research we know that organisational performance may indeed result from a matching of strategy, internal structures and systems, and external conditions. We should acknowledge that the correlations found were sometimes quite low, and that the connection with a strong theoretical model that maps the process of design, implementation and use of these systems is poor or even absent. There is however a general consensus that the design of management control systems should be tailored to support a business strategy towards superior performance (Langfield-Smith, 1997) and we expect that this will be valid for EMA as well.

The implementation of EMA, or environmental management in general, is often advocated by arguing that this would contribute in different ways to the development of competitive advantage (Braakhuis et al., 1995). It seems reasonable to assume that organisations which understand their environmental costs are better equipped to control these costs. There are many case studies that describe how environmental management accounting techniques can contribute to the identification of profitable options for pollution prevention (Bartelomeo et al., 2000; Korean Ministry of Environment and World Bank, 2001). It seems obvious that companies which face increasing environmental costs whilst operating in markets where price competition prevails, will be particularly more interested in collecting environmental cost information. Although these companies are likely to apply environmental management accounting techniques, it is unknown how effective the alternative various EMA techniques are. To analyse the relationship between business strategy and EMA, we will explore whether, and to what extent, theories on the relationship between business strategy and management control systems are applicable and useful.

One of the critical remarks by Langfield-Smith (1997), in a review of research on the relationship between management control systems and strategy, relates to the problem of defining and measuring strategy. This criticism is shared by Chapman (1997), who points to the limitations in the contingency view of accounting, that complex concepts like strategy have been reduced to a limited set of stereotypes such as Porter's model of alternative competitive strategies. According to Porter, to compete effectively a firm must derive its competitive advantage in one of two ways: by product differentiation to provide customer satisfaction from factors such as superior quality, product flexibility, delivery and product design; or by low-cost production, which allows the firm to compete by offering its products at lower prices than competitors. While Porter contended that a firm should choose between competing on either product differentiation or low price, firms may in fact focus on a variety of combinations of product differentiation and low-price strategies.

In relating environmental management accounting practices to business strategy, we can now use the more balanced approach to measuring business strategy developed by Chenhall and Langfield-Smith (1998). Respondents were asked to indicate on a 7-point Likert scale their strategic priorities as developments over the last three years, from no emphasis (scored as 1) to great emphasis (scored as 7), from the following options:

- provide high quality products
- low production costs
- provide unique product features
- low price
- make changes in design and introduce new products quickly
- make rapid volume and/or product mix changes
- provide fast deliveries
- make dependable delivery promises
- provide effective after-sale service and support
- product availability
- customise products and services to customers' needs

We hypothesise that:

> *Organisations that emphasise low production cost and price will be more advanced in the allocation of environmental costs.*

To measure the effectiveness of environmental accounting systems, one could easily be tempted to select measures of environmental performance. The major argument for allocating environmental cost is however economic: that it would help organisations to identify profitable options for pollution prevention. Also, we are relating environmental cost allocation to business strategy rather than to environmental strategy. We will therefore measure effectiveness by asking the respondents whether environmental measures have contributed towards realising any of the competitive strategies listed above. As contingency theory states that a matching of strategy and management control systems is related to organisational effectiveness, we hypothesise the following:

> *Organisations that have both a strategic focus on low production costs and are advanced in the allocation of environmental costs, will be more effective in realising competitive advantages through environmental management.*

From the environmental cost statistics for Dutch industry (CBS, 2000), we know that while the cost of environmental measures and levies on average tend to stabilise at around 1% of sales, environmental costs are considerably higher for sectors such as chemicals, steel, oil and utilities. It seems logical to assume that it makes more sense for companies with relatively high environmental costs to implement accounting systems to control these costs. For our first, exploratory study we will restrict ourselves to the analysis of only those organisations with high environmental costs.

## 22.5. The institutional approach

Clearly, the contingency approach is not oblivious to the external conditions of firms, but explicitly acknowledges them. The development of EMA was strongly influenced by the role of organisations such as accounting associations and governmental agencies. But for a better understanding of their role in the adoption of EMA, an institutional perspective may be rich in providing insights into the effects of the organisational field of a firm.

DiMagio and Powell (1983) define this field as 'those organisations that, in the aggregate, constitute a recognised area of institutional life'. Although these institutional approaches may increase the understanding of the relationship between EMA and business policy by pointing to other factors that are external to firm itself, the management theories address the setting in which a firm operates. This was not the case with classical management theories, which attempted rather to derive universal prescriptions that indicate the best way to manage and structure an organisation (Emmanuel and Otley, 1985). A main strand of management theory was scientific management, but this lost the attention of those who design and analyse the implementation of management accounting systems and techniques, in favour of open system approaches that seek to understand the activities of organisations by including the context of the environment as a set of contingencies to which organisations adapt and accommodate themselves. Although the factors, both internal and external to the organisation, are identified as having an impact on the choice of an appropriate organisational structure and hence on management accounting systems and techniques, an overall theoretical foundation is needed to understand the relationship between the firm's strategy and its accounting systems and techniques.

The insights from institutional theory add to understanding the interaction between the context of an organisation and its behaviour, but make little contribution to understanding the process of design, implementation and use of management accounting systems and techniques. The context in which accounting tools develop can be characterised by the existence of highly formalised institutional bodies that try to set boundaries around the area that can be addressed as management accounting. Accounting is a field that constantly undergoes change, especially management accounting which focuses on supporting the information needs of an organisation's management. The financial sector and accounting institutions show that they face difficulties in determining its identity. It is noticeable that although accounting associations such the European Accounting Association and the National Association of Accountants in the United States define their working field or come up with recommendations on accounting practices, academic accounting articles show the dynamics in its field.

Accounting seems to be a form of bricolage, an activity whose tools are largely improvised and adopted to the tasks and materials at hand (Miller, 1998). With respect to the increase in information needs that are related to the environmental problems and challenges that firms are confronted with, management accounting systems and techniques seem to undergo changes. Figures of environmental costs, and information on the financial consequences of environmental investments, is to be presented to the management of almost all types of firms. The accounting profession acknowledges this new area of information needs that has to be addressed, which is reflected in a variety of activities in the professional accountancy bodies. These activities range from practical accounting activities at a firm level up to more strategic research activities. Perhaps the most striking was the *Federation des Experts Comptables Europeens* (FEE) who brought the European accounting bodies together and gave the environment a prominent place on their research agenda. Also, environmental management accounting has gained in importance in the political domain as an instrument to help to meet environmental policy objectives. The European Commision (DG XII, Human Dimension of Environmental Change) has sponsored research that focussed on the implications of environmental awareness for accounting (Bartolomeo et al., 1999), and in 1999 the United Nations Division for Sustainable Development (UN-DSD) established a working group to explore the role of governments in stimulating the development and use of management accounting systems and techniques for environmental management within firms.

In order to understand and explain different developments in management accounting, institutional theory offers a fruitful perspective. Miller (1998) focussed upon general trends in behaviour in the financial sector and accounting practices and theory, referring for example to the increasing appeal that is being made within management accounting to a wide range of non-financial measures, including all kinds of control information (inventory levels, defect and rework rates, etc.). Miller classifies this as a trend since it has been embedded in management accounting practices for over 10 years. He also focuses on the important role of firms' external stakeholders in the development and use of new accounting techniques, which is illustrated by a reference to the rise of discounting techniques that were adopted by firms because of their use by the financial sector. However, this author does not provide any framework to map and explain the development and adoption of any accounting systems or technique in a more general sense.

The institutional perspective is shaped by theories such as transaction cost economics (Williamson, 1975, 1986), evolutionary economics (Nelson and Winter, 1982; Espen-Anderson, 1994) and the work of North (1989) on institutional economics, which try to explain organisational changes and take the institutional context of economic phenomena into consideration. A recent application of this institutional perspective by Boons et al. (2000) established a conceptual framework to understand the process of change in Dutch industrial firms as a result of ecological pressures. They show the poor empirical basis to apply such a institutional theoretical approach on changes that occur with respect to management accounting systems and techniques. Figure 22.1 summarises this model with its variables and relations. The thick arrows indicate the link between the general variable 'ecological pressure' and 'organisational change'. Ecological pressure is regarded as the socially constructed 'image' of the natural environment that influences the organisation (Hannigan, 1995). The trajectories (A and B) are the aggregate result of actions of individual and/or of corporate social actors, whose beliefs and desires (Elster 1983) are influenced by the process of problem definition and its outcome (arrow 1). The actions of these social actors influence the process of problem definition (arrow 2). The actions of social actors have effects on the ecological system (arrow 3). This conceptual model aims to identify organisational changes more broadly than only those related to management accounting.

Organisational change is mapped at four levels of analysis (Argyris and Schoen, 1978; and Van den Broeck, 1994). At the *operational* level the main focus is on what people actually do, how they act and react within the organisation. At the *model* level the main focus is on the formal representation of the organisation; on the discursive, cognitive aspects of organisational behaviour. At the *coalition*, or power level, the main focus is on the dominant and subordinate coalitions within and around the organisation, on who has the initiative, on who is in control and who is not. Lastly, there is a *value* level at which the main focus is on the sense of mission – of direction and of what is important within the organisation, which include ideas, visions and motivations. Mechanisms are identified that explain these organisational changes at the four different levels.

The theoretical perspective presented above was applied to interpreting existing data. A secondary analysis was conducted by examining existing research reports over a ten-year period (1985–1995) in the Netherlands. It was concluded that in capital budgeting, an important area within management accounting, organisational changes are widespread in the Netherlands although research into this topic is difficult to find for the period up to 1995. Research in the United States has accelerated research on the topic of capital budgeting in the Netherlands (Boons et al., 2000). For the period 1986–1995, organisational changes with regard to capital budgeting are fundamental only to a limited extent. Some of these changes refer to the operational level (for example the involve-

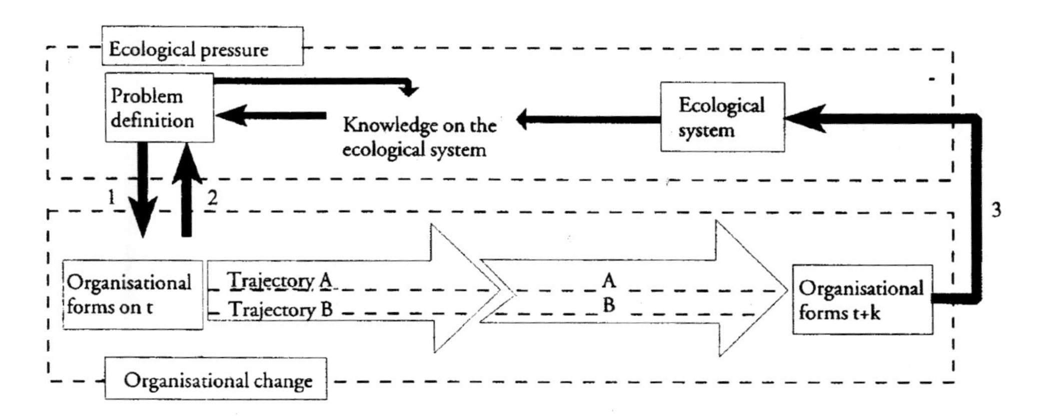


Figure 22.1. Conceptual model (Boons et al., 2000).

ment of an environmental co-ordinator in the capital budgeting process). Also, at the model level, organisational changes occurred that are mapped by research related to capital budgeting (for example the development of environment-oriented accounting systems). Although organisational changes at external stakeholders, such as accountants are well documented, this has not yet resulted in assessments of the impacts of these organisational factors within industry itself. However, no conclusions can be drawn on the effectiveness of the accounting systems that facilitate the capital budgeting with respect to dealing with environmental pressure. Nevertheless, it can be concluded that the organisational field had an impact on the development of EMA in the Netherlands (Boons et al., 2000) and probably has on the further development and adoption. Therefore the following hypothesis can be generated:

> *The acknowledgement of environmental costs as a parameter for decision-making depends on the organisational field that creates a concept for capturing environmental costs in the mindset of management.*

## 22.6. Empirical feedback

To address the three hypotheses that were derived from the contingency and institutional approaches, we aim at a empirical research that in the longer term goes beyond case studies. However, in order to identify the appropriate variables at a company level and in the organisational field, a case study will be conducted here, in a structured way. The data in the case study is derived from the external environmental reports and from a structured closed and open interview with the environmental manager, the manager responsible for marketing strategy, and the controller. Often the environmental manager contacted the controller and the marketing manager to obtain the answers to the questions that we as researchers asked.

### 22.6.1. *Case study at TI*[1]

The company TI belongs to the chemical sector. The firm has sites in The Netherlands, Germany, Argentina and China, with its main plant being located in Vlissingen where the case study takes place. TI employs about 900 people of which more than 400 work at the Dutch plant, which produces phosphates as a material for foodstuff, medicines, pesticides, paint, clothing, drinking water and cleaning agent.

#### *Corporate strategy*

In its environmental report the board formalises its environmental goal as follows: 'With regard to the environment the emphasis is shifting steadily towards sustainable development. TI aims to provide solutions to support this trend'. To achieve this goal, a number of quantified objectives are formulated. For example, the firm aims to produce 20% of its products from renewable raw materials within the past 5 years (similar objectives were also formulated in previous years). From the interviews, the general corporate strategy was characterised according to the following aspects. For these aspects (strategic priorities), the respondent scaled the emphasis that the firm had given to them in last three years,

---

[1] The name of the company is fictitious in order to keep the real name secret.

Table 22.1. Emphasis on different aspects of strategies.

| | 1 (no emphasis) | 2 | 3 | 4 | 5 | 6 | 7 (high emphasis) |
|---|---|---|---|---|---|---|---|
| Provide high quality products | • | • | • | • | • | X | • |
| Low production costs | • | • | • | • | • | X | • |
| Provide unique product features | • | • | • | • | • | X | • |
| Low price | • | • | • | • | • | X | • |
| Make changes in design and introduce new products quickly | • | • | • | • | X | • | • |
| Make rapid volume and/or product mix changes | • | • | • | X | • | • | • |
| Provide fast deliveries | • | • | • | X | • | • | • |
| Make dependable delivery promises | • | • | • | • | X | • | • |
| Provide effective after-sales service and support | • | • | • | X | • | • | • |
| Product availability | • | • | X | • | • | • | • |
| Customise products and services to customers needs | • | • | • | • | X | • | • |

ranging from no emphasis (a score of 1) to high emphasis (a score of 7). Table 22.1 presents the findings for TI.

### *The role of environmental management*

It is noticeable from Table 22.1 that low-cost production is scaled as an important strategic priority. However, a closer look at the role that environmental management plays in enhancing low-cost production shows that this role is only limited. This insight was gained by asking about the importance of environmental management with respect to the different strategic aspects (see Table 22.2).

### *Cost control*

Environmental management, by which environmental costs can be controlled, is not so much integrated into the regular cost accounting systems. This is also illustrated by the development of a separate environmental cost calculation technique by which environmental impacts are valued according to the shadow pricing technique. Using this technique, different emissions can be weighted and compared with each other in the process to achieve its environmental objectives. However, when dealing with the strategic issue of producing at a low production cost, monetarized environmental impacts (based on the shadow pricing technique) are excluded from the cost concept. Managing production costs is not perceived as a strategic issue that is strongly enhanced by environmental management. This does not mean that the management of TI was not provided with information on waste management costs, energy costs, water costs and wastewater costs. On the contrary, information on these types of cost items is available to the management. For example, waste costs, wastewater costs are specified up to the level of departments, and

Table 22.2. The importance of environmental management with respect to the different aspects of strategies.

| | 1 (no emphasis) | 2 | 3 | 4 | 5 | 6 | 7 (high emphasis) |
|---|---|---|---|---|---|---|---|
| Provide high quality products | • | • | • | X | • | • | • |
| Low production costs | • | • | X | • | • | • | • |
| Provide unique product features | • | • | • | • | X | • | • |
| Low price | • | • | X | • | • | • | • |
| Make changes in design and introduce new products quickly | • | X | • | • | • | • | • |
| Make rapid volume and/or product mix changes | • | • | • | X | • | • | • |
| Provide fast deliveries | • | X | • | • | • | • | • |
| Make dependable delivery promises | • | X | • | • | • | • | • |
| Provide effective after-sales service and support | • | • | • | • | • | X | • |
| Product availability | • | • | • | X | • | • | • |
| Customise products and services to customers needs | • | • | X | • | • | • | • |

energy and water usage costs are specified even up to product level. For all these costs, past trends are monitored and management is informed about the expected future trends. Waste water abatement costs and waste management costs are dealt with as overhead and not allocated to the products, although waste management costs are allocated to specific processes in order to control these costs. This is a logical consequence of achieving low production costs, which is ranked as an important strategic priority. However, the concept of environmental costs is very much linked to issues such as external reporting, communication with the financial sector, and take-overs, and not so much to reducing production costs. Another cost concept is used for dealing with the concept of controlling production. External parties such as the national statistics office (CBS), the external accountant, banks, insurance companies and research institutes are typical stakeholders who wish to receive environmental cost information addressing the term environmental costs. To deal with these requests, environmental costs are defined as 'additional costs that are made to accomplish projects that aim for environmental performance improvements'.

## 22.7. Comments on the confrontation between theoretical perspectives and empirical insights

From the case study at TI it appears that at this organisation, with an emphasis on low production costs and price, the four selected cost items are dealt with by using fairly advanced procedures to allocate these costs. Waste costs, wastewater costs are specified up to the level of departments and energy and water usage costs are specified even up to product level. Whether this is related to achieving a competitive advantage by means of environmental management is questionable. TI regards environmental management as of only 'less importance' in achieving low production costs. Environmental management has its own calculation technique for deriving figures on environmental impacts and the monetary value of these, and these are not generated to control the costs covered by the term production costs.

The control of production costs is structured by decision-makers within the company (the overruling perspective of the decision-maker). The concept of environmental costs however, is highly influenced by external factors (such as external parties like national statistic office (CBS), the external accountant, banks, insurance companies and research institutes). Also, the calculation technique for attaching a monetary value to environmental impacts is designed by an external research institute. Based on a remark in the external environmental report the hypothesis that environmental costs are acknowledged as a parameter for decision making is underpinned. The report states: 'If an independent organisation . . . has provided input and monitored a proposal, it is much more likely to be taken seriously'. The case study shows that the organisational field was of importance while creating a concept for capturing environmental costs in the mindset of the management of TI.

## 22.8. Conclusions for further research

From the contingency approach, two hypotheses were derived, which are underpinned by the TI case study. Coming from the institutional approach another hypothesis was formulated, which similarly was also underpinned by the TI case study. The empirical basis for drawing conclusions that can be generalised (at a meso or macro level) has to be broadened, but this was not the aim of this paper which was to understand better the adoption of EMA. For this, two theoretical approaches were followed, and with regard to their appropriateness it is concluded that each has its own benefit. The contingency approach focuses on the perspective of a decision-maker who structures the accounting systems and techniques. The costs to achieve low production costs are defined and specified according to internal rules set by the management. Institutional theory helps in understanding the process of defining environmental costs and in explaining the mechanisms that explain the adoption and use of specific accounting systems and techniques for environmental costs. This adoption takes place at different levels within a firm. With respect to the TI-case the following was noted:

- at an *operational* level different accounting systems and techniques are implemented and used to calculate and allocate environmental costs;
- at a *model* level the systems and techniques that are used to calculate environmental costs are formalised;
- at a *coalition* level it is noticed that decision-makers who are in control use the figures derived from the calculation techniques. Independent research institutes play an important role in underpinning these figures;
- at a *value* level it is stated that environmental costs stimulate the integration of environmental concern into the firm's mission. However, there is not necessarily a direct link between a strategic priority such as lower production costs, and environmental management, because environmental costs can be excluded from the terminology being

used by the management. For example, the term 'production costs' may be not linked with environmental costs. Although these different cost categories may be based on the same use of resources (for example, the input of labour or materials), the cost figures related to environmental activities are not regarded as production costs.

The mechanisms that explain changes in the above-mentioned levels are not fully recorded in this limited study. For this, further research is necessary to understand the process by which EMA is adopted. Institutional theory may be used beside contingency theory to understand the factors that affect this important change towards more sustainable enterprises.

# References


AFNOR (1996a), NF EN ISO 14001 *Systèmes de management environnemental. Spécifications et lignes directrices pour son utilisation*

AFNOR (1996b), NF ISO 14004 *Systèmes de management environnemental. Lignes directrices générales concernant les principes, les systèmes et les techniques de mise en œuvre*

Argyris, C., and D.A. Schön (1978), *Organizational Learning: A Theory of Action Perspective*, Addison-Wesley: Reading

Arnold, J., and R. Scapens (1981), 'The British Contribution to Opportunity Cost Theory', in: M. Bromwich and A. Hopwood (eds.), *Essays in British Accounting Research*, London: Pitman

Arnold, J. (1980), 'Budgets for Decisions', in: J. Arnold, B. Carsberg, and R. Scapens, *Topics in Management Accounting*, Oxford: Philip Allan

Asa, S., and U. Wennberg (1998), *Continuity, Credibility and Continuity: Key challenges for corporate environmental performance: a report commissioned by the European Environment Agency*, The International Institute for Industrial Environmental Economics, Lund University, Sweden

Bartolomeo, M., M. Bennett, J.J. Bouma, P. Heydkamp, P. James, F. de Walle and T. Wolters (1999), *Eco-Management Accounting*, Dortrecht/Boston/London: Kluwer Academic Publishers

Bartolomeo, M., M. Bennett, J. Bouma, P. Heydkamp, P. James and T. Wolters (2000), 'Environmental Management Accounting in Europe: Current practice and future potential', in: *The European Accounting Review*, Vol. 9, No. 1, 31–52

Batschari, A. (1995), 'Maschinenbau und Umweltschutz in Deutschland', in: Verband Deutscher Maschinen- und Aanlagenbau e.V. (VDMA) (ed.), *Maschinen- und Anlagenbau im Zentrum des Fortschritts*, Frankfurt a. M., pp. 159–168

Baxter, W.T. (1938), 'A Note on the Allocation of On-cost between Departments', in: *The Accountant*, November 5, pp. 633–636

Baxter, W.T., and A.R. Oxenfeldt (1961), 'Costing and Pricing: The cost accountant versus the economist', in: *Business Horizons*, Winter

Behrendt, S., C. Jasch, M. Peneda, and H. van Weenen (1997), *Life Cycle Design: A manual for small and medium-sized enterprises*, Berlin: Springer Verlag

Bennett, M., and P. James (1996), 'Environmental Related Management Accounting in North America'. In: C. Tuppen (ed.), *Environmental Accounting in Industry: A Practical Review*. London: BT.; James, P., and M. Bennett (1994). Environment-related Performance Measurements in Business. Ashridge Management Research Group.

Bennett, M., and P. James (1996), 'Environmental Related Management Accounting in North America', in: C. Tuppen (ed.), *Environmental Accounting in Industry: A Practical Review*, London: BT

Bennett, M., and P. James (1998a), 'The Green Bottom Line', in: M. Bennett and P. James (eds.), *The Green Bottom Line: Environmental Accounting for Management – Current Practice and Future Trends*, Sheffield: Greenleaf Publishing, pp. 30–60

Bennett, M., and P. James (eds.) (1998b), *The Green Bottom Line: Environmental Accounting for Management – Current Practice and Future Trends*, Sheffield: Greenleaf Publishing

Bennett, M., and P. James (1999), 'Key Themes in Environmental, Social and Sustainability Performance Evaluation and Reporting', in: M. Bennett and P. James (eds.), *Sustainable Measures – Evaluation and Reporting of Environmental and Social Performance*, Sheffield: Greenleaf Publishing, pp. 29–74

Bennett, M., and P. James (1999a), *Environment-related Management Accounting in North America and its Implications for UK Companies*, Working paper, Ashridge Management College & Wolverhampton Business School

Bennett, M., and P. James (1999b), *Sustainable Measures: Evaluating and Reporting on Environmental and Social Performance*, Sheffield: Greenleaf Publishing

Bezou, E. (1997), *Système de management environnemental*, AFNOR, Paris

Blonk, H. (assisted by C. Dutilh) (2000), *Duurzaamheidsindicatoren voor de primaire sector, Een verkenning van het werkveld als basis voor dialoog*, Blonk Milieu Advies, voor Stichting DuVo, Rotterdam

BMU and UBA (1996) (German Ministry of the Environment and Federal Department of the Environment), *Handbook of environmental cost accounting, Munich* (only in German)

Boons, F., L. Baas, J.J. Bouma, A. de Groene and K. Le Blansch (2000), *The Changing Nature of Business*, Utrecht, The Netherlands: International Books

Bouma, J.J., and T. Wolters (1998), *Management Accounting and Environmental Management: A Survey among 84 European Companies*, Zoetermeer: EIM and Erasmus University

Bouma, J.J., and T. Wolters (eds.) (1999), *Developing Eco-Management Accounting: An International Perspective*, Zoetermeer: EIM

Bouma, J.J. (1999), 'The Expectations of Stakeholders on Environmental Information', in: *Continuity, Credibility and Comparability*, IIIEE Communication 1999:3, pp. 65–72

Bouma, J.J., and T. Wolters (1999b), 'Synopsis and Outlook', in: J.J. Bouma and T. Wolters (eds.), *Developing Eco-Management Accounting: An International Perspective*, Zoetermeer, NL: EIM

Bouma, J.J., M.H.A. Jeucken, L. Klinkers (2001), *Sustainable Banking: The Greening of Finance*, Sheffield: Greenleaf Publishing

Boyer, R., and J.P. Durand (1998), *L'après fordisme*. Paris: Syros – Alternatives économiques

Braakhuis, F.L.M., M. Gijtenbeek and W.A. Hafkamp (1995), *Milieumanagement: Van Kosten naar baten*, Alphen aan den Rijn: Samsom H.D. Tjeenk Willink

Broas, C., J. Ulhøi and H. Eriksen H. (1999), *Virksomhedsøkonomiske konsekvenser af miljøledelse* (The Economic Consequences of Environmental Management in Companies), Copenhagen: Danish Environmental Protection Agency

Brouck, H. van den (1994), *Learning Management*, Lannoo: Tielt (in Dutch)

BTI Consulting Group (1999), *Market Opportunities in the Environmental Management Systems Market*. Boston: BTI

Buchanan, J., and G. Tullock (1975), 'Polluters' Profits and Political Response: Direct Controls Versus Taxes', in: *American Economic Review*, Vol. 65, March, pp. 14–22

Bullinger, H.-J., and G. Jürgens (1999), 'Betriebliche Umweltinformationssysteme als Grundlage für den Integrierten Umweltschutz', in: H.-J. Bullinger, G. Jürgens and U. Rey (eds.), *Betriebliche Umweltinformationssysteme in der Praxis*, Stuttgart: Fraunhofer IRB-Verlag

Burritt, R.L., and S. Welch (1997), 'Accountability for Environmental Performance of the Australian Commonwealth Public Sector', in: *Accounting, Auditing and Accountability Journal*, Vol. 10, No. 4, 1997, pp. 532–561

Burritt, R.L. (2001a), 'Voluntary Agreements: Effectiveness Analysis – Tools, Guidelines and Checklist', in: P. ten Brink (ed.), *Environmental Agreements: Process, Practice and Future Use*, Sheffield, UK: Greenleaf Publishing, 2001

Burritt, R.L. (2001b), 'Application of Effectiveness Analysis – the Case of Greenhouse Gas Emissions Reduction', in: P. ten Brink (ed.), *Environmental Agreements: Process, Practice and Future Use*, Sheffield, UK: Greenleaf Publishing, 2001

Burritt, R.L., T. Hahn and S. Schaltegger (2002), 'An Integrative Framework of Environmental Management Accounting – Consolidating the Different Approaches of EMA to a Common Framework and Terminology', in: M. Bennett, J.J. Bouma and T. Wolters (eds.), *Environmental Management Accounting: Informational and Institutional Developments*, Dordrecht: Kluwer Academic Publishers

Burritt, R.L., and K. Gibson (1993), 'Accounting for the Environment', in: *Australian Accountant*, pp. 17–21

Canadian Institute of Chartered Accountants (1993), *Environmental Costs and Liabilities: Accounting and Financial Reporting Issues*

Capillon, A. (1992), 'Utilité et spécificité de l'approche de l'environnement par l'agronome', *Cahiers Agricultures*, 1:113–122

CBS (2000), *Milieukosten van Bedrijven 1997*, Voorburg/Heerlen: CBS

Cerf, M., and M. Sebillotte (1997), 'Approche cognitive des décisions de production dans l'exploitation agricole', in: *Econ.rurale*, 239:11–18

Chacón, R., M. Danse and T. Wolters (1999), *Sustainable Coffee in an International Supply Chain: A pilot for Costa Rica and the Netherlands*, Paper presented at the Eighth Annual Conference of the Greening of Industry Network, Chapel Hill, USA (in conference cd-rom)

Chambers, R.J. (1966), *Accounting, Evaluation and Economic Behavior*, Houston, Texas: Scholars Book Co.

Chapman, C.S. (1997), 'Reflections on a Contingent View of Accounting', in: *Accounting, Organizations and Society*, 22:189–205

Chenhall, R.H., and K. Langfield-Smith (1998), 'The Relationship between Strategic Priorities, Management Techniques and Management Accounting', in: *Accounting, Organizations and Society*, 22:243–264

Coase, R.H. (1938), 'Business Organisation and the Accountant', in: *The Accountant*, 1 October,17 December

Coenenberg, A.G. (1997), *Jahresabschluß und Jahresabschlußanalyse. Grundfragen der Bilanzierung nach betriebswirtschaftlichen, handelsrechtlichen, steuerrechtlichen und internationalen Grundsätzen*, 16th rev. and ext. ed., Landsberg/Lech

Cohen, S., and S. Kamieniecki (1991), *Environmental Regulation Through Strategic Planning*, Boulder, San Francisco, Oxford: Westview Press

Committee on Industrial Performance Metrics (1999), *Industrial Performance Metrics: Challenges and Opportunities*, National Academy of Engineering, National Research Council, National Academy Press, Washington. The Executive Summary is at http://www.nap.edu/html/ind-env-perf-met

Danse, M., and T. Wolters (2001), *Sustainable Production Methods in the Coffee Chain as a Strategy to Gain Competitive Advantage: With special reference to a Costa Rican case*, Paper presented at the Ninth Annual Conference of the Greening of Industry Network, Bangkok, Thailand

Darré, J.P. (1996), *L'invention des pratiques dans l'agriculture*, Paris: Khartala

Devienne, S., and B. Montel (2000), *Changes in Dutch Dairy Farming due to New Environmental Policies*, Working notes, not published

DiMaggio, P.J., and W.W. Powell (1983), 'The Iron Cage Revisited: Institutional isomorphism and collective rationality in organizational fields', in: *American Sociological Review*, 48:147–160
Ditz, D.; J. Ranganathan, and R.D. Banks (1995). *Green Ledgers: Case Studies in Corporate Environmental Accounting*. Washington DC: World Resource Institute
Ditz, D., J. Ranganathan, and R.D. Banks (eds.) (1995), *Green Ledgers: Case Studies in Corporate Environmental Accounting*, Baltimore: World Resources Institute
Dockès, A.C., J. Capdeville and A. Farrugia (1995), *Diagnostic des risques de pollution des eaux dans les élevages: la méthode DeXeL*, Renc.Rech.Ruminants, 2:362
DOE (Washington State Department of Ecology) (1992a), *Guidance Paper. Economic Analysis for Pollution Prevention*, Olympia, WA.
DOE (Washington State Department of Ecology) (1992b), *Success Through Waste Reduction*, Volume I, Olympia, DOE
DOE (Washington State Department of Ecology) (1992c), *Success Through Waste Reduction. Proven Techniques for Washington Businesses*, Volume II, Olympia, DOE
DOE (Washington State Department of Ecology) (1993), *Success Through Waste Reduction. Pollution Prevention Planning in Washington State Businesses*, Olympia, DOE
Dold, G. (1997), *Computerunterstützung der produktbezogenen Ökobilanzierung*, Wiesbaden: Gabler Verlag
ECOMAC (1996), 'EIM Small Business Research and Consultancy', *Synreport: Eco-management accounting as a tool of environmental management (the Ecomac project)*
ECOMAC (1998), *Three Case Studies on Environmental Management Accounting in Germany*, Rotterdam: Erasmus Centre for Environmental Studies
Edwards, R.S. (1938), 'The Rationale of Cost Accounting', in: A. Plant (ed.), *Some Modern Business Problems*, London
Elster, J. (1983), *Explaining Technical Change*, Cambridge: Cambridge University Press
Elwood, H., and S. Case (2000), 'Private Sector Pioneers: How companies are incorporating environmentally preferable purchasing', in: *Greener Management International*, Issue 29, Spring, 70–94
Emmanuel, C., and D. Otley (1985), *Accounting for Management Control*, Berkshire: Van Nostrand Reinhold
EPA (Environmental Protection Agency) (1995), *An Introduction to Environmental Accounting as a Business Management Tool: Key Concepts and Terms*. Washington DC: EPA
EPA (US Environmental Protection Agency) (1995b), *Environmental Accounting Case Studies: Green Accounting at AT&T*, Washington DC: Office of Pollution Prevention & Toxics
Epstein, M.C. (1996), *Measuring Corporate Environmental Performance*, Chicago: Irwin Professional Publishing
Epstein, M.J. (1996), *Measuring Corporate Environmental Performance: Best Practices for Costing and Managing an Effective Environmental Strategy*, Chicago, USA: Irwin
Espen-Anderson, S. (1994), *Evolutionary Economics*, London: Pinter Publishers
Etterlin, G., P. Hürsch, and M. Topf (1992), *Ökobilanzen – ein Leitfaden für die Praxis*, Mannheim
European Commission (2001), *Commission Recommendation of 30 May 2001 on the recognition, measurement and disclosure of environmental issues in the annual accounts and annual reports of companies*. Official Journal of the European Communities, 13.6.2001, I.156/33–42

European Environmental Agency (1999), *Guideline for Defining and Documenting Data on Possible Environmental Protection Measures*, Technical Report No. 27

Fabrycky, W.J., and J.H. Mize (1991), *Life-Cycle Cost and Economic Analysis*, New Jersey: Prentice Hall

Fichter, K., T. Loew, and E. Seidel (1997), *Betriebliche Umweltkostenrechnung*, Berlin u.a.: Springer

Fichter, K., T. Loew, and E. Seidel (1997), *In-house Environmental Cost Accounting – Methods and Practice-oriented Further Development*, Berlin, Heidelberg (only in German)

Fischer, H., and R. Blasius (1995), 'Environmental Cost Accounting', in: BMU and UBA (pub. 1995) (German Ministry of the Environment and Federal Department of the Environment): *Handbook of Environmental Controlling*, Munich, 1995, pp. 439–457 (only in German)

Fischer, H., C. Wucherer, B. Wagner, and C. Burschel (1997), *Environmental Cost Management (Umweltkostenmanagement)*, Munich: Hanser

Fischer, H. (1997), 'Management-tools for Cutting Costs by Relieving Environmental Stress', in: H. Fischer et al. (pub. 1997), *Environmental Cost Management – cutting costs through tried and tested environmental controlling*, Munich, 1997, pp. 1–27 (only in German)

Glenna Ford (2000), *Managing Environmental Information in the 21st Century: The Environmental Knowledge Management System* at http://www.greenware.ca/downloads/down_files/ekm_article.pdf

Freeman, H. (ed.) (1990), *Hazardous Waste Minimization*, New York: McGraw-Hill

Galbraith, J. (1973), *Designing Complex Organizations*, Reading: Addison-Wesley

Garrison, R.H., and E.W. Noreen (2000), *Managerial Accounting*, Boston, USA: Irwin, McGraw-Hill

Giraldi, G. (1999), 'A Guideline for Accounting for Waste as a Business Management Tool: Environmental accounting from an Australian perspective', in: J.J. Bouma, and T. Wolters (1999), *Developing Eco-Management Accounting: An International Perspective*. Zoetermeer: EIM

Giraldi, G. (1999), 'A Guideline for Accounting for Waste as a Business Management Tool: Environmental accounting from an Australian perspective', in: J.J. Bouma and T. Wolters (1999), *Developing Eco-Management Accounting: An International Perspective*. Zoetermeer, NL: EIM

Global Environmental Management Initiative (1994), *Finding Cost-Effective Pollution Prevention Initiatives: Incorporating Environmental Costs into Business Decision Making – A Primer*

Gotsche, B. (1995), *Wertschöpfungsanalyse der deutschen Stahlindustrie*, Wuppertal

Gould, J.R. (1974), 'Opportunity Cost: The London Tradition', in: H. Edey, and B.S. Yamey, *Debits, Credits, Finance and Profits*, London: Sweet and Maxwell, pp. 91–107

Gray, R.H., J. Bebbington and D. Walters (1993), *Accounting for the Environment*, London: Paul Chapman Publishing Ltd.

Gray, R., D. Owen, and C. Adams (1996), *Accounting and Accountability: Changes and Challenges in Corporate Social and Environmental Reporting*, London: Prentice Hall Europe

Günther, E., and A. Sturm (2000), 'Environmental Performance Measurement', in: *Die Betriebswirtschaft* (DBW)

Hahn, R.W. (1989), 'Econonomic Prescriptions for Environmental Problems: How the Patients Followed the Doctor's Orders', in: *Journal of Economic Perspectives*, Vol. 3, No. 1, 95–114

Hahn, R.W., and A.M. McGartland (1989), 'The Political Economy of Instrument Choice: An Examination of the US Role in Implementing the Montreal Protocol', in: *Northwester University Law Review*, Spring, Vol. 83, No. 3, 592–611

Hahn, R.W. (1990), 'The Political Economy of Environmental Regulation: Towards a Unifying Framework', in: *Public Choice*, Vol. 65, 21–47

Hannigan, J.A. (1995), *Environmental Sociology. A Social Constructionist Perspective*, London: Routledge

Hessisches Ministerium für Wirtschaft, Verkehr, und Landesentwicklung (1999), (Hessen Ministry for the Economy, Traffic, and Development). *Flow Cost Management, Cutting Costs and Raising Ecological Efficiency through Materials Flow Orientation in Cost Accounting (a guide)*, Wiesbaden (in German)

Hinterberger, F., and M.J. Welfens (1996), *Warum inputorientierte Umweltpolitik?*, in: J. Köhn & M.J. Welfens, Neue Ansätze in der Umweltökonomie, Marburg

Hirshleifer, J. (1980), *Price Theory and Applications*, 2/e, Prentice Hall

Holmark D., P.M. Rikhardsson and H.B. Jørgensen (1995), *The Annual Environmental Report: Measuring and Reporting Environmental Performance*, Price Waterhouse

Horngren, C., and G. Foster (1987), *Cost Accounting: A Managerial Emphasis*, 6th ed., Englewood Cliffs, NJ: Prentice-Hall

IASC (International Accounting Standards Committee) (1995), *International Accounting Standards 1995*, London: IASC

IFAC (1998), *Environmental Management in Organizations. The Role of Management Accounting*, Financial and Management Accounting Committee, International Federation of Accountants, Study #6, New York, March

IISD (1996), *Global Green Standards*. IISD, Winnipeg, Canada, p. 97

IKARUS (2000), Internet catalogue company environmental management systems, URL: http://www.lis.iao.fhg.de/ikarus/, Fraunhofer-IAO, Stuttgart

Information Society Forum (2000), *A European Way for the Information Society*

International Standards Organisation (ISO) (1997), ISO 14040: *Life Cycle Assessment – Principles and Framework*

James, P., and M. Bennett (1994), *Environment-related Performance Measurement in Business*. Ashridge Management Research Group

Japan Accounting Association (2000), *Development and Construction of Environmental Accounting*, Japan Accounting Association (in Japanese)

Japanese Ministry of Environment (2000), *A Guideline for Adopting Environmental Accounting System*, March

JEMAI (2000), *A Report on Environmental Accounting*, Japan Environmental Management Association for Industry (JEMAI)

JICPA (2000), 'A Framework of Environmental Accounting', *JICPA Journal*, February (in Japanese), Japanese Institute of Certified Public Accountants

Johnson, H., and R. Kaplan (1987), *Relevance Lost: The Rise and Fall of Management Accounting*, Cambridge MA: Harvard Business School Press

Jürgens, G., and J. v. Steinaecker (1999), *Aufbau eines betrieblichen Stoffstrom-management auf Basis von SAP R/3-Daten, Fachtagung 'Environmental Management Accounting and the role of Information Systems*, EMAN – Eco-Management Accounting Network, 10.12.1999, Wuppertal

Jürgens, G., C. Liedtke, and H. Rohn (1997), *Zukunftsfähiges Unternehmen (2), Beurteilung des Öko-Audits im Hinblick auf Ressourcenmanagement in kleinen und mittleren Unternehmen. Eine Untersuchung von 13 Praxisbeispielen*, Wuppertal Paper Nr. 72. Wuppertal: Wuppertal Institut für Klima, Umwelt und Energie

Kaplan, R.S. (1984), 'The Evolution of Management Accounting', *The Accounting Review*, LIX (3), pp. 390–418

Kaplan, R., and D. Norton (1996), *The Balanced Scorecard*. Cambridge MA, Harvard Business School Press

Katajajuuri, J.-M., T. Loikkanen, K. Pahkala, J. Uusi-Kämppä, P. Voutilainen, S. Kurppa, P. Laitinen, H. Mikkola, T. Kivinen and S. Salo (2000), *Developing Data Production on Environmental Impacts into The Quality Management of Finnish Farms – Case: Life Cycle Assessment on Barley Production*, Espoo 2000, VTT Research notes 2034 (in Finnish)

Keeler, T., and G. Schiefer (1997), *Computer Aided Environmental Control to Support Environmental Management Systems in Agricultural and Food-industrial Production Chains*. 1st European conference for information technology in agriculture, Copenhagen, 15–18 June 1997

Kempf, M., and P. Chotteau (1996), *Aux Pays-Bas, les contraintes environnementales justifient la poursuite de l'intensification laitière*, Renc. Rech. Ruminants, 3:271–274

Klinkers L., W. van der Kooy and H. Wijnen (1999), 'Product-oriented Environmental Management Provides New Opportunities and Directions for Speeding up Environmental Performance', *Greener Management International*, Issue 26, Summer, 91–18

Knight, P. (1994), 'What Price Natural Disasters?', *Tomorrow*, pp. 48–50

Kokubu, K. (2000), *Environmental Accounting*, Shinsei-sha Tokyo (in Japanese)

Kokubu, K., E. Nashikoka and R. Daikuhara (2001), *Environmental Accounting Practices in Japan*, Kobe University, Graduate School of Business Administration, Discussion Paper No. 2001.7

Kokubu, K., A. Noda, Y. Ohnishi and T. Shinabe (2001), *Determinants of Environmental Report Publication*, Kobe University, Graduate School of Business Administration, Discussion Paper No. 2001/4

Korea Accounting Institute (2001), *Accounting Standard for Environmental Costs and Liabilities*, Seoul

Korean Ministry of Environment and World Bank (2001), *Environmental Accounting Systems and Environmental Performance Indicators*, World Bank-Korea Environmental Management Research Final Report 2001, POSCO Research Institute, Seoul, Korea

Korean Ministry of Environment and World Bank (2001), *Environmental Accounting Systems and Environmental Performance Indicators*, March

Krasowski, H., and J. Friedrich (1998), 'Society of Automotive Engineers Publication P-339, Technical Paper 982227', in: *Proceedings SAE Total Life Cycle Conference*, Graz, Austria

Krcmar, H., G. Dold, H. Fischer, M. Strobel and E. Seifert (2000), *Information systems for environmental management – the reference model 'ECO-Integral'*, Munich, Vienna, 2000 (only in German)

Kreps, D.M. (1990), *A Course in Microeconomic Theory*, Harvester Wheatsheaf

Kriegbaum, H. (1995), 'Konjunkturzyklus – Schicksal des Maschinenbaus', in: Verband Deutscher Maschinen- und Anlagenbau e.V. (VDMA) (ed.) (1995): *Maschinen- und Anlagenbau im Zentrum des Fortschritts*, Frankfurt a. M. 1995, pp. 49–60

Landais, E., and J.P. Deffontaines (1990), 'Les pratiques des agriculteurs', in: *Modélisation systémique et système agraire*, Paris: INRA, pp. 31–51

Langfield-Smith, K. (1997), 'Management Control Systems and Strategy: A Critical Review', *Accounting, Organizations and Society*, 22:207–232

Le Menestrel, M., and F. Planes (1996), Rationality and legitimacy. Legitimacy requirements for collective decision-making based on instrumental rationality, actes du colloque

*Ecologie, Economie et Société*, Université de Versailles Saint-Quentin en Yvelines – 23/25 mai 1996 – tome2

Le Moigne, J.L. (1974), *Les systèmes de décision ans les organisations*, Paris: PUF, p. 244

Le Moigne, J.L. (1984), *La théorie du système général*, Paris: PUF, p. 320

Lee, B.W. (1996), 'Studies on Concept and Methodology of Environmental Accounting', *Korean Accounting Education and Studies*, Seoul: Parkyoungsa, pp. 169–196

Lee, B.W. (1997), *Environmental Management*, Seoul: Bibong Publishing Co.

Lee, B.W. (1998), *Studies on International Trends of Environmental Accounting and Policy Options*, Seoul: Korea Chamber of Commerce & Industry

Leibenstein, H. (1966), 'Allocative Efficiency versus X-Efficiency', *American Economic Review*, No. 56, 392–415

LfU and Ministerium für Umwelt und Verkehr Baden-Württemberg (1999), *Materials Flow and Energy Flow Management, Ecological Efficiency through Sustained Reorganisation (a guide)*, Karlsruhe, October (only in German)

Lied, W. (1999), 'Energie- und stoffstromorientiertes Umweltcontrolling als Basis eines betrieblichen (Umwelt-)Informationssystems', in: H.-J. Bullinger, G. Jürgens and U. Rey (eds.), *Betriebliche Umweltinformationssysteme in der Praxis*, Stuttgart: Fraunhofer IRB-Verlag

Liedtke, C., R. Nickel, H. Rohn and U. Tischner (1995), *Öko-Audit und Ressourcenmanagement bei der Kambium Möbelwerkstätte GmbH*, Wuppertal

Liedtke, C., T. Orbach and H. Rohn (1998), 'Umweltkostenrechnung', in: U. Lutz, K. Döttinger and K. Roth, *Springer LoseblattSystem Betriebliches Umweltmanagement: Grundlagen – Methoden – Praxisbeispiele*. Berlin, Heidelberg, New York

Llerena, D. (1996), *Integration of Environmental Issues in the Firm: Learning processes and coordination*. Working paper du BETA, n∞9608

Loew, T., and G. Jürgens (1999), 'Flußkostenrechnung versus Umweltkennzahlen – Was ist das richtige Instrument für das betriebliche Umweltmanagement?', in: *Ökologisches Wirtschaften*, Ausgabe 05–06/1999, München

Loikkanen, T. (1999), 'Innovation System and Sustainable Development – Towards an Innovation Driven Sustainable Development', in: Schienstock and Kuusi (eds.), *Transformation Towards A Learning Economy, The Challenge for The Finnish Innovation System*, SITRA 213, Helsinki

Loikkanen, T., and M. Hongisto (2000), *The Integration of Sustainable Development and Innovation Activities*, IEE Reports 05/00, VTT Chemical technology, 138 p. (in Finnish)

Loikkanen, T., H. Mälkki, Y. Virtanen, J.-M. Katajajuuri, J. Seppälä, J. Leivonen and A. Reinikainen (1999), *Life Cycle Assessment as A Support for Decision-Making in Companies and Administration – State of The Art and Development Needs*, Technology Survey 68/99, Technology Development Centre of Finland (TEKES) (in Finnish)

Mansfield, E. (1997), *Applied Microeconomics*, WW Norton and Co., New York

Mayer, M. (1999), 'Ressourcenmanagement auf Basis von SAP R/3', in: Hockerts, Kai; Hamschmidt, Jost; Dyllick and Thomas: *Prozesskostenoptimierung durch integriertes Ressourcenmanagement*, IWÖ-Diskussionsbeitrag Nr. 74, St. Gallen, 1999

Miller, P., (1998), 'The Margins of Accounting', in: M. Callon (ed.), *The Laws of the Markets*, Blackwell Publishers/The Sociologican Review, Oxford, pp. 174–193

MOE, *Environmental Performance Indicators for Businesses (Fiscal Year 2000 Version)* at http://www.env.go.jp/en/eco/epi2000.pdf

MOE/EAJ (1999), *Grasping Environmental Cost: A Draft Guideline for Evaluating Environmental Cost and Publicly Disclosing Environmental Accounting Information (Interim Report)*, Environment Agency Japan

MOE/EAJ (2000), *Developing an Environmental Accounting System*, Environment Agency Japan. (This document is downloadable at Report & Publication section of the MOE's website: http://www.env.go.jp/en/org/org-index.html)

Nelson, R., and S. Winter (1982), *An Evolutionary Theory of Economic Change*, Cambridge (Mass): Harvard University Press

North, D.C. (1989), 'Institutional Change and Economic History', in: *Journal of Institutional and Theoretical Economy*, 145, pp. 238–245

O'Dell, C., C. Jackson Grayson Jr. (1998), *If Only We Knew What We Know: The Transfer of Internal Knowledge and Best Practice*, The Free Press

OECD (1999), Proceedings of the OECD International Conference on PRTRs, env/jm/mono(99)16/ part1 and env/jm/mono(99)16/part2, Paris: OECD

Opschoor, J.B. (1991), 'Economic Instruments for Controlling PMPs', in: H. Opschoor and D. Pearce, *Persistent Pollutants: Economics and Policy*, Dordrecht: Kluwer Academic Publishers

Parker, L. (1999), *Environmental Costing: An Exploratory Examination*, Australian Society of Certified Practising Accountants, February

Pearce, D., and R.K. Turner (1998), *Economics of Natural Resources and the Environment*, 12th printing, New York: Harvester Wheatsheaf

Petersen, J.S. (1999), 'Documentation of Quality in Primary Production', in: *Regulation of Animal Productions in Europe*, KTBL Arbeitspaper 270:179–182

Porter, M., and C. van der Linde (1996), 'Green and Competitive: Ending the Stalemate', in: R. Welford and R. Starkey (eds.), *Business and the Environment, The Earthscan Reader*, Earthscan Publications Ltd.

Porter, M.E. (1980), *Competitive Advantage*, Boston, MA: Harvard University Press

POSCO Research Institute (ed.) (2001), *Environmental Accounting Systems and Environmental Performance Indicators*, Seoul: Korean Ministry of Environment and World Bank

Preimesberger, C. (no year), 'Deckungsbeitrag und Stofffluss von Maerktleistungen als ökonomisch-ökologisch Managementinformation', in: A.H. Malinksi (ed.), *Betriebliche Unweltwirtschaft. Grundzüge und Schwerpunkte*, Sonderdruck

PricewaterhouseCoopers in cooperation with the Aarhus School of Business (2000), *Environmental Management Accounting: What Why and How*. Copenhagen: PricewaterhouseCoopers. Available by contacting: van@pwc.dk

Rautenstrauch, C. (1999), *Betriebliche Umweltinformationssysteme - Grundlagen, Konzepte und Systeme*, Springer, Berlin Heidelberg

Rey, U., G. Jürgens and A. Weller (1998), *Betriebliche Umweltinformationssysteme – Anforderungen und Einsatz*, Stuttgart: IRB-Verlag

Ricoh (2000), *Ricoh Group Environmental Report 2000*

Rosemann, M., and R. Schütte (1997), 'Grundsätze ordnungsmäßiger Referenzmodellierung', in: J. Becker, M. Rosemann and R. Schütte (eds.), *Entwicklungsstand und Entwicklungsperspektiven der Referenzmodellierung*. Proceedings zur Veranstaltung vom 10.03.1997, Arbeitsbericht Nr. 52 des Institutes für Wirtschaftsinformatik der Westfälischen Wilhelms-Universität Münster, S. 16–33

Schaltegger and Sturm (1990), 'Ökologische Rationalität', in: *Die Unternehmung*, Nr. 4, 273–290

Schaltegger, S., and R. Burritt (2000), *Contemporary Environmental Accounting*, London: Greenleaf

Schaltegger, S., and K. Müller (1997), 'Calculating the True Profitability of Pollution

Prevention', *Greener Management International*, Special edition on Environmental Management Accounting (edited by M. Bennett and P. James), Spring, pp. 86–100

Schaltegger, S. (1999), 'Bildung und Durchsetzung von Interessen zwischen Stakeholdern der Unternehmung. Eine politisch-ökonomische Perspektive', in: *Die Unternehmung*, Vol. 53, Nr. 1, 3–20

Schaltegger, S., and R.L. Burritt (2000), *Contemporary Environmental Accounting – Issues, Concepts and Practice*, Sheffield UK: Greenleaf Publishing

Schaltegger, S., T. Hahn and R. Burritt (2000), 'Environmental Management Accounting. Overview and Main Approaches', in: M. Kreeb and E. Seifert (eds.), *Environmental Management Accounting*, Witten/Herdecke, Germany

Schaltegger, S., K. Müller and H. Hindrichsen (1996), *Corporate Environmental Accounting*, Chichester: Wiley

Schaltegger, S., and C. Stinson (1994), 'Issues and Research Opportunities in Environmental Accounting' (discussion paper 9124; Basel: Wirtschaftswissenschaftliches Zentrum [WWZ])

Schaltegger, S. and A. Sturm (1992/1994/2000), *Ökologieorientierte Entscheidungen in Unternehmen. Ökologisches Rechnungswesen statt Ökobilanzierung. Notwendigkeit, Kriterien, Konzepte*, Bern: Haupt. (can be downloaded: www.uni-lueneburg.de/umanagement under 'publications')

Schaltegger, S. and A. Sturm (1998), *Eco-Efficiency by Eco-Controlling*. Zürich: Vdf

Schaltegger, S., and A. Sturm (1994), *Ökologieorientierte Entscheidungen in Unternehme*, 2. Auflage. Bern: Haupt

Schaltegger, S., and C. Stinson (1994), *Issues and Research Opportunities in Environmental Accounting*, Basel. WWZ Discussion Paper No. 9124

Schaltegger, S., T. Hahn and R.L. Burritt (2000), *Environmental Management Accounting – Overview and Main Approaches*, Lueneburg: Center for Sustainability Management at the University of Lueneburg (www.uni-lueneburg.de/umanagement)

Schaltegger, S., T. Hahn and R.L. Burritt (2001), *EMA-Links – The Promotion of Environmental Management Accounting and the Role of Government, Management and Stakeholders*, Lueneburg: Center for Sustainability Management at the University of Lueneburg (www.uni-lueneburg.de/umanagement)

Schaltegger, S., T. Hahn and R.L. Burritt (2002), 'Environmental Management Accounting: Overview and Main Approaches', in: M. Bennett, J.J. Bouma and T. Wolters (eds.), *Environmental Management Accounting: Informational and Institutional Developments*, Dordrecht: Kluwer Academic Publishers

Schaltegger, S., K. Müller and H. Hindrichsen (1996), *Corporate Environmental Accounting*, London: John Wiley & Sons

Scheer, A.-W. (1995), *Wirtschaftsinformatik. Referenzmodelle für industrielle Geschäftsprozesse*, Berlin et al.: Springer Verlag

Scheide, W., M. Strobel, S. Enzler, R. Pfennig and H. Krcmar (2001), *Flow Cost Accounting in practice – ERP-based solutions of the ECO-Rapid-Project*, Proceedings of the Conference 'Sustainability in the Information Society' 2001, Environmental Informatics 2001, 15th International Symposium, Informatics for Environmental Protection, October 10th to 12th, 2001, ETH Zurich

Schmidheiny, S., and F.J.L. Zorraquin (with the WBCSD) (1998), *Financing Change, The Financial Community, Eco-Efficiency, and Sustainable Development*, The MIT Press

Schmidt-Bleek, F. (1998), *Das MIPS-Konzept: Weniger Naturverbrauch – mehr Lebensqualität durch Faktor 10*, Munich

Schmidt-Bleek, F. (1994), *Wieviel Umwelt braucht der Mensch? MIPS – Das Maß für ökologisches Wirtschaften*, Berlin, Bale, Boston

Schmidt-Bleek, F., and U. Tischner (1995), *Produktentwicklung, Nutzen gestalten – Natur schonen*, Vienna

Schroeder, G., and M. Winter (1998), 'Environmental Accounting at Sulzer Technology Corporation', in: M. Bennett and P. James (eds.), *The Green Bottom Line: Environmental Accounting for Management – Current Practice and Future Trends*, Sheffield: Greenleaf

Serageldin, Ismail, Tarig Husain, Joan Martin-Brown, Gustavo Lopez Ospina, and Jeanne Damlamian (eds.) (1998), *Organizing Knowledge for Environmentally and Socially Sustainable Development: Proceedings of the Fifth Annual World Bank Conference on Environmentally and Socially Sustainable Development*

Siemens AG (2000), *Solutions for the cities of tomorrow*

Siemens AG (2000), *Environmental Report 2000 'Our Constant Quest for Greater Sustainability'*

Society of Automotive Engineers (1998), *Proceedings SAE Total Life Cycle Conference*, Graz, Austria

Stichting DuVo (2000), *Begin van een dialoog*, Rotterdam

Stichting DuVo (1999), *Duurzaamheid in de Voedingsmiddelenketen*, Rotterdam

Strobel, M., B. Wagner and J. Gnam (1999), 'Flußkostenrechnung bei der Firmengruppe Merckle-Ratiopharm', in: H.-J. Bullinger, G. Jürgens, U. Rey (Hrsg.), *Betriebliche Umweltinformationssysteme in der Praxis*, Stuttgart: Fraunhofer IRB-Verlag

Strobel, M. (2001), *Systemic Flow Management, Flow-oriented Communications as a Perspective for Corporate Development in Both Ecological and Economic Terms*, Augsburg: IMU (only in German)

Strobel, M., and S. Enzler (2001), 'Flussmanagement – Kostensenkung und Umweltentlastung durch einen materialflussorientierten Managementansatz', in: *UmweltWirtschaftsForum*, 9. Jg., Heft 2-2001, S. 54–60

Sturm, A. (2000), *Environmental Performance Measurement*, Dresden: University of Technology

Technical Report 54 (2001), *Business and the Environment: Current trends and developments in corporate reporting and ranking*, European Environment Agency, downloadable from http://reports.eea.eu.int/Technical-report-No-54

Teffène, O., M. Rieu, J. Dagorn, P. Mainsant, H. Marouby and F. Porin (1998), 'Trente ans d'évolution du secteur porcin français', in: *Journées Rech. Porcine en France*, 30:133–152

Tietenberg, T. (1996), *Environmental and Natural Resource Economics*, 4th Edition, New York: Harper & Collins

Toyokeizai (2001), 'Current Trends of Environmental Management,' *Tokeigeppo*, April (in Japanese)

Tuppen, C. (ed.) (1996), *Environmental Accounting in Industry: A Practical Review*, London: British Telecom

UNDSD (2000), *Improving Governments' Role in the Promotion of Environmental Managerial Accounting*, Meeting Document for the First Expert Working Group, Washington DC

UNEP (1998), *Voluntary Industry Codes of Conduct for the Environment*, Technical Report # 40, United Nations Environment Programme, Paris: France

UNEP (1998), *UNEP Financial Institutions Initiative 1998 Survey*, United Nations Environment Programme/PWG, Geneva

United Nations Conference on Trade and Development (1999), *Accounting and Reporting for Environmental Costs and Liabilities*

Uno, K., and P. Bartelmus (eds.), *Environmental Accounting in Theory and Practice*, 1998, Kluwer Academic Publishers; S. Simon and J. Proops (eds.), Greening the Accounts, 2000, Edward Elgar

US EPA Environmental Accounting Project (1995), *An Introduction to Environmental Accounting as a Business Management Tool – Key concepts and terms*, Washington, DC, June, EPA 742-R-95-001 (or as download from http://www.epa.gov/opptintr/acctg)

Varian, H. (1986), *Intermediate Microeconomics*, Norton

Veen, M.L. van der (2000), 'Environmental Management Accounting', in: A. Kolk, *Economics of Environmental Management*, Harlow: Financial Times Prentice Hall

Vornholz, G. (1999), 'Branchenperspektiven 1999', in: *Sparkasse*, 116th vol., no. 1, 1999, pp. 40–42

Wagner, B., and M. Strobel (1999), 'Cost Management with Flow Cost Accounting', in: J. Freimann (1999), *Tools for Successful Management – A compendium for in-house company practice*, Wiesbaden, pp. 49–70 (only in German)

Walden, W.D., and B.N. Schwartz (1997), 'Environmental Disclosures and Public Policy Pressure,' in: *Journal of Accounting and Public Policy*, Vol. 16, No. 2, pp. 125–154

WCED (1987), *Our Common Future*. Oxford, Oxford University Press

White, A.L., M. Becker and J. Goldstein (1991), *Alternative Approaches to the Financial Evaluation of Industrial Pollution Prevention Investments*, Boston MA: Tellus Institute

Williamson, O.E. (1975), *Markets and Hierarchies*, London: The Free Press

Williamson, O.E. (1986), *The Economic Institutions of Capitalism*, London: The Free Press

Wolters, T. (1996), *Eco-Management Accounting as a Tool of Environmental Management: A Background Document*. Presented at the 5th Annual Conference of the Greening of Industry Network, November 24–17 1996, Heidelberg

Wolters, T., and J. Hoeben (2000), *Milieubewust inkopen, Verkenning van ontwikkelingen bij het MKB*, Zoetermeer: IOO bv

Wübbenhorst, K.L. (1984), *Konzept der Lebenszykluskosten*, Darmstadt: Verlag für Fachliteratur

Zehbold, C. (1995), *Lebenszykluskostenrechnung*, Wiesbaden: Gabler

DATE DUE

MAY 31 2007

APR 29 2007

SEP 30 2011

SEP 28 2015

MAY 31 2019

GAYLORD

PRINTED IN U.S.A.